A WAR OF CHOICE

The British in Iraq 2003–9

JACK FAIRWEATHER

JONATHAN CAPE
LONDON

Published by Jonathan Cape 2011

2 4 6 8 10 9 7 5 3 1

First published in Great Britain in 2011 by
Jonathan Cape
Random House, 20 Vauxhall Bridge Road,
London SW1V 2SA

www.vintage-books.co.uk

Addresses for companies within The Random House Group Limited can be found at:
www.randomhouse.co.uk/offices.htm

The Random House Group Limited Reg. No. 954009

A CIP catalogue record for this book is available from the British Library

ISBN 9780224089586

The Random House Group Limited supports The Forest Stewardship Council (FSC®), the leading international forest certification organisation. Our books carrying the FSC label are printed on FSC® certified paper. FSC is the only forest certification scheme endorsed by the leading environmental organisations, including Greenpeace. Our paper procurement policy can be found at www.randomhouse.co.uk/environment

Typeset in Granjon by Palimpsest Book Production Limited, Falkirk, Stirlingshire
Printed and bound in Great Britain by Clays Ltd, St Ives plc

For my parents, Rufus and Cherry

'The king must first wage only obligatory wars. What is an obligatory war? It is a war against the seven nations, the war against Amalek, and a war to deliver Israel from an enemy who has attacked them. Then he may wage authorised wars, which is a war against others in order to enlarge the borders of Israel and to increase his greatness and prestige.'

Maimonides, *Mishneh Torah, c.*1170–1180

Tim Russert: 'Do you believe the war in Iraq is a war of choice or a war of necessity?'
President George Bush: 'I think that's an interesting question. Please elaborate on that a little bit. A war of choice or a war of necessity? It's a war of necessity.'

'Meet the Press with Tim Russert', NBC, 7 February 2004

'The freest human being is not one who acts on reasons he has chosen for himself, but one who never has to choose.'

John Gray, *Straw Dogs*, 2002

NOTE ON THE TEXT

Quotations attributed to individuals are from their own recollections, except where footnotes indicate another source. Subsequent quotations from the same individual without footnotes indicate that the same source is being used.

I have sought to protect the identities of British Special Forces personnel and Secret Intelligence Service agents using pseudonyms, except if they are no longer serving and have given their permission to be identified.

Several individuals cited in the book received honorary titles during the period covered. To avoid confusion and unnecessary explanation, I have refrained from using titles.

For Arabic words and place names I have sought to use the most common transliterations. I have used the suffix al- in surnames when giving an individual's full name. When only referring to the surname I have dropped the suffix for convenience.

Maps xii

Prologue The Birthday Party 1

Chapter 1 Strategic Communication 11

Chapter 2 Basra Mon Amour 23

Chapter 3 Occupation 27

Chapter 4 The Emerald City 37

Chapter 5 Breadbasket 43

Chapter 6 Death in the Marshes 51

Chapter 7 Ministry of Culture 65

Chapter 8 Our Man in Baghdad 75

Chapter 9 Neocolonialists 85

Chapter 10 Uprising 99

Chapter 11 Danny Boy 115

Chapter 12 Out of Iraq 127

Chapter 13 Police 135

Chapter 14 Fartosi 147

Chapter 15 Get Elected 161

Chapter 16 Transition 173

Chapter 17 A Proposal 183

Chapter 18 Don't Mention Iraq 187

Chapter 19 Iran 195

Chapter 20 Jamiat 207

Chapter 21 Helmand 223

Chapter 22 Pike Force 237

Chapter 23 New Coin, Old Rope 247

Chapter 24 The Sheriff 261

Chapter 25 Breaking Point 269

Chapter 26 Sinbad 279

Chapter 27 A Marked Man 285

Chapter 28 Lambo 289

Chapter 29 Palermo 301

Chapter 30 The Deal 309

Chapter 31 On the Run 319

Chapter 32 Charge of the Knights 327

Epilogue 341

Dramatis Personae 351

Notes 361

Bibliography 399

Acknowledgements 407

Index 411

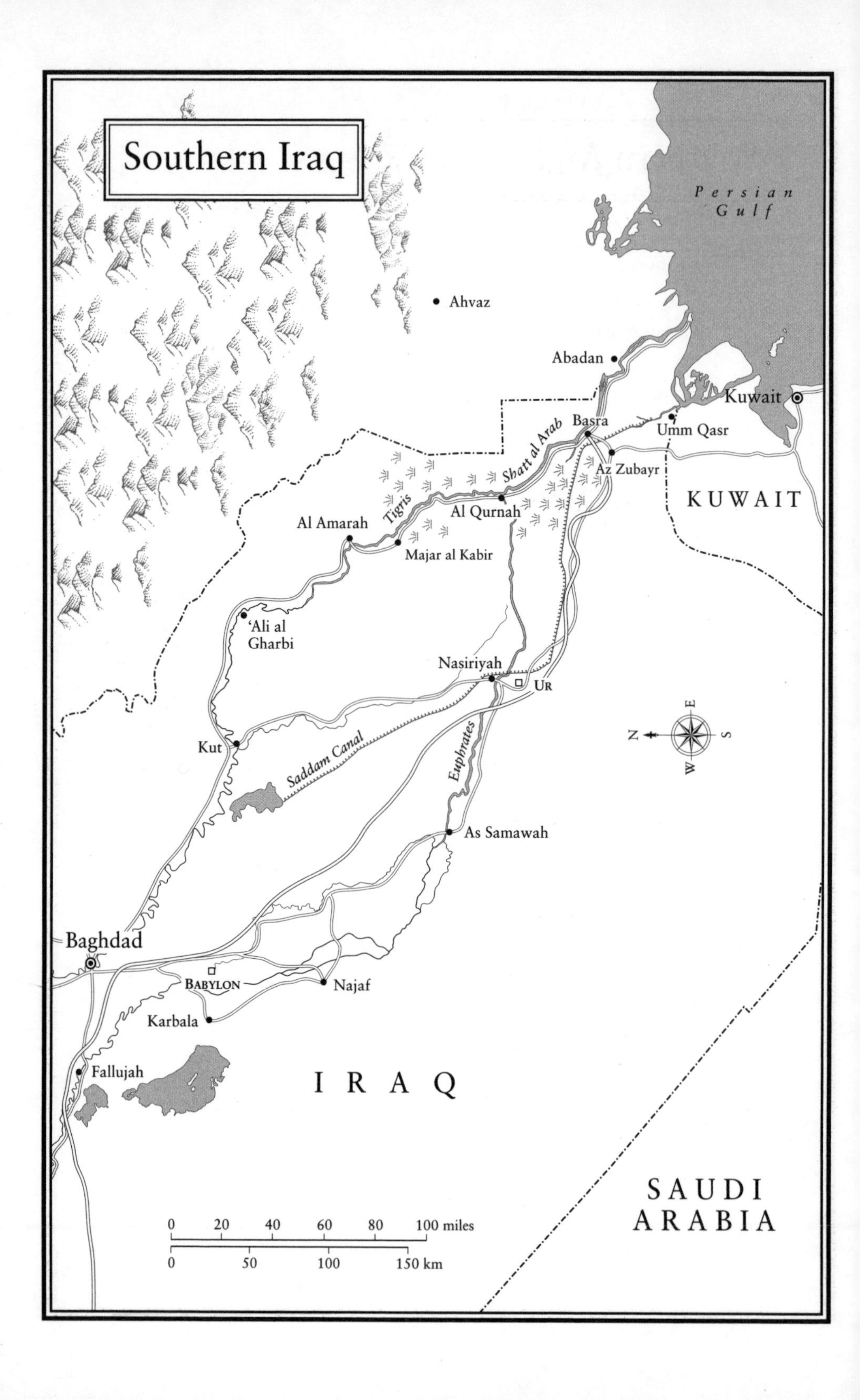
Southern Iraq
Persian Gulf
Ahvaz
Abadan
Kuwait
Basra
Shatt al Arab
Umm Qasr
Az Zubayr
KUWAIT
Tigris
Al Qurnah
Al Amarah
Majar al Kabir
'Ali al Gharbi
Nasiriyah
Ur
N
E
S
W
Kut
Saddam Canal
Euphrates
As Samawah
Baghdad
Babylon
Najaf
Karbala
Fallujah
IRAQ
SAUDI ARABIA
0
20
40
60
80
100 miles
0
50
100
150 km

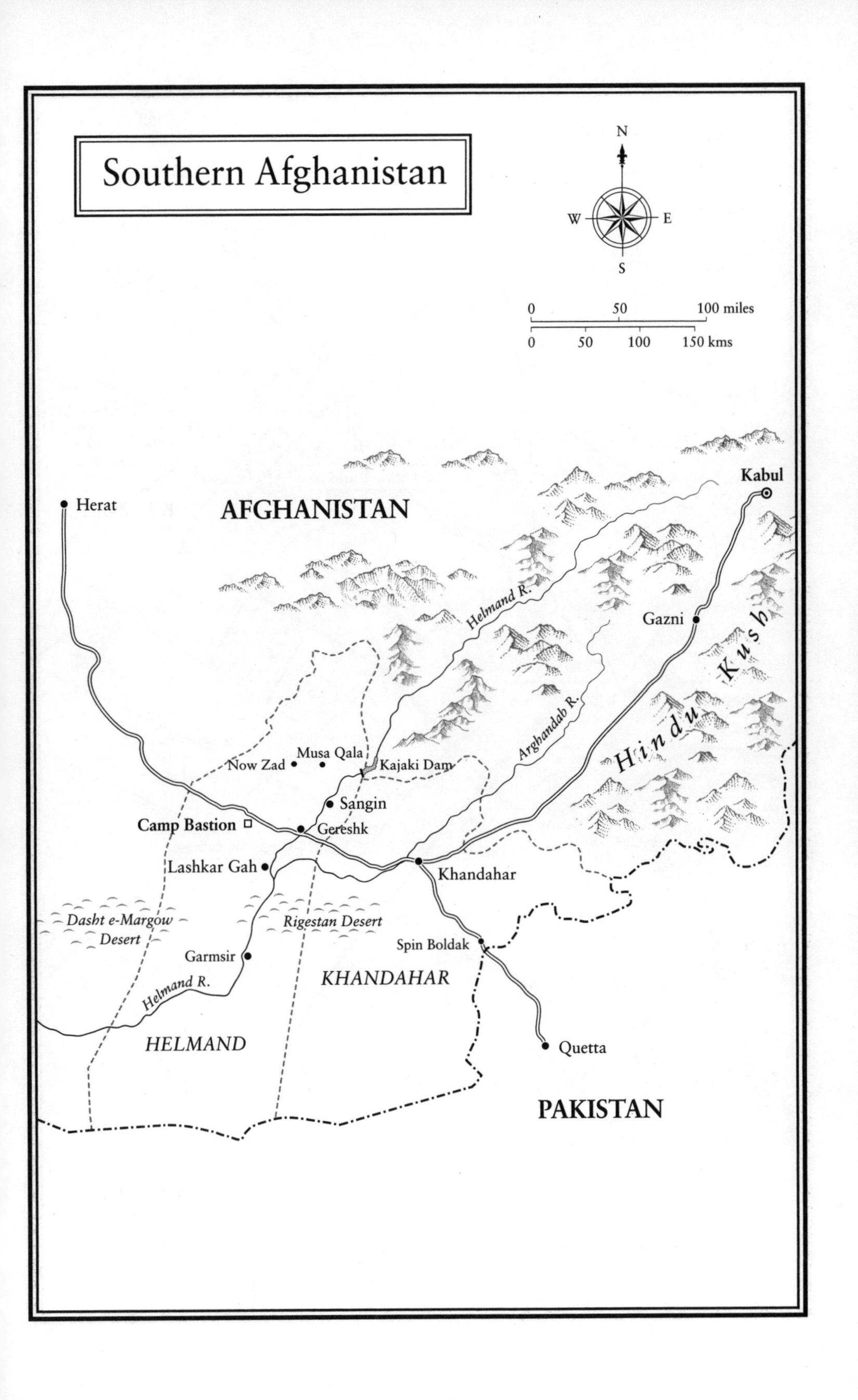

Southern Afghanistan
N
W
E
S
0
50
100 miles
0
50
100
150 kms
AFGHANISTAN
Herat
Kabul
Gazni
Hindu Kush
Helmand R.
Arghandab R.
Musa Qala
Now Zad
Kajaki Dam
Sangin
Camp Bastion
Gereshk
Lashkar Gah
Khandahar
Dasht e-Margow Desert
Rigestan Desert
Spin Boldak
Garmsir
KHANDAHAR
Helmand R.
HELMAND
Quetta
PAKISTAN

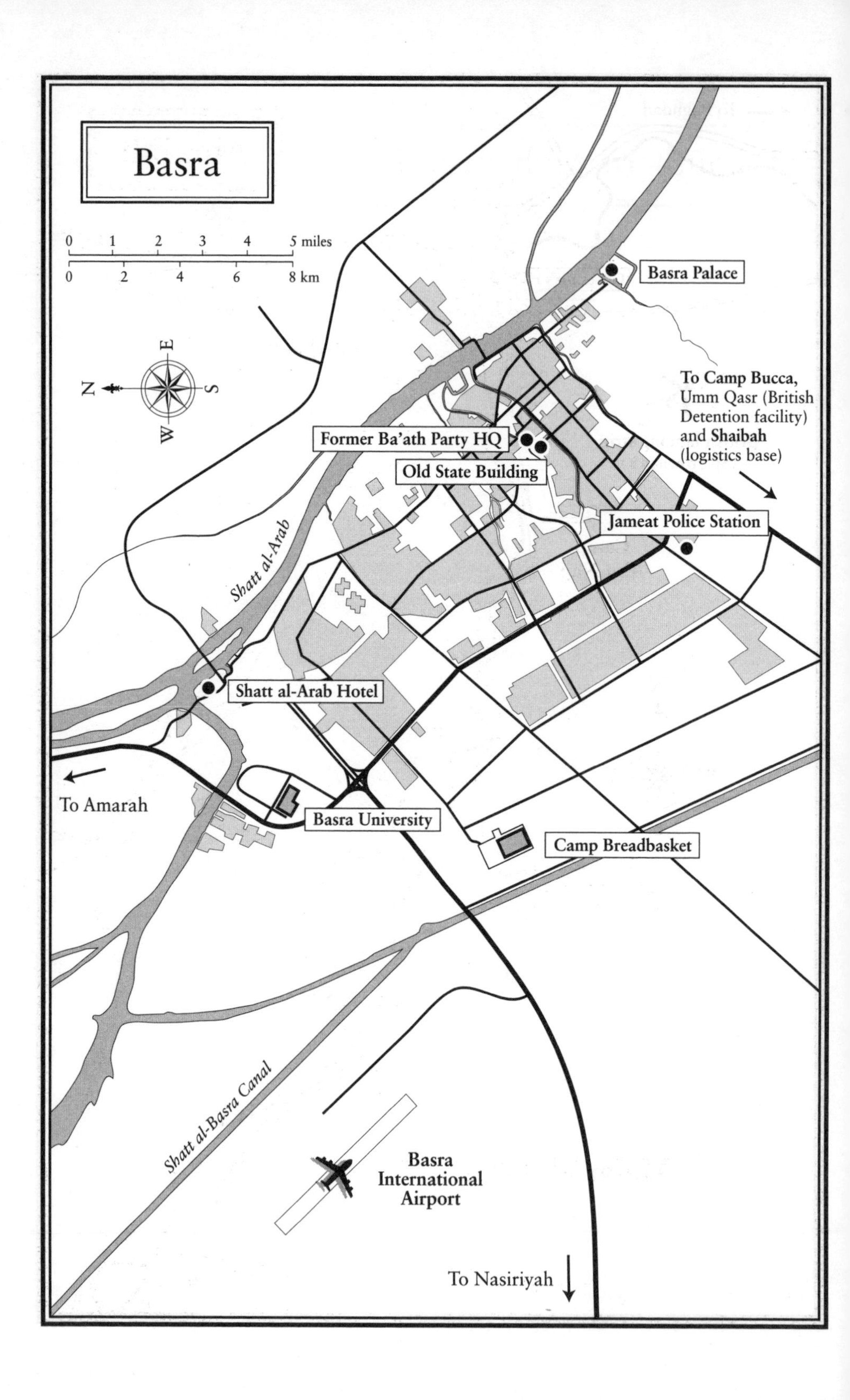

Basra
0 1 2 3 4 5 miles
0 2 4 6 8 km
N
E
S
W
Basra Palace
To Camp Bucca, Umm Qasr (British Detention facility) and Shaibah (logistics base)
Former Ba'ath Party HQ
Old State Building
Jameat Police Station
Shatt al-Arab
Shatt al-Arab Hotel
To Amarah
Basra University
Camp Breadbasket
Shatt al-Basra Canal
Basra International Airport
To Nasiriyah

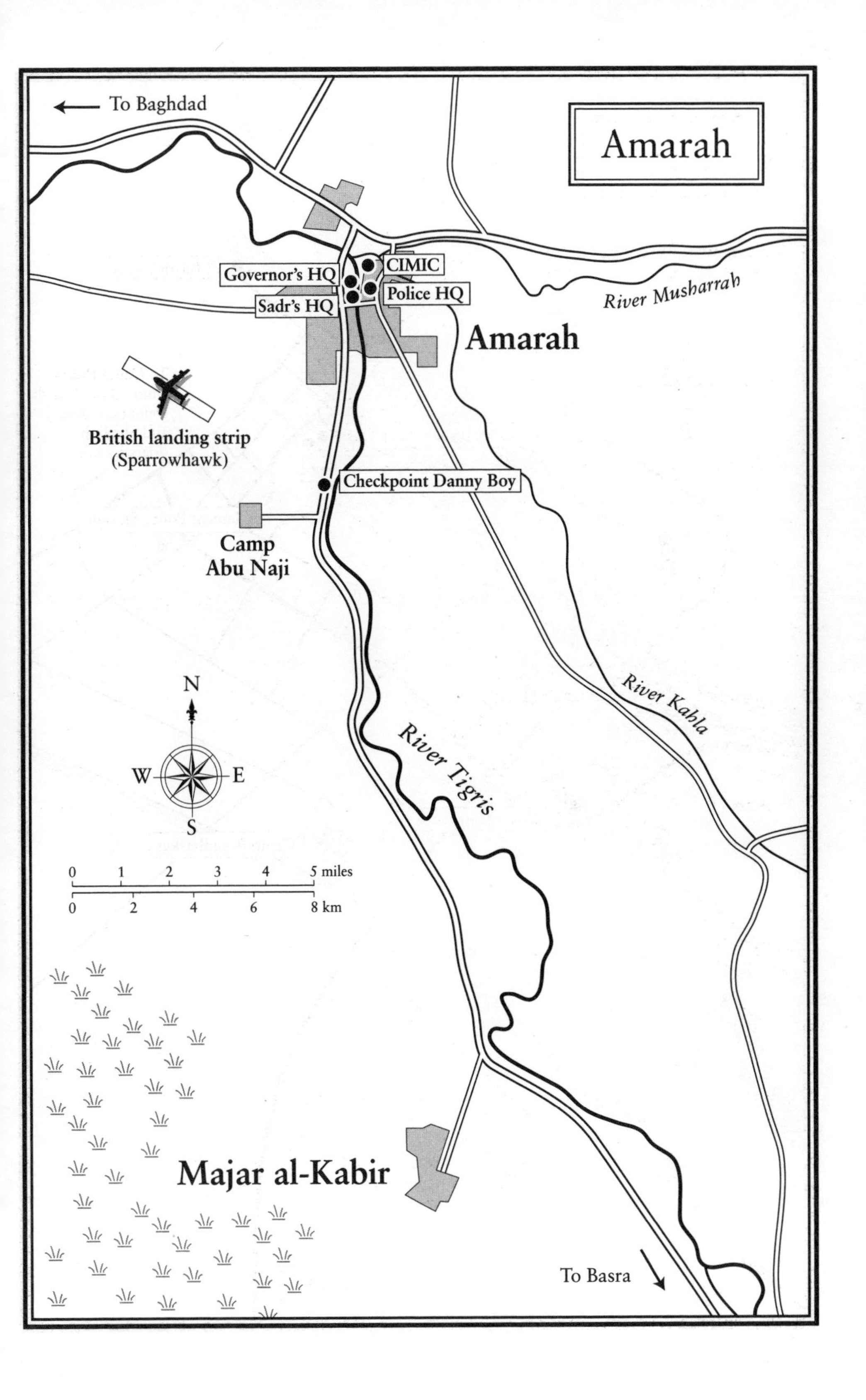

Amarah
To Baghdad
CIMIC
Governor's HQ
Police HQ
Sadr's HQ
River Musharrah
Amarah
British landing strip
(Sparrowhawk)
Checkpoint Danny Boy
Camp
Abu Naji
River Kahla
River Tigris
N
W
E
S
0 1 2 3 4 5 miles
0 2 4 6 8 km
Majar al-Kabir
To Basra

Prologue

THE BIRTHDAY PARTY

BASRA'S POSTAL SERVICE WAS STILL not running, which meant the invitations would have to be hand-delivered. Ahmed, the consulate's office manager, was dispatched into the city with two hundred of the stiff white cards. The florid lettering on each read, 'You are kindly invited to attend the Queen's Birthday Party at Basra Palace. Date April 21st 2006.'[1]

Consular staff had compiled an impressive guest list to celebrate the royal birthday: southern Iraq's most notable sheikhs, imams, security officials and politicians. Many owed their position and wealth to British patronage, although few would publicly defend Her Majesty's Government's presence. Iraqis working inside the consulate whispered among themselves that some of the guests were known to have organised attacks against the British army.[2] But a well-attended event was important to British diplomats like James Tansley, the consul general, who anxiously texted the tribal chiefs he knew, encouraging them to attend. Pulling off the Queen's Birthday Party was one of the annual rigours of life in the field for diplomats, but this year's required special dedication. Three years after the invasion, Britain's empire in southern Iraq was teetering.

This country's greatest nation-building exercise since the colonial era stretched from the dense marshes and war-ravaged palm groves of the Iranian border to the vast expanse of the Arabian desert, covering an area larger than Ireland. Over four million Iraqis looked to British guidance on electricity, water, healthcare, education and security. Despite the insistence by UK diplomats that authority lay in the hands of Iraqis, the 8,000 British troops stationed in southern Iraq suggested otherwise. With power had come the expectation among Iraqis that the UK would transform the region. Britain's failure to deliver since the US-led invasion of Iraq in 2003 had stoked popular resentment and a growing insurgency.

The military was feeling the strain. Two months earlier, the *News of the World* had released a video of British soldiers beating Iraqi rioters in the Marsh Arab town of Amarah.[3] In response, protesters took to the streets, UK bases were hit with nightly mortar and rocket attacks and in Amarah a roadside bomb killed two British soldiers. In addition, the local council in Basra, an elected body of suspected militia sympathisers, dodgy businessmen and clerics, was testing the promise of a smooth transition to democratic rule. The UK was relying on the council to take control of the city so that British troops could begin withdrawing. However, after the video emerged, the council had stopped speaking to the British altogether.

In his daily reports to London Tansley played down the attacks, knowing that the media was likely to blow the violence out of proportion. His job was to reassure Number 10 that progress was still being made. When a rocket slammed into the ground thirty metres from his office but failed to detonate, he made light of the situation: 'If that had gone off I certainly would have felt it.'

Tansley looked the part of a cautious bureaucrat – plump and softly spoken – but he shared that peculiar streak of romance that marks out many Arabists at the FCO. As a young diplomat posted to Baghdad in the 1980s, he had taken a taxi to the front line of the murderous Iran–Iraq war, and had been the first Western diplomat into liberated Kuwait in 1991. He had left government for business school and a brief stint working for the Australian Cricket Board before the lure of adventure and the FCO's pension scheme brought him back. An ambassadorship was not out of the question, especially as he had supported the invasion, unlike many in the FCO. He had seen Saddam Hussein's tyranny first hand and felt Britain should not stand idly by while an entire nation suffered in fear and poverty.

However, his rose-tinted reports to London were becoming more difficult to write as the violence increased. A fortnight before the Queen's Birthday Party was due to take place, rockets were slamming nightly into the British headquarters. Tansley had discussed withdrawing non-essential staff – almost half of the fifty-strong team – from the consulate with

Dominic Asquith, the Iraq director at the FCO in London. Asquith was the doyen of the department. Great-grandson of the prime minister who had authorised the British invasion of Iraq during the First World War, he embodied the gentlemanly values that many in the FCO sought to preserve. He also had a first-class political brain and a self-deprecating manner that won him loyalty among his subordinates. Ultimate responsibility for non-military British personnel in Iraq rested with him.

'Shouldn't you be getting out?' Asquith asked over the phone.[4]

Tansley argued that the rockets were being fired by a handful of militants and that to retreat would only empower them. Both men knew the smell of British defeat was in the air. The war was growing in unpopularity in the UK, and the politicians were turning their attentions towards Afghanistan. Even as they planned the party in Iraq, the first troops were landing in Helmand, Afghanistan. Asquith mentioned that Peter Ricketts, one of the FCO's most influential mandarins and an Iraq sceptic, was due to be in Basra the week of the party.[5] Tansley felt a successful event, with a strong Iraqi showing, would send the message that the British were still in control of southern Iraq.

ON THE BALMY SPRING AFTERNOON OF THE party, the Iraqi guests queued up outside the concrete guardhouse to be frisked. In the spirit of the event, Imam Ali al-Musawi had brought a two-metre-square cake, garishly decorated with a digitally inscribed image of Queen Elizabeth II. At the gate already-skittish Filipino guards had panicked at the possibility that the cake contained a bomb, but Musawi and the cake were eventually cleared to enter. The leader of the Ismaili Shia sect in Basra, Musawi was a long-standing favourite of the British, who had rewarded him with many lucrative rebuilding contracts, only some of which had been completed. Tansley smiled diplomatically as the tasteless rendering of the Queen was brought through the giant arched entranceway of the compound and placed proudly on the table of kebabs and hummus that marked the focal point of the party on the consulate lawn.

A contrast to Saddam's regal home in Baghdad, currently serving as the American headquarters, the British centre of operations consisted of a dozen villas which the Iraqi dictator had built in 1988 to host an Arab summit that never took place. The dictator had spent little time in the Shia south, where he had viciously crushed several uprisings over the years.

Compared to the rest of Basra, a once elegant city that Saddam had reduced to a squalid, overcrowded cesspool, the villas were luxurious, boasting neo-Palladian columns and marble floors. Divided from each other by twisting canals and ornate bridges, they evoked the salubrious air of a Florida golf resort. Beyond a fringe of palm trees the swollen Shatt al-Arab river sent cool breezes through the palace windows as it ran the thirty-four miles towards the Persian Gulf. After the 2003 invasion, the villas had been converted into offices for the use of diplomats and Iraqi staff, while nearby portacabins, interspersed with concrete blast walls and bomb shelters, were used for housing. Next to the villas, the portacabins looked like a Butlins holiday camp, although they were air-conditioned with en-suite bathrooms. A phalanx of golf buggies provided transport within the compound, which was where most of the British diplomats stayed.

Musawi stood beaming beside his cake as the rest of Basra's elite arrived. Sheikhs in gold-trimmed cloaks were greeted by Tansley and a crowd of besuited diplomats and contractors clutching wine glasses. Tansley was aware that Islam forbade alcohol, but such a vital British custom could be discreetly observed without offending anyone. Furthermore, Muzahim al-Tamimi, a garrulous Iraqi sheikh who had ingratiated himself with the British, welcomed a drink. As he declared to the crowd, surely the whole point was to offend the unsophisticated 'turbans', whose piety did not represent the moderate majority. To emphasise the point, Tamimi had brought his own supply of Johnnie Walker whisky.

Peter Ricketts, a clever man dulled by the endless machinations of Whitehall, commented to Tansley, 'I don't think I've ever seen an embassy where everyone is so relaxed on the Queen's Birthday.'[6] Tansley laughed a little nervously.

The party was in full swing when the sun dipped below the line of palm trees, and the mood suddenly changed. By some accounts, phones began to ring, and groups of Iraqis hastily departed. Ole Jepsen, a sixty-one-year-old Danish contractor advising on southern Iraq's tomato and date crops, remembered talking to the deputy head of Basra Council as he received a call and left without a word.[7] Tansley would later insist that the party was actually winding down, and that his guests had left for evening prayers. However, shortly after their departure, the first Katyusha rocket salvo slammed into scrubland just outside the compound walls, the explosion rippling through the ground. The Katyusha rocket system was designed by the Soviets during the Second World War to deliver a barrage of rockets – each one containing up to fifty pounds of explosives – from the back of a truck. Iraq's militias had developed a more ad hoc system, in which individual rockets were propped up against a rock and pointed in roughly the right direction.

The remaining crowd broke for cover as the air-raid siren wailed. The drill was well established: make for the nearest building, in this case the concrete veranda of the consulate office, which offered some protection. Whether due to alcohol or collective fright at being caught out in the open, the diplomats raced for the meagre cover, jostling past one another, women tottering in high heels and men doing their best to look civil. Rachel Schmeller, a US State Department representative in Basra, recalls the FCO's privately hired British security guards in their sunglasses and bullet-proof jackets, waving machine guns and hurrying along those in the crowd still trying to appear unflustered.[8]

'We were trying not to spill our drinks,' recalled Jepsen.

The remaining Iraqis made a beeline for the gates. Only Tamimi remained at the buffet, sipping his whisky, inured to the explosions after a few drinks and the experience of living through three decades of Iranian, Iraqi and American bombardments.

'I wish every party could end like this,' he joked.[9]

Under the veranda the diplomats nervously traded past experiences of rocket attacks, but as the adrenaline faded, awkward silence set in. Half an hour later, when the all-clear sounded, Tansley hurried with Ricketts to the helicopter landing pad. Just as a Lynx helicopter approached to whisk away the senior diplomat another rocket salvo landed, forcing the helicopter away.

'Someone inside the compound tipped the militants off about the helicopter's arrival,' Tansley speculated. A shaken Ricketts finally left just before midnight.

The next day Tansley conferred with Dominic Asquith in London and the first diplomats began packing their bags to leave. The partial evacuation of the palace had begun, and with it the beginning of the end of the British occupation of southern Iraq. Even Tansley, for all his sanguine reports, had to confront the consequences of what had just happened.

'What are we doing here?' he wondered.

* * *

In 1999, seven years before the Queen's Birthday Party in Basra, Tony Blair gave a speech in Chicago outlining his views on when one country could justly intervene in the affairs of another. Genocide, he told Chicago's Economic Club, merited external intervention; sometimes armed force was 'the only means of dealing with dictators'. His views came at the end of a decade that had seen bloody wars in the Balkans and Rwanda flare out of control as the international community watched. Many in Britain saw Blair's remarks as deliberately provocative, coming from a Labour politician whose party was known for its pacifism during the 1980s. Since coming to power in 1997, Blair had demonstrated his willingness to use force for a just cause. In 1998 he had confronted Saddam Hussein during the Desert Fox bombing campaign after the Iraqi dictator had stopped complying with UN weapons inspections. The month before Blair's Chicago speech he had succeeded in pushing the US and NATO to intervene in Kosovo. Arguably, his greatest success lay just ahead, in 2000, when a small British force intervened decisively in Sierra Leone's civil war.

Blair did not take such missions lightly. As he told his audience in Chicago, leaders needed to be sure of their case and to have exhausted all diplomatic options. Military operations should be well planned, with a recognition that 'we cannot simply walk away once the fight is over'. And to declare war a prime minister needed to show that Britain's national interests were at stake – although in Blair's interconnected view of the world that was a fairly broad definition. Blair's Chicago speech was widely praised back home. Such mature reasoning was rare among world leaders, but then Blair had always had a gift for turning strategy into fine words. Having laid out his doctrine for intervention, he appeared well placed to judge when Britain should next deploy its forces.

It was not just the prime minister who was ready for the challenges of the twenty-first century. Since 1997 New Labour had revolutionised the country's approach to world affairs, with Robin Cook as foreign secretary famously promising to add an 'ethical dimension' to foreign policy. That claim unravelled almost immediately when it transpired that the Foreign and Commonwealth Office had known about an arms deal between Sandline International, a private British security firm, and the deposed president of Sierra Leone, in contravention of a United Nations embargo. However, New Labour persevered with its plans, setting up the Department for International Development to administer how foreign aid would be spent.

Aid money, which had traditionally been a tool of the FCO to exert British influence, was now in the hands of a new ministry whose stated objective was the elimination of 'world poverty'.[10] Furthermore, the DFID would focus on building the capacity of poor nations to design, implement and run their own projects. It was an approach that drew on the best practice of the development community, which stated that struggling nations were better served in the long run by training people rather than by giving them handouts. Led by the strong-willed Clare Short, the DFID was soon drawing international praise for its work.

The Ministry of Defence was also undergoing a transformation. Over the previous decade, success in the counter-insurgency campaign in Northern Ireland and the nation-building missions in the Balkans and Sierra Leone had primed the military for the complexities of the post-Cold War world. In the early 1990s morale in the armed forces had hit rock bottom following Conservative cuts to the MOD's budget. Blair aspired to please the armed forces and overcome his party's anti-military reputation by redefining its mission. The 1998 Strategic Defence Review offered one of the first budget increases in years and a new mandate for the military: rather than focus on the defence of Europe, the military would prepare for expeditionary warfare and rapid deployment into war zones and humanitarian crises.

The September 11 2001 attacks on the World Trade Center in New York forced Blair to further refine his thinking about war. Along with US President George Bush, he came to believe radical Islam posed an existential challenge to the values of liberal democracy and needed to be contained before it struck again. The idea of taking pre-emptive action against terrorist organisations or rogue states before they could attack the West marked a radical shift from his earlier philosophy, although on the surface, the subsequent campaign in Afghanistan to overthrow the repressive Taliban regime was in the mould of his earlier triumphs in that it evoked humanitarian goals as much as military ones. British troops played a key role during the US-led operation, and Blair was prominent in efforts to create an international coalition to support both the war and subsequent nation-building efforts.

Then in December 2001, after the apparent success in Afghanistan, President Bush telephoned Blair to raise the issue of 'regime change' in Iraq. Saddam Hussein had never convinced United Nations weapons inspectors that he had disposed of his biological and chemical arsenal after the First Gulf War. Bush argued that the threat of Saddam using weapons of mass destruction could no longer be tolerated. Blair agreed with Bush's assessment, although he preferred to focus on the need to disarm the dictator, preferably through the UN. But he assured Bush that 'if [regime change] became the only way of dealing with this issue, we were going to be up for that'.[11]

By the spring Blair's foreign policy adviser David Manning had reframed British policy towards Iraq. Current UN sanctions were failing, a Cabinet Office paper commissioned by Manning explained. The Oil-for-Food programme had been hopelessly corrupted by the Iraqi regime, and China and France were pushing for a resumption of normal relations. One proposal was for tougher sanctions, but the US had 'lost faith in containment', the paper warned. Another suggestion was regime change through the support of a coup by opposition groups, although such efforts had failed in the past. The paper concluded, 'the use of overriding force is the only option that we can be confident will remove Saddam and bring Iraq back into the international community'.[12]

On 2 April 2002 Blair invited Chief of the Defence Staff Admiral Michael Boyce and his senior planner to Chequers, the prime minister's country retreat in Buckinghamshire. Jack Straw also attended. In a few days Blair was due to meet Bush at the president's ranch in Crawford, Texas. The Chequers meeting was an opportunity to discuss the change in Iraq policy beyond the prime minister's inner circle. Blair already knew it would be difficult to sell a war against Iraq to the public, although he felt the moral case for tackling Saddam was strong. As he wrote to his chief of staff, Jonathan Powell, 'The persuasion job on this seems very tough . . . Yet from a centre-left perspective, the case should be obvious. Saddam's regime is a brutal, oppressive military dictatorship.'[13] The mood at Chequers was confident without being bullish. Iraq would be Blair's fifth campaign in almost as many years, and Blair had grown comfortable with the military. Powell warned those gathered against complacency, referring back to Blair's Chicago speech and the justification for intervening in the affairs of another country. The Americans needed to be told not to 'rush into anything, move at a deliberate pace and, above all, build a coalition'.[14] No strong objections were raised.

* * *

This book is the story of what happened after that fateful meeting, of the giddy plunge into war on false premises and the extraordinary occupation of Iraq that followed.

On the face of it, the concept of a 'war of choice' sounds dubious.[15] Given the likelihood of death and destruction, what conscientious leader would willingly send his forces to war? Back in the twelfth century, the Jewish scholar Maimonides identified the distinction between an 'obligatory' war of survival and what he termed an 'authorised' or voluntary war for the aggrandisement of a country or its leader. Obvious contrasts lie between Britain's confrontation with Nazi Germany during the Second World War and the UK's invasion, along with France and Israel, of Egypt during the 1956 Suez crisis. However, the historian Lawrence Freedman, a supporter of liberal intervention who drafted sections of Tony Blair's Chicago speech, has argued that this dichotomy is misleading. All wars, whether of necessity or choice, involve decision-making at some level, and to suggest that some might be thrust upon us suggests an abnegation of responsibility. As Freedman writes, 'The issue is not whether we now have choices in war whereas before our decisions were made for us. There is always choice. The issue is whether, in the complex and fluid international system, we can frame the debate in a way that does it justice. The issue is not the fact of choice but the quality of choice.'[16]

Focusing on the quality of Blair's decision-making can be illuminating, especially in the months leading up to the invasion, when he, more than any other figure, drove the country to war. But conflicts do not end with the defeat of the opposing army, as Iraq was to so painfully show. Guiding the country through the occupation of southern Iraq was to prove no less important or demanding. In the messy aftermath of war, the decisions taken on Blair's soda or in drab Whitehall meeting rooms had little impact on the battlefields of Iraq. The real decisions that were to determine the fate of the British occupation of southern Iraq were often taken by soldiers, diplomats and contractors on the ground.

This struggle captures Britain at the start of the twenty-first century. When Britain last occupied Basra, ninety years earlier, the country stood at the height of its imperial glory. Although it is unfashionable to espouse Victorian attitudes, many still idealise notions of thrift and hard work, tradition and the stiff upper lip, and find it hard to reconcile those ideals with the current culture of self-entitlement and instant gratification: frozen dinners and celebrity news, ASBOs and online shopping. These contrasting aspects of British culture were on display in Iraq and later Afghanistan. Some thought that the country's imperial history in the Middle East conferred an innate ability to nation build and that a swift victory and pull-out were possible; others knew that rebuilding would be perilous, but once committed to the task would not leave, even at the cost of their own lives. This book describes the quality of their choices, and nature of their sacrifice.

Chapter 1

STRATEGIC COMMUNICATION

In February 2003, a month before the war began, Major General Tim Cross sent a worried memo to Whitehall from the British embassy in Washington DC. Dispatched to work on the post-war plan for Iraq, he had made the alarming discovery that there wasn't one. Cross was not a man to mask his feelings behind the dry language of officialdom. At fifty-two, he was at the height of his powers: handsome and ruddy, with bushy eyebrows that loomed over sensitive eyes and a downturned mouth. Where his more flashy peers excelled at manoeuvre warfare, the Cold War proving ground for officers destined for high command, Cross had made his name in logistics. As the UK's logistics commander in the Balkans he had been in charge of delivering humanitarian aid, a role that nicely married his Christian faith with the soothing application of numbers to people, weapons and supplies.[1]

Cross was the first person the defence chiefs turned to when Whitehall heard about an organisation the Americans were setting up to coordinate post-war planning. On 20 January 2003 President Bush had ordered the creation of the Office for Post-War Iraq. Until then, the issue of what to do with Iraq after the invasion had not been high on the agenda in either London or Washington. The problem had begun, at least for the British, ten months earlier when Tony Blair met George Bush at his ranch in Texas.

The two-day conference at Crawford was dominated by the Middle East peace process and a nuclear stand-off between India and Pakistan. But away from the advisers and minute-takers, the two leaders bonded over their response to the post-9/11 world. If America wasn't aggressive,

Bush believed, the country risked looking weak. Blair did not question the president's decision to target Iraq and offered his support. Given Saddam's history, Blair believed it was only a matter of time before the Iraqi dictator acquired new weapons of mass destruction, if he had not done so already.[2] By supporting an American confrontation with Iraq, Blair also saw the opportunity to position Britain at the heart of US decision-making with a clout not enjoyed since the days of empire.[3]

That meeting in Crawford was the moment to exert maximum influence over the Americans and ensure that post-war planning for Iraq was an integral part of the discussions. Blair did set three conditions for British involvement, but they reflected his preference for grand strategic thinking and preoccupation with the media.[4] Firstly, he urged Bush to engage in the Middle East peace process. The transformation of the region might start with Iraq, but it would only end with a deal between the Israelis and Palestinians. The second condition was for Bush to build his case for war in the United Nations. Blair feared the international damage that unilateral action by America would cause. He wanted UN weapons inspectors, who had withdrawn from Iraq in 1998, to return and ensure that Saddam Hussein had disarmed. The final condition was for enough time to convince the public of the need to go to war. Early polling in the UK had confirmed Blair's hunch that a majority was opposed to the invasion.[5]

There was already considerable scepticism within the British Cabinet, as Blair discovered the day after leaving Crawford. He gave a speech at the George Bush Senior Presidential Library, outside Houston, reaffirming his support for combating terrorism and threatening rogue nations with regime change 'if necessary and justified'. For Foreign Secretary Jack Straw, a former criminal lawyer, Blair's use of the phrase 'regime change' was alarming. Under the United Nations' founding charter, which both the UK and US had signed in 1945, no country can invade another unless acting in self-defence or with the authorisation of the UN Security Council.[6] No matter how vile Saddam's regime was, Straw knew that overthrowing the dictator without the Security Council's endorsement would undermine the entire international system established after the Second World War.

To avoid that calamity a UN resolution was needed, and that would only be obtained by proving Saddam Hussein had failed to disarm and continued to pose a threat. Straw had seen British and American intelligence alleging that Iraq had WMDs. In July he warned a Cabinet meeting that the evidence was 'thin'.[7] A week later Attorney General Peter Goldsmith wrote that current intelligence suggested there 'would not be any ground for considering Iraqi use of WMD as imminent'.[8]

However, Blair rarely felt constrained by the opinions of his Cabinet, preferring those of a small clique of advisers whose views chimed with his own.[9] Jonathan Powell, one member of this 'sofa government', advised Blair, 'The specific state of Saddam's WMD was simply not the essence of our concern. What concerned us was the threat he could potentially pose.'[10] Such nuanced arguments were often unnecessary with Blair, who often appeared more interested in seeking Bush's approval than assessing the details. At the end of July Blair wrote a private note to Bush that began, 'You know, George, whatever you decide to do, I'm with you.'[11]

In early September, following a second meeting between the two leaders, Blair secured a diplomatic victory when President Bush told the UN General Assembly he would work with the Security Council to disarm Iraq. However, the awkwardness of the Anglo-American stance was soon apparent. In order to be seen to give UN weapons inspections a chance, Britain could not appear to be planning for the war's aftermath.[12] Preparations were limited that autumn, with Blair's team more concerned about making the case for war. In September the government released a dossier of intelligence to reveal the threat posed by Saddam Hussein, including the claim, later proved to be false, that Iraq could deploy weapons of mass destruction in 45 minutes. The one government agency that could handle the complex planning needed to rebuild Iraq, the Department for International Development, was excluded from initial discussions for fear that the department's left-leaning officials would leak details to the media.[13] That left the small team of British military planners, dispatched to the US military's Central Command in late September, to raise the issue of post-war planning with the Americans. Tim Cross, a member of the team, did not get a satisfactory answer, but he was confident that 'somewhere in the bowels of the Pentagon someone was working on this'.[14]

Cross was already running into his own problems posed by the secrecy of the planning. Military options for British involvement ranged from air support to a 20,000-strong invasion force. Admiral Michael Boyce, Blair's senior military adviser, initially favoured a limited force entering northern Iraq through Turkey. In order to keep British options open, and with the army pushing for a role, Boyce opted for a division-sized force.[15] The largest deployable unit of the British army, a division usually consists of around 15,000 men drawn from several brigades, which, in turn, are comprised of up to a dozen battle groups formed around a single regiment. When assessing the challenges of the world after the Cold War, the 1998 Strategic Defence Review had envisioned a considerably smaller force for overseas expeditions, but at a stretch the invasion force would be ready in six months.

The problem was that Number 10, fearing negative publicity, was preventing the Ministry of Defence from making vital preparations like ordering equipment from suppliers.[16] Among the items needed were 37,000 sets of body armour, requested by the MOD on 13 September 2002 but only ordered in November after the passage of a UN resolution ordering Saddam Hussein to allow the return of weapons inspectors or face 'serious consequences'. The delay was to have fatal results during the invasion, when a thirty-year-old British soldier without appropriate body armour was shot and killed in Basra.[17]

The contradictions within the British approach were further exposed on Christmas Eve 2002, when the military was shocked to learn that the Turkish government had refused permission for them to use the country as a launch pad into Iraq. The British ambassador to Turkey, Peter Westmacott, might have warned military planners that Turkish consent was unlikely, given the massive opposition across the Muslim world the threat of war was creating. Unfortunately Westmacott was kept in the dark about the military plans, with the result that when the Turkish parliament voted against involvement in the war on 24 December the news was a bombshell for the planners.[18]

The resulting scramble to deploy British troops through Kuwait set back planning and meant there was even less time to think about what would happen after the invasion. Thus the defence chiefs were only too pleased to hear about the new Office for Post-War Planning that Bush was setting up. In January, the Chief UN weapons inspector, Hans Blix, had recently issued a lukewarm report on Iraqi compliance, making war likely. The following month, Cross was sent to Washington to meet the retired American general running the operation, Jay Garner. Cross knew him from the First Gulf War, during which Garner had organised relief efforts in northern Iraq. With his thick steel-rimmed glasses and soft fleshy features, the American general resembled a mild-mannered farmer from the Midwest. He was not a great communicator, and grew nervous when asked to think beyond military parameters. A man ill-suited to be the public face of the reconstruction effort in Iraq, thought Cross, who was shocked to find Garner in a half-empty office suite downtown, clearly struggling with a staff of only forty. Most of his team consisted of retired officers with whom Garner had served; great field commanders in their day but hardly the men to penetrate the murky politics of the Washington Beltway or envision Iraq's future.

To his credit, Garner was aware of the problems and had scheduled a meeting at the Fort McNair military base in Washington for 21 February. He invited anyone in government with an interest in Iraq to come. It was Cross's first introduction to the poisonous relations between the US Departments of State and Defense. Prior to the creation of Garner's office, the State Department had run post-war planning. Among the attendees at the conference was Tom Warrick, a mid-level diplomat who had spent the previous eleven months interviewing Iraqi exiles and experts on the region to compile a thirteen-volume report on post-war realities entitled *The Future of Iraq Project*.[19] It warned of looting and the need for slow and concrete steps to restore Iraqi sovereignty. Few in the Pentagon read the report because US Defense Secretary Donald Rumsfeld took the opposite view: he believed there would be a quick military campaign and then a seamless handover of authority back to the Iraqis.[20]

'Basically the American plan was we don't need a plan,' noted Cross. 'They thought they could win the military campaign, stand up an Iraq government and then dramatically reduce their force levels – all within just a few months.'

When Garner asked Warrick to join his reconstruction team he was summoned to Rumsfeld's office and told, 'You've got to let him go, Jay. He's not one of us. He's not a man who thinks like us.'[21]

Cross was surprised to see Garner being pushed around. 'He could have resigned or pushed back,' said Cross. 'But at the end of the day he was a soldier and he was going to follow orders.'

Cross was under similar pressure to pretend that post-war planning wasn't really necessary. Despite repeated requests, he had received little guidance from Whitehall on the UK's position. When Garner asked Cross to become his deputy in charge of 'international coalition building' and to run the organisation's offices in south-eastern Iraq, the only direction he received from London was an urgent memo from the FCO's deputy Middle East director, Dominic Chilcott, telling him categorically not to accept the south-eastern role, which might bring with it British responsibility for the region after the invasion and a hefty bill for reconstruction. Chilcott was running the newly created Iraq Planning Unit,[22] with the aim of co-coordinating British thinking on Iraq. The small size of his team meant this became little more than a secretariat for the Cabinet Office, or a Cabinet 'post office' as one member dismissively called it.[23] 'What it never became was the planning headquarters of the British perspective on post-war Iraq,' noted Cross.

The unit was conspicuously situated in the Foreign and Commonwealth Office, as Clare Short, the head of the Department for International Development, had grown increasingly hostile to Britain's involvement. Although the ban on DFID's participation in planning had been lifted, the combative Member of Parliament from Birmingham Ladywood was not one to forget a slight. On 13 February Tony Blair had finally convened a Cabinet meeting to discuss post-war Iraq and Britain's possible role in the south. Ministers were informed that Blair expected the British effort to be 'exemplary'.[24] Short had a simple answer for him: show me the money. The Treasury had yet to indicate what additional funds would be available for any reconstruction effort, and she was not prepared to take money away from the genuinely 'poor and needy' in other countries.[25]

Cross knew Clare Short well from their work together in the Balkans and had had dinner with her before departing for Washington. He sympathised with her position – as a committed Christian he only liked to fight when he was convinced of the justness of the cause – what he could not understand was why her department wasn't doing more to help when she had given her conditional support for the war. While the FCO was sending two highly capable diplomats to work for him, Short could only provide one official from DFID's New York office on a part-time basis. Short responded that until there was United Nations approval for the invasion she was not going to expose her staff to accusations of breaking international law. A large protest in London on 18 January had demonstrated growing public resentment over the rush to war. Short, who had resigned from the Labour front bench over the First Gulf War, was torn. 'I can't be seen to be helping you,' she informed Cross.[26]

Filing his cables home every night from the embassy in Washington,

Cross felt an odd sense of loneliness. 'What is the UK's foreign policy here? What are we trying to achieve? What resources are we prepared to put into this?' he asked without getting a reply. He had the distinct impression that Whitehall was stalling, hoping that either war could be averted, or that the UN would step in and take charge of Iraq.[27]

Cross's next meeting destroyed any illusion that the UN would support the war. In late February he and Garner visited UN headquarters in New York. US Secretary of State Colin Powell had recently addressed the General Assembly, holding up an empty test tube to warn of the dangers posed by a small dose of anthrax. The speech coincided with massive anti-invasion demonstrations around the world. On Blair's insistence, both the UK and US were pushing for a second resolution, this time explicitly authorising a war. The prospects looked bleak. Many on the UN Security Council felt they were being railroaded by the Anglo-Saxons. The discovery that a British government dossier arguing the threat posed by Saddam had been lifted from a postgraduate thesis hardly helped.[28]

Cross understood the UN's discomfort. He had worked with enough United Nations representatives in the Balkans to realise many in the organisation saw themselves as glorified aid workers and viewed the military with distaste. But even he was surprised by the unfriendly reception they received from Louise Frechette, a former Canadian ambassador to the UN who had set up an ad hoc group to look at the possibilities for UN involvement in a post-conflict Iraq.[29] She was clearly in no mood to negotiate. Flanked by a dozen aides, Frechette laid into Garner: 'Firstly, we can't give you anybody to work with because we can't be seen to be endorsing the fact there may be a war,' she said.[30] Her second point was delivered with icy satisfaction: 'You had better understand that under international law you will be responsible for what follows. If what follows is hundreds of thousands of people displaced, hungry, thirsty, maybe having been hit with chemical and biological weapons, there's going to be a major humanitarian catastrophe. And you, Garner, are going to be responsible for dealing with them.'

Cross duly informed London that the prospect of UN involvement in Garner's organisation was slim. Tony Blair continued to insist publicly that international backing was feasible, but behind the scenes his advisers had begun to confront the prospect of Britain going to war without a new UN resolution explicitly authorising an invasion. That meant the earlier resolution from November 2002, which threatened 'serious consequences' if Saddam did not comply with weapons inspections, might have to suffice as legal cover.

The decision over whether that resolution did indeed provide UN backing for an invasion now lay in the hands of Attorney General Peter Goldsmith. Since he had expressed his doubts about the case for war over the summer, Goldsmith had largely been sidelined by Number 10. Blair's advisers were keen to avoid an early judgement from Goldsmith that might box in the government's options. Their fears were confirmed when Goldsmith offered his first impressions in January that the nebulous phrase 'serious consequences' did not sanction an invasion.[31] But as the US pushed aggressively for a quick decision from the UN on whether Saddam Hussein had complied with weapons inspections, and French opposition to a second resolution mounted, it was clear to Number 10 that Goldsmith would have to change his mind.

At the end of January Jeremy Greenstock, the British representative at the UN, flew to London to argue Goldsmith around. Greenstock told the attorney general that the case for going to war rested on a twist of language and inside knowledge of the negotiations. France had wanted the first resolution to insist Britain and America seek further UN authorisation if they wanted to go to war. But all France had got was the promise that the Security Council would 'consider the situation' before an invasion. 'The French understood that they had failed to achieve the need for a second resolution,' insisted Greenstock.

Goldsmith was not convinced. The French concession could not be considered without also looking at what Britain and the US had given up – explicit language in the first resolution authorising them to invade Iraq. But a month later Goldsmith changed his mind following discussions in Washington with senior American officials, including National Security Adviser Condoleezza Rice. Like Blair before him, he seemed awestruck by proximity to power. The senior legal adviser to the US National Security Council John Bellinger later boasted, 'We had a problem with [Goldsmith] who was telling us it was legally doubtful under international law. We straightened him out.'[32] The American arguments were little different to Greenstock's – they insisted that the French knew they had 'lost' the argument when they signed up to the first resolution – but this time Goldsmith *was* convinced. As he confessed to the Iraq Inquiry, he did not think to check with the French as to what their opinion was.[33]

'I used a test which I quite frequently use when I'm having to advise on difficult matters, which is to say, "Which side of the argument would you prefer to be on?" and I took the view I would prefer to be on the side of the argument that said a second resolution wasn't necessary,' reflected Goldsmith.[34]

On 7 March 2003 Goldsmith issued a thirteen-page document suggesting that a second resolution might not be needed, though war without one risked Britain's indictment before an international court. The same day Jack Straw flew to New York to hear Hans Blix give his latest report on weapons inspections. Since his own doubts the previous summer over British intelligence, Straw had wavered for and against the war. On the flight he read the draft of Blix's report, which he found a 'sensational' chronicle of Saddam Hussein's deceit and evasiveness since the 1990s and a powerful case for taking on the dictator. Blix later expressed his surprise that Straw saw his report as evidence for the need to overthrow Saddam. 'I don't think anyone else took it as sensational. It was reporting on concealment and obstruction in the 1990s, but not much more than that.'[35] His own presentation to the Security Council suggested Saddam Hussein was increasing cooperation.

The uncomfortable truth was that Blair's determination to support the US was twisting his advisers in knots and backing the government into a corner. It was unclear whether he had enough parliamentary support to go to war without a second resolution, but if Britain did not go to war, Blair would suffer a massive blow to his credibility. The government could fall either way. Clare Short had already sensed the desperate atmosphere at Number 10, where Alastair Campbell, the communications director, was threatening and cajoling wavering ministers like herself. In an interview with the journalist Andrew Rawnsley broadcast on 10 March she threatened to resign without a UN mandate for war, declaring, 'the whole atmosphere of the current crisis is deeply reckless, reckless for the world, reckless for the undermining of the UN in this disorderly world, which is wider than Iraq, reckless with our government, reckless with his [Blair's] own future, position and place in history'.

As British and American troops streamed into Kuwait ahead of the invasion, the mounting pressure created further diplomatic twists and turns. The same evening as Short's outburst, French President Jacques Chirac announced he would use France's veto at the Security Council to block the passage of any further resolution authorising an invasion of Iraq. Campbell immediately seized on Chirac's threat as an explanation for Britain's failure to get a second resolution.[36]

But the relief at Number 10 did not last long. The following day US Secretary of Defense Donald Rumsfeld, expressing his frustration with international diplomacy, told a press conference that the US was happy to 'go it alone without Britain'. For a prime minister so anxious to appease his American counterparts, the comments struck a raw nerve. Furthermore, Rumsfeld's claim was palpably false. The proposed British deployment to Iraq had grown to 40,000, almost a third of the American-led effort, which had been dubbed the Coalition. Admiral Boyce believed it would take weeks for the US to replace them, pushing the timetable for war into the dangerous summer months.[37] Blair went 'bonkers about it' noted Campbell. 'He couldn't believe how the US kept fucking things up.'[38]

But other UK government advisers saw Rumsfeld's comments as a final chance for Britain to wriggle out of the war. Michael Jay, permanent undersecretary at the FCO, pushed this view with Jack Straw. Britain had taken its influence with the Americans as far as it could and was now risking its international reputation by slavishly clinging to the Americans.[39] Straw, vacillating once again, presented this view to the prime minister on 12 March, but Blair would not back down. He knew his foreign secretary was a party man who believed that loyalty should triumph in any battles of conscience he might be having. Straw later told the Iraq Inquiry if he had resigned then he could have stopped the war.[40]

Neither was Clare Short prepared to jeopardise the government. In a decision that would ultimately damage her reputation, Blair convinced her not to resign with promises of progress on the Middle East peace process and that the UN would take charge of post-conflict reconstruction, although neither were in his power to deliver.[41] Short was mollified but still rebellious. At a 17 March Cabinet meeting, at which Peter Goldsmith presented a single-page statement confirming the war's legality, Short punctured the sense of relief among ministers. 'That's extraordinary,' she declared sarcastically. 'Why is [this judgment] so late? Did you change your mind?'

Given Blix's favourable report on Saddam Hussein's compliance with inspections, Goldsmith had actually needed an assurance from Blair himself that Iraq had failed to disarm in order to reach his verdict, an extraordinary case of the prime minister acting as judge and jury.[42] Short was eventually shouted down by other ministers. 'Everything was very fraught by then and they didn't want me arguing,' she recalled.[43] In the end only Robin Cook, the former foreign secretary, embraced the chill of life outside government, announcing his resignation in protest against the war the same day, along with Jack Straw's deputy legal adviser Elizabeth Wilmhurst.

On 18 March Tim Cross arrived in London to report on the shambolic state of US post-war planning. Jay Garner had finally had enough of the 'Beltway bandits', as he called Washington politics, and in mid-March decided to deploy his headquarters to Kuwait. His thinking on Iraq was still rudimentary, consisting of a three-pronged approach focusing on humanitarian response, re-establishing Iraq's ministries and forming a new government.

London itself was nervous ahead of another 'Stop the War' rally. Cross found few in the capital ready to engage with his concerns. At the British military's command centre in the leafy London suburb of Northwood officers were immersed in the invasion plan. In Whitehall, civil servants still thought war could be averted, despite the prospect of a second UN resolution disappearing the previous week.[44] Cross managed to get a meeting at Number 10 with Alastair Campbell in a last-ditch effort to gain attention for his reports. Campbell was starting to show the strain of the past few weeks. The communications director had signed off on the 'dodgy dossier' of claims about Saddam Hussein's weapons programme, opening the government to the charge that they were manipulating the case for war. His usual brand of dark humour verged on the gallows.[45]

Campbell's eyes glazed over as Cross ran through Garner's sketchy plan for establishing stability in Iraq after the invasion. Then Cross hit upon a theme that got the spin doctor excited: the fact that Garner's press team consisted of just one naval reservist. 'He thinks that he can just issue a press release once a day; even I know he'll never get away with it,' Cross told Campbell.

'Wait right there,' he replied. This was just the sort of basic mistake that Campbell had seen in Kosovo. One of his finer achievements was setting up a media centre in Brussels, modelled on New Labour's election war rooms which could respond to the demands of the twenty-four-hour news cycle. He hurried out of the room to return a few minutes later with Blair.

The prime minister was obviously running on adrenaline, noted Cross. His shirtsleeves were rolled up and he leaned forward as Cross spoke, listening intently. His government would fall if he could not get enough votes from the parliamentary Labour Party. But Blair also stood at the zenith of his power, the world statesman who had put Britain at the heart of the American war on terror and who was now marching boldly with his allies to confront the enemies of Western liberalism.

'This is definitely something we can do,' he said.[46] Blair also heard Cross's concerns about the lack of post-war planning but did not respond.

Campbell had just the man to transform Garner's approach to the media: a young diplomat and Arabist called Charles Heatley, who had caught Campbell's eye on secondment to the Cabinet Office. In his early thirties, roguishly good-looking with thick black hair and a neat goatee, Heatley was to play a remarkable role in Iraq's future. But for now he was just one of a few British diplomats dispatched to Kuwait to prepare for the aftermath.

Cross left the meeting feeling pensive. The next day he flew to Kuwait, where Garner was orchestrating his team from a luxurious Hilton chalet overlooking the tepid waters of the Persian Gulf. Back in the field, Garner reverted to being the excellent operational commander he was, marshalling tents and supplies for the humanitarian disaster that the UN had warned him about. But Cross was far from reassured. Once the dark and unpredictable forces of war had been unleashed, he doubted whether Garner's meagre resources would cope. A few days later Britain plunged into battle.

Chapter 2

BASRA MON AMOUR

When Haider Samad heard the first bombs falling, he scrambled onto his rooftop to watch the night sky flaring with explosions. From listening to the BBC World Service on his radio he knew what was happening. The nights in late spring still contained a chill, but he lay on the roof, too excited to sleep, until the soft heat of the early morning sent him downstairs. The British army was coming. When he tried to picture where the soldiers came from he could see nothing beyond the unpaved alley outside his house and the piles of garbage where the black crows fed.[1]

For most of his adult life Haider had been under house arrest. In 1992 the Ba'ath Party had stripped him and his family of their identification papers after his father, Muhammad, a prominent lawyer, was arrested for allegedly contacting exile groups during the First Gulf War. Haider never knew the truth of the charges. He was only fourteen years old at the time, at an age when his father still loomed over his life. Muhammad was a flowery talker who liked to evoke the city he had known as a young man.

In the late 1950s Basra was swinging. While the capital Baghdad drew the modernists and malcontents, Basra offered an easy blend of portside bars and love poetry. Nightclubs along Al-Watan Street, like Matilda's Bar, drew a steady brotherhood of drinkers who would load up on gin and tonics, a vestige of British colonial rule, and stagger off to old Basra, where Madam Hasiba ran a brisk trade in her brothels.[2] Or Muhammad might wander down the street that followed the meandering creek connecting the city's canals to the corniche and the dark expanse of the Shatt al-Arab. There was usually some poet ready to recite a snatch of doggerel for a few dinars. On the opposite bank lay the dense palm groves where Haider's grandfather had worked as a farmer.

In the 1980s the war against Iran took its toll on the city. Only five miles from the border, Basra was hit by nightly artillery salvos. Rubble choked the canals, forming stinking ponds. The date palms were replaced by trenches and shallow graves. At first the city preserved its edgy vitality, but one effect of the war was to strengthen the Ba'ath Party's grip. The nightclubs closed down, and more arrests followed soon after.

The last time Haider saw his father, two secret police officers were dragging him through the front door as his mother clung desperately to his shirtsleeve. The family was given an hour to pack up its possessions and the house was seized. Haider spent the next ten years in a slum on the northern fringes of the city, which had swollen in size after the war and Saddam Hussein's brutal reprisals against the Shia population. He grew into a troubled young man, thin and cerebral, inheriting his mother's delicate features, with an aquiline nose and dark almond eyes that held the world at a distance. His home was a single-storey concrete building facing a street with an open sewer and a line of houses either half built or falling down. Haider completed high school and enrolled in medical college, only to be kicked out when the dean discovered his missing paperwork and family history. The police subsequently arrested Haider for army desertion, a false charge as his college place had exempted him. A year in prison, confined in a small windowless room, manacled to other prisoners, broke his health. After his release he rarely left his house.

Haider did have one dalliance. He had sneaked out of the house to attend a family wedding. The raucous music and bright lights were overwhelming. He stood to one side and found himself watching the partygoers with Nora al-Sadoon, a shy and demure girl several years younger than him, who shared his outsider's view of the world. They poked fun at the other wedding guests before exchanging telephone numbers at the end of the evening – a risky business in Iraq, where any contact between a man and woman outside the family can condemn a woman as a whore. But after a few phone calls the relationship petered out. Nora's father was a Sunni who had a good job at the electrical plant. He would never consent to his daughter marrying a suspected enemy of the regime.

Haider lost contact with Nora, and the brief friendship only seemed to heighten the emptiness of his life. He was not religious, but in his darkest moments he found an inner conviction that matched the radical sermons he heard via the loudspeakers at his local mosque every Friday. In the 1990s Ayatollah Muhammad Sadeq al-Sadr transformed the Shia faith in Iraq, rejecting the quietism of an earlier generation in favour of religious fervour and political activism, which he expounded at his weekly sermons. After failing to control him, Saddam Hussein had Sadr and two of his sons assassinated in 1999, but the movement lived on in a network of preachers.

Haider listened to their calls for the blood of martyrs and the rise of a new order, but mostly he watched from his rooftop as the world drifted by unchanging: the family down the street heading off to school, all eight of the children in a line behind the mother's black *abaya*; the local Ba'ath Party informer smoking cigarettes in the shade of the doorway opposite; the smell of cumin and dried lemons drifting up from the kitchen. Haider longed for a return to his father's world.

The day British tanks entered the city, he stayed at home as the other young men streamed downtown. He heard later about the looting from his brother, who had been in the crowd that stormed the Ba'ath Party headquarters, ripping out office equipment and burning files. Haider had seen the smoke from his rooftop, barely able to suppress his excitement. The following morning he dressed in his best brown velour tracksuit. In his pocket he had a small piece of paper on which he had scribbled the British national anthem and a rapturous poem on freedom, which he had styled after one of his father's favourite love songs. 'Democracy is my love, a subtle balm for all my wounds, let me embrace you.' Suitably armed, he set off cautiously for the nearest British base.

Chapter 3

OCCUPATION

As he navigated his Land Rover through the crowds in Basra, Major Chris Parker observed with foreboding the gangs of Iraqis tearing the city apart. For the past six weeks he had been camped out in the desert with brigade headquarters, living on a few hours' sleep and bad coffee, and rarely straying from the maps and chatter of the radios in the command tent. At thirty-two Parker had already racked up seven tours in almost as many years. His last mission to Kosovo had given him experience of rebuilding a society shattered by war, but that paled in comparison to his latest job. Parker was driving through the wreckage of a city of 1.8 million that he was now charged with running.

By the time Parker arrived in Iraq, he had been the 7th Armoured Brigade's chief of staff for two years.[1] Only officers destined for high command got such a job, and Parker acted the part. He was tall and handsome, with a square jaw and green eyes that rarely betrayed his thoughts. He had originally joined the RAF at eighteen and started a degree in aeronautical engineering, only to be told he was 'crap at flying'. Some young men might not have recovered from that knock, but Parker did not lack confidence. He ditched the degree and signed up with the Royal Hampshire Regiment, where he had found his calling not among machines but as a commander of men.[2]

His colleagues complained about his intensity, but it was Parker's job to bully and batter his way through the hourly situation reports, on the lookout for backsliding. The staff at brigade headquarters had worked together for a year running the NATO mission in Kosovo, but they knew Iraq was going to be a tour like no other. Parker found it easy to tune out the media circus and UK anti-war protests. After nine months of training the only sensation he felt when the bombs started falling was a deep thrill at the prospect of combat. Colonel Tim Collins, one of the brigade's battle-group commanders, captured the mood on the eve of battle when he addressed his men: 'Those who do not wish to go on that journey, we will not send. As for the others, I expect you to rock their [the Iraqis'] world. Wipe them out if that is what they choose. But if you are ferocious in battle remember to be magnanimous in victory.'

Under the invasion plan, British forces were to hold the Americans' south-eastern flank as they charged for Baghdad. After the initial advance by US Marines on 22 March, Parker's brigade took over American positions around Basra as the night sky above the city flashed with explosions and the distant rumble of gunfire. The city's western suburbs were bordered by the Shatt al-Basra, a man-made canal spanned by five bridges. Cautious about entering Basra, the British launched nightly raids into the city from the bridges to weaken the defenders. Parker's commanding officer, Brigadier Graham Binns, let him manage some of the operations. An old-fashioned commander, Binns lacked the taste for politics that marked out those destined for the top jobs at the MOD. Binns knew his limitations, as any lad from a Hull grammar school might among the double-barrelled surnames at Sandhurst. He did not approve of fraternisation between officers and other ranks, but beneath the bluff exterior he was a warm supporter of his men.

On 5 April, in the early evening, the British Secret Intelligence Service informed Parker that they had located Saddam Hussein's southern commander, Ali al-Majid, in a villa on the eastern bank of the Shatt al-Arab. Known as Chemical Ali, Majid was responsible for gassing the Kurds at Halabja and crushing the Shia uprising in 1992. He was high on the US most-wanted list of regime insiders and an American airstrike was hastily called in. A spy plane observed bodies being carried from the remains of the building and a convoy of white trucks speeding up the road to Baghdad.[3] Parker woke up Binns, who immediately grasped the significance of the news: the city's Ba'athist leadership had fled. The Scots Dragoon Guards were planning a raid into the city that night. Binns asked the SDG company commander, Captain James Fenmore, to hold their ground after the raid instead of withdrawing. When Fenmore announced he had reached downtown with little resistance, Binns ordered in the brigade. Basra had fallen.

The next morning Binns rode into Basra with the commander of British land forces in Iraq, Major General Robin Brims. A cloud of dust hung over the city, giving the already tan-coloured concrete of the slums on the outskirts a ghostly air. As the officer commanding 1st Armoured Division, Brims had overseen the campaign's strategy, leaving much of the operational heavy lifting to his subordinates. As they neared the city centre, the officers' convoy was surrounded by Iraqis – young men in jeans and cheap nylon shirts, older women in ballooning black tents that left only their faces visible.

Binns studied their faces for a sign of the elation he felt. He had anticipated losing sixty British soldiers taking Basra; instead there had been just three fatalities.[4] Some of the Iraqis made V signs, or gave the thumbs up; others ignored the soldiers and focused on ransacking government buildings. British forces made some attempt to control the crowds, but efforts were isolated, and the crowds of looters too great. The British defence minister, Geoff Hoon, claimed they were 'redistributing that wealth among the Iraqi people' and flippantly told Parliament that he regarded such behaviour as 'good practice'. The realisation was slow to dawn that the hospitals, schools and government buildings being ransacked would shortly have to be rebuilt with British aid, and that lives hinged on the result.

On the outskirts of the old city Binns called a stop and clambered onto his tank with Brims to soak up the atmosphere.[5] Most officers develop a taste for military history as they serve, and both officers were familiar with the last time the British had taken Basra. During the First World War a British-led force had captured Basra from the Turks after the latter sided with Germany. The British aim was initially to protect the oil fields of southern Iran, but the ease of Basra's fall and the subsequent push up the Shatt al-Arab to Qurnah meant the possibility of taking Baghdad and opening a new front in the war. The British advance ended in disaster when the commanding officer Major General Charles Townshend and his men were besieged a hundred miles south of the capital in the city of Kut, their advance exposed as ill judged and undermanned.[6]

British plans in 2003 were more modest. UK forces would be rapidly reduced in Iraq from 40,000 to a brigade-size force of several thousand. The MOD was acutely aware of the cost of deploying a division.[7] During the invasion the Treasury picked up the £1 billion bill, but if the British occupation dragged on the MOD risked having to cover further costs from its operational budget. The military did not intend to pick up the tab for reconstructing the country. Drawing down troop numbers, however, raised the awkward question of what came next. Senior generals like Brims had been asking since January, only to be told by London, 'Don't worry; there is a plan.'[8] Realising he was being fobbed off, Brims eventually drew up a two-page brief on what his officers might expect after the fighting. Basically, they were going to have to make it up as they went along.[9]

That was how Major Parker ended up with the job of running the city. Brigade Commmander Binns wanted to focus on establishing a local council, leaving his headquarters to work out what to do. The day after Basra's 'liberation', Parker drove into the city to set up his headquarters in Saddam's palace. After the poverty outside, the place looked like Disneyland. The brigade had selected the largest of the villas for its base, reached via a bridge over a short waterway. The front of the building had a high portico held up by columns that split a wide staircase leading into the marbled interior. Royal Marines, preparing to go home, were busily looting the place, carrying out sofas and beds. Parker gave them an earful, but they carried on regardless.

Parker's mind was already racing to the challenges ahead. Lieutenant Colonel Nick Ashmore, a resourceful Royal Horse Artillery commander, had created 'The Basra Province Roadmap' – a PowerPoint presentation that broke down British responsibilities into key areas with benchmarks to measure progress to the end state: 'stable, self-governing Iraq competing in international trade. No WMDs.' It amounted to the first detailed plan for the British occupation.

At the evening situation report Parker ran through the presentation with his fifteen-man team. Establishing law and order was clearly the priority, but he did not have enough troops to police the city. All his men could do was fire a few warning shots, and pack the worst of the looters off to a recently built British detention facility outside the city. They needed to get the police back onto the street and start work on restoring basic services. The brigade's medical officer, James Pattison, would be in charge of running five hospitals and nineteen healthcare clinics. The paymaster, Major Ian Jaggard-Hawkins, was to run the multi-million-dollar banking system, and the education officer, Major Brian Elliot, who usually helped squaddies get a couple of GCSEs, was to run the school system.

'It's just like running Portsmouth,' Parker told them. 'We can do this.'

Meanwhile, Binns was making progress in creating the local council. The SIS had a shortlist of candidates drawn from its pre-war spy network in Iraq. As his forebears had discovered after expelling the city's Turkish administrators in 1916, there were few capable leaders to take their place. Back then the British had turned to tribal leaders, who appeared to embody the aristocratic ideals of many of the officer class.[10] This time round they would settle on Muzahim al-Tamimi, a sheikh from a largely Sunni town outside Basra with a taste for Ray-Bans and Johnnie Walker's, and well-known to the SIS. When Binns met Tamimi he judged the man to have enough legitimacy and a strong enough personality to keep the religious leaders included on the council in check.[11] Few gave much thought to the potential consequences of putting a well-known Sunni in charge of a Shia city like Basra, and no one had adequate contacts to gauge local views on Tamimi.

On 11 April Binns convened the new council at the palace. Tamimi and the other sheikhs and imams sat around a conference table; British officers lined the walls. Among them stood Haider Samad, the young Iraqi with high expectations of the occupation. A squadron commander, Major Matthew Botsford, had recently hired him to work as an interpreter in Iraq. Haider recognised Tamimi at once. He was notorious for being on Saddam Hussein's payroll, and was known locally as 'Saddam's Sheikh'. Why would the British select him? Haider wondered. But he wasn't going to openly question his new masters.

'We've come not as an invading army, but with the offer of friendship and the promise of democracy,' Binns declared to a smattering of applause and inscrutable looks from the Iraqis. Undeterred, Binns ushered the council onto the steps outside for a photograph in the amber hues of the setting sun. He allowed himself a moment's satisfaction. Baghdad had fallen on 9 April. The looting had dwarfed anything seen in Basra. In contrast, a modicum of calm was returning to the port city.

The contentment did not last long. Haider was not the only Iraqi to take note of Tamimi's appointment. The day after Binns announced the members of the council, several hundred protesters gathered outside the palace waving placards denouncing Tamimi as a Ba'athist and demanding elections. 'Let us choose our leaders,' one placard pointedly declared. Trouble was also flaring at the British-run detention facility. Several truckloads of prisoners a day were arriving at the compound, a walled enclosure covering half a square mile of desert. The task of building the prison had fallen to Lieutenant Colonel Gil Baldwin of the Queen's Dragoon Guards. Division had originally requested a specialised battalion to deal with detainees, but in the rush to deploy there hadn't been time to prepare such a force. Baldwin was to build a temporary prison until the Americans could sort out a longer-term facility.

A forceful, ambitious officer, Baldwin knew that rear-echelon work wasn't exactly the field to make one's reputation. He had worked with the International Committee of the Red Cross in the Balkans on issues like transitional justice and was one of the few officers in the British army who knew how to apply the Geneva Conventions on human rights to war zones. From what he was seeing at the detention facility, he knew that Britain was coming dangerously close to breaking them.[12]

Baldwin did not have enough men to process the detainees as they arrived. Deserting soldiers, suspected Iraqi militia, looters and anyone else unlucky enough to be arrested were dumped at his gate throughout the day and night, some of whom had been hooded and manacled for hours. Baldwin knew that hooding was a controversial practice. In 1973 the Heath government had banned its use on detainees at the height of the troubles in Northern Ireland, along with stress positions, starvation, noise and sleep deprivation. How hooding became prevalent in Iraq can be traced back to a legal ruling given by Attorney General Peter Goldsmith before the war. In early 2003, the question of what to do with potential detainees had provoked a debate within government over whether the 1998 Human Rights Act, which brought UK law into line with the European Convention on Human Rights, should apply to British forces overseas. In consultation with Christopher Greenwood, a Queen's Counsel and professor at the London School of Economics, Goldsmith decided that a war like Iraq should have its own 'special law' and that British forces were not bound by either piece of legislation.[13]

The army's legal team was relieved that there would be no last minute need to translate the onerous human rights laws into standard operation procedures. Instead, British forces would rely on the Geneva Conventions, as they had in the past. They too banned hooding but there was an understanding in the conventional military and among lawyers like Greenwood that under the Geneva Conventions hooding could be used in extreme circumstances for brief periods when sensitive locations might be exposed.

Within the Special Forces and intelligence community, however, this subtle legal distinction had ceased to matter, if it had ever applied. Hidden from public scrutiny at the Defence Intelligence and Security Centre at Chicksands, Bedfordshire, military interrogators were taught to strip their subjects naked, put them in stress positions, and scream abuse and threats at them until they were cowed into submission.[14]

How the MOD could promote such abuse, so clearly in violation of national and international law, is unclear. One factor may have been pressure from the US. Since 9/11 there had been growing readiness to use 'enhanced interrogation techniques' such as water-boarding in the name of extracting information from suspected al-Qa'eda members. Such abusive behaviour was further legitimised by Washington's belief that many of those detained should not be treated as prisoners of war, and therefore not subject to the basic human rights guaranteed by the Geneva Conventions. President Bush contributed to the malaise with the aggressive, dehumanising rhetoric of the 'war on terror', often aped by British and American soldiers. With military interrogators in Iraq under pressure to deliver evidence of Saddam Hussein's weapons of mass destruction that would justify the war, and no clear plan for how to deal with detainees, the situation was ripe for widespread abuse.

The British government still had ample warning. The army's top lawyer in Iraq, Lieutenant Colonel Nicholas Mercer, felt the military should be applying the highest possible standards to Iraqi detainees. Since January 2003 he had been pushing for the army to use the European Convention on Human Rights in Iraq. On 24 March, as the prospect of a British occupation loomed, he raised the matter once again with the British military headquarters in Northwood but received a terse response from its legal adviser, Rachel Quick. She argued the convention did not apply during the war, or subsequent occupation. 'If the [attorney general] and Prof Greenwood are wrong on this advice, perhaps you could put yourself up to be the next attorney general!' she exclaimed.[15]

A dapper lawyer in his early forties, Mercer had pale blue eyes and a bookish air. Before Iraq, Mercer had approached his work assuming that the British army always stood on the right side of the law and that his job was merely to provide the legal backdrop, a virtuous circle that appealed to his scruples. He was not going to be fobbed off by Quick. An alarming report from Baldwin had alerted him to conditions at the detainment facility, which was already holding over 3,000 prisoners after just a week. To alleviate the overcrowding Baldwin had created a military tribunal to make quick decisions on who should be held at the camp or released. Mercer feared the tribunals might be illegal under international law.

On 27 March Mercer visited Baldwin's barbed-wire enclosure. He was shocked to discover forty detainees at the main gate, hooded, hands cuffed behind their backs and kneeling in the hot sun. Inside, the mass of prisoners looked like abused cattle, herded together in giant pens, with no food or water in sight and little shelter.

After witnessing one tribunal, which confirmed his worst fears about the lack of due diligence, Mercer confronted Baldwin in his grimy office. 'This facility is a moral disgrace,' he yelled.

Baldwin could not have agreed more, but was not going to be held responsible for the mess. 'Why doesn't headquarters get off its arse and do something about it,' he erupted. His tribunals were the only way to work through the backlog of cases at the camp.

Mercer shot back that they lacked a jury or judge, and this was just the sort of summary justice Saddam Hussein would have favoured. The only legal thing to do, Mercer said, was bring over a British judge to try cases, following a model the Australians had used in East Timor and described by the UN as in 'accordance with the highest human rights standards'.

Baldwin listened incredulously to Mercer's plans. What he needed were quick solutions. The lawyer had only caught a glimpse of the problems at the camp. Worse still was the behaviour of a team from the Joint Service Interrogation Team, an SIS offshoot trained at Chicksands that was manned by special forces who reported directly back to Northwood. They used a pen near the main holding facility where hooded detainees were held for hours in stress positions. The prisoners that Mercer had seen at the main gate were also waiting to be questioned by the interrogation unit. Such treatment of detainees, often in view of other prisoners, was clearly agitating those within the camp. Baldwin had already asked the interrogators to stop, only to be told they were following Intelligence Corps doctrine. When he pushed further, he was told to mind his own business.[16]

Baldwin and Mercer sparred for several more minutes about military tribunals before the lawyer left to raise his own concerns with the special forces' interrogators about the prisoners he had seen outside the main gate. He was also brushed off. Returning to headquarters Mercer told Major General Brims that the use of hooding and stress positions were in breach of international law. Yet his proposals to bring out a British judge and apply the European Convention on Human Rights to the handling of detainees did not get far in London. He was informed that his plans might be fine for 'individuals locked up on a Saturday night in Brixton', but not appropriate for a 'bit of looting' in Iraq.

At the detainment facility Baldwin was now facing a full-scale riot, as angry Iraqis lobbed rocks, shoes and whatever else lay to hand at the British soldiers guarding them. It was clear to Baldwin that things had to change. He picked up the phone and called some of his contacts at the International Committee of the Red Cross to tip them off about the disastrous state of the camp. Baldwin was aware that by doing so he would be making serious enemies at headquarters and probably jeopardising his career, but he knew he was right. The ICRC, it turned out, had already recorded allegations of abuse from detainees they had interviewed, and had raised their own concerns with the British military.

Two days later, on 29 March, an ICRC delegation met Major General Brims and Mercer in Baldwin's office. The ICRC had issued a formal complaint and the potential for a public relations disaster was clear. Brims asserted that the British were looking after the prisoners adequately. When Mercer tried to speak he was told by a senior political adviser, who had flown out from London to supervise the meeting, to keep his mouth shut. 'I was so appalled by the attempts of the United Kingdom to justify the conduct of the UK that I walked out,' recalled Mercer.[17] He tracked down a member of the ICRC team after the meeting and listed the problems he had seen.

Back at headquarters, Mercer pushed the general to take action. A rattled intelligence officer showed Mercer a document that he claimed said that hooding was mentioned in army guidelines. However, on closer examination there was no such reference.[18]

Brims still appeared reluctant to take action, until his media adviser Colonel Chris Vernon raised the dangers of adverse press coverage.[19] That evening Brims' chief of staff issued a verbal order banning hooding, although it would be months before the British military headquarters in Northwood confirmed the decision. Meanwhile riots continued to flare at the facility. Mercer was relieved to hand over custodianship of the facility to the Americans a few weeks later, but the issue of the treatment of Iraqi prisoners wasn't going away.[20]

Chapter 4

THE EMERALD CITY

FOR THE FIRST TEN DAYS AFTER BAGHDAD'S FALL, Major General Tim Cross was stuck in the five-star Hilton Hotel in Kuwait. Garner's plans for post-war Iraq were rapidly coming undone: the humanitarian crisis he had prepared for had not occurred, and as for setting up Iraqi ministries quickly, they were watching on CNN as looters tore them down. Garner was anxious to get to Baghdad, but the retired general was being fobbed off by the military, a reminder, if one was needed, of the low esteem with which the Pentagon viewed their efforts.

They eventually got a flight on a blacked-out C-130 transport plane in the early hours of 19 April. Arriving at Saddam Hussein's Republican Palace in Baghdad with Garner and the others, Cross glimpsed against the night sky one of the four giant busts of the dictator that crowned the palace's corners. In the gloomy interior they searched for military cots where they could sleep. It wasn't the confident entrance Cross had wanted for his team, which had been named the Office of Reconstruction and Humanitarian Aid. The country needed someone to fill the power vaccuum and Cross urged Garner to think of himself as the 'viceroy of Iraq'. But Garner refused. 'The Iraqis are going to be running this place soon,' he insisted.

The next day the two men toured Baghdad. Garner sat beside Cross in the back of their Toyota Land Cruiser, squinting out of the windows and occasionally taking off his baseball cap to mop the sweat from his brow. A milky haze hung over the city, broken by the blackened hulks of ministry buildings. There were still hundreds of looters on the streets, and both men carried revolvers. They spun around Al-Tahrir Square, where

Saddam's statue had stood, before passing the grimy office blocks lining Al-Sadoon Street.

Baghdad had once led the Arab world's embrace of modernity. In 1958, the year the British-backed monarchy was overthrown, famed American architect Frank Lloyd Wright visited Iraq and submitted a sweeping modernist plan for the capital. His blueprints did not survive the king, but the impulse to tear down the old appealed to the succession of centralising and increasingly totalitarian military leaders which culminated in Saddam Hussein. Baghdad had the air of a half-abandoned Soviet experiment, a city of tower blocks riddled with bullets, abandoned ceremonial parks and a hinterland of hulking factories all coloured the egalitarian grey of the desert.

At the palace Garner did his best to motivate his staff, grown to over a hundred. ORHA's plan was to attach a small team of Western advisers to each of Iraq's ministries, but with violence still flaring, Garner deemed it too dangerous to send out his men without a military escort, and those were difficult to get. With concrete barricades hastily erected around the palace and surrounding villas, an area dubbed the Green Zone, Garner's advisers had little idea what was happening in the rest of Baghdad. They spent their days in Saddam Hussein's old ballroom, sitting at plywood desks hastily built by engineers, trying to send emails on slow internet connections or standing outside in the blistering heat to use satellite telephones that required a clear view of the sky to work. With no air conditioning or running water, the place had a warm and fetid smell, like a men's locker room, and an air of quiet desperation.

One of the few officials who refused to be confined to the Green Zone was the young spin doctor Charles Heatley, who was now ORHA's 'strategic media adviser'. After flying in from Kuwait with Cross and Garner, Heatley found the claustrophobia of the palace self-defeating, and had arranged to have a BMW motorbike he had owned in Jordan delivered. He spent his evenings speeding through the streets of Baghdad, his head wrapped in an Arabic *shemah*. His destination most evenings was the Hamra Hotel, a journalist hang-out with kitsch 1970s style apartments and a frenetic party scene beside the pool. There he revelled in challenging the gloomy prognosis of correspondents, who spent their days on the city streets talking to Iraqis.

'Is it just me or are they fucking up?' said Peter Foster, a *Daily Telegraph* correspondent who had gone to Oxford University with Heatley.[1]

'Come on, Peter, you don't expect me to answer that,' he replied.

Heatley would return at first light, when the air was still cool, and the Tigris looked like molten lead. Only a few Iraqis were on the streets at that time. Some stared, but for the most part Heatley felt like he was flying over the city as it stirred. Crossing over the July 14th Bridge he could see American tanks guarding the Assassins' Gate, a massive sandstone entrance to the palace complex. After flashing his diplomatic credentials at the American guards, he drove past the parking lot filled with SUVs and trucks, and LZ Washington, the helicopter pad, already buzzing with flights. He was at the heart of the American war machine and liked the feeling of power.[2]

Heatley was not surprised to learn at the end of April that Washington was replacing Garner with a proper viceroy. Recognition was finally dawning on the White House that events on the ground were spiralling out of control. Although Garner had tried to warn of the dangers of a power vacuum, he now became the scapegoat. The Pentagon was scrambling for an alternative. Its initial plan had been to put in charge Iraqi exile leader Ahmed Chalabi, who had not been inside the country for years, but Iraqis were making clear at private meetings in the Green Zone and in demonstrations outside the palace that they viewed Chalabi as an outsider and CIA stooge. US Defense Secretary Donald Rumsfeld, recognising the need to establish some legitimacy for whoever took over, was won round to the argument that they needed to 'slow things down' and bring more Iraqis into the discussion.[3]

In London Number 10 hoped that the US would now consider a United Nations lead on reconstructing Iraq. At a 7 April meeting at Hillsborough Castle Tony Blair had coaxed Bush into suggesting the United Nations would play a 'vital' role in Iraq, but the illusion did not last long. The UN was still reluctant to get involved and Rumsfeld had no desire to surrender control. He wanted an American-run administration and had just the man to do it: Paul Bremer, a former ambassador to Holland who had left the State Department twelve years before to work for the former secretary of state Henry Kissinger's strategic consulting firm and then a crisis management firm. Rumsfeld had met him in the 1970s, when Bremer had worked as a special assistant to Kissinger and Rumsfeld was then in his first term as secretary of defense. Bremer had little experience of Iraq or the Middle

East, but he was considered an anti-terrorism expert, whose Christian faith and neoconservative views on exporting democracy made him popular with the right-wing crowd in the White House.[4]

To London Bremer's appointment was a welcome sign that the US was taking charge of the situation. Bremer would head an American-run administration called the Coalition Provisional Authority, which would replace Garner's Office of Reconstruction and Humanitarian Aid. No timetable for handing over control to Iraqis was announced. At the same time the UN Security Council was prepared to grant the occupation an air of legitimacy by passing a resolution giving Britain and America joint custody of Iraq.[5] It was a far cry from the British plans for UN involvement of just a few weeks before, but London viewed any agreement between the US and the UN as major progress.

The scale of the destruction in Baghdad was also beginning to dawn on Number 10. The FCO's political director John Sawers, a dapper former spy, had arrived in the Iraqi capital as Blair's personal envoy. He delivered a withering assessment of the 'unbelievable mess' in Garner's administration.[6] Bush reassured Blair over the telephone that Bremer would get 'on top of things', little realising how radical the new US administrator was to prove.[7]

Five days before his departure from Washington, Bremer attended a briefing at the Pentagon given by Douglas Feith, a fervent neoconservative who ran the influential Office of Special Plans. He advocated cleansing the country of all links to Saddam's regime. The US commander Tommy Franks had already outlawed the Ba'ath Party. Feith wanted to go further and disband the Iraqi army and purge the government of those with mid- and high-level Ba'ath Party membership, disregarding the reality that Saddam had forced most government officials to join his party.[8] Feith's plans were supported by exiled Iraqis, whose hatred of the former regime had grown over the years and who were themselves jockeying for government jobs.

Feith's views struck a note with Bremer, who saw post-war Germany and its denazification as a model for how to deal with Iraq. After the drift under Garner, he felt that Iraq needed a bold statement of intent to galvanise the reconstruction process. On 16 May, shortly after his arrival, Bremer announced the 'de-Ba'athification' of Iraqi society. A week later he issued his next order, disbanding the army. Tim Cross attended both news conferences. His worst fears were coming true. Not only was Bremer's team back-briefing that Garner had been a failure, but the new laws threatened to alienate half the country. The army was the country's second largest employer, and disbanding it put thousands of angry young men onto the streets, while the proposal to sack the top four cadres of the Ba'ath Party would strip ministries of their most able employees.[9]

Cross sent a furious memo to London, where David Manning, the prime minister's foreign policy adviser, was raising the first doubts about American strategy within Blair's inner circle.[10] A slight and softly spoken man in his mid-fifties, Manning had the air of a parson with none of the sanctimony. Having been plucked from the ambassador's job in Israel to serve in Number 10, Manning played an important role in bringing clarity to complex international issues after 9/11. Manning supported the war, and although he had been shocked by the post-war chaos, had not doubted that the US would 'come good'.

Bremer's decisions looked like grave blunders. Blair, however, had not been privy to the decision-making process and concluded there was little he could do to reverse them. When he and President Bush next spoke on the phone the subject did not arise. Blair's attitude was, 'So if this is what the Americans want to do, how can we work with it?'[11] He told Peter Mandelson at the time that post-war Iraq was 'chiefly America's responsibility, not ours'.[12]

As Manning glimpsed the scale of the unfolding disaster, Cross was preparing to leave Iraq at the end of June a disillusioned man. His services were clearly not required at the CPA, and he had been offered a job briefing the UN, to build support for the effort in Iraq. On his final day he scheduled a round of farewell media interviews. The first was with Foster, the *Daily Telegraph* correspondent. All the disappointment and frustration of the past few weeks welled up as he laid into the US effort, calling it 'chaotic' and a 'shambles'. It was a harsh but accurate portrayal.

'I'd never been on the losing side before, and I didn't like how it felt,' he recalled. The story made the front page the next morning. Charles Heatley got an earful from Alastair Campbell, but the Americans had the final say. A few days later Cross was informed by Northwood that his services would not be required at the UN. The US Department of Defense had vetoed his appointment.[13]

Chapter 5

BREADBASKET

Major Chris Parker, the British officer in charge of running Basra, believed he had found an answer to the security problem. In late April he had met Rear Admiral Muhammad Jawad, commandant of Basra's naval academy. Jawad's sailors were the latest government employees to picket the palace demanding their salaries. Teachers, engineers, doctors and university professors had all gone without for over a month and were getting desperate. Their heated protests presented Parker's fledgling administration with the most serious challenge to law and order since the looting began. The brigade's paymaster, Major Ian Jaggard-Hawkins, had recovered $30 million (£19 million) from Basra's central bank, which he intended to use to pay 180,000 government employees. The difficulty was that most ministry records had been destroyed in the looting. Jaggard-Hawkins was working long shifts at a computer compiling names from whatever lists he could find.

Parker was expecting a tirade from Jawad when they met; instead he found a calm and articulate officer who had spent a year at Dartmouth Naval College and spoke perfect English. He had also brought a CD with him containing the names and salaries of his employees, which he had made before the invasion in anticipation of looting. Parker saw the opportunity at once. So far, he had struggled to find enough Iraqi police to man a station let alone patrol the city, and those who had reported for work were by and large the most corrupt. Why didn't Jawad organise his men to patrol Basra's docks and shipyards?

Parker's new brigade commander, Adrian Bradshaw, sent the idea up the chain of command, but received the answer that no police strategy would be available until next year.[1] A week later Parker took matters into his own hands, and put 188 of Jawad's men on the streets, each wearing a baseball cap emblazoned in English and Arabic with the words 'Basra River Force'. With a single turbine in Basra's power station working again, the water treatment plant flowing, and most of the city's schools and health clinics reopened, Parker felt the city was getting back on its feet.

Parker's tangible success contrasted with the still nebulous strategy for British forces in Iraq. By midsummer over half of the 40,000-strong invasion force had withdrawn. A force of 12,000, headquartered in Basra, would remain for at least the next six months. Some in the military wanted to get out quickly and minimise Britain's role in what was obviously going to be an expensive and long-term nation-building project. Others argued for expanding the British presence beyond southern Iraq to Baghdad, where the major decisions over the country's future would be made, and UK forces could have greatest effect.

One officer calling for an expanded mission was the SAS commander in Baghdad, Richard Williams. 'We don't want to create a British enclave in the south. It's not Basra that is important, it is Iraq,' he argued.[2] Williams proposed sending the Parachute Regiment to Baghdad to start an intensive police-training programme. He estimated they could put 5,000 Iraqi officers on the streets in a week, if the UK moved quickly.

Back in London, General Mike Jackson, the new head of the army, supported the proposal. A gravel-voiced bruiser with hooded eyes and a deeply creased face, Jackson was viewed by the media as the embodiment of British grit and determination. Jackson's more urbane and sophisticated boss, Michael Walker, took the opposite view. Three thousand 'nice smiling Paras' would not transform Baghdad overnight, Walker argued.[3] The focus had to be on getting troops out of the country. Keep the military in Basra, Walker decided, and make it a model for the Americans to follow.

As the British debated their role, the Americans were about to make it clear that they would act alone and call the shots. When reports of Parker's Basra River Force arrived in Baghdad, the US response was immediate and negative. Three days after setting up the unit, Parker got a call from US headquarters in Baghdad. He was accused of operating a unit that should have been disbanded with the rest of Saddam's military.

'You are in breach of CPA Directive Number 1,' the American officer informed Parker, referring to Bremer's decree disbanding the army.[4]

'What the hell are you talking about?' Parker shouted down the phone. The priority was securing the streets, not playing politics.

He ignored the call, but a few days later he got the order again, this time from British divisional headquarters. Bremer had apparently complained to Washington. The US administrator was increasingly irked by laudatory reports in the media about Britain's softly-softly approach, which was usually contrasted with the aggressive behaviour of the Americans. Bremer told his aides he thought the Brits were not taking a firm enough grip of the city, which should start with purging the military of Saddam loyalists. This time there was no dodging the order.[5]

'This is going to set back rebuilding weeks or months,' he warned his headquarters. Not only had his embryonic police force been scrapped, but Parker had lost credibility with an influential Iraqi leader and made enemies of his men. 'Did some of those guys who were sacked join the insurgency? I wouldn't be surprised,' Parker reflected after the sackings.

Meanwhile, the looting continued across Basra, as did the British strategy of rounding up suspected gangs and throwing them into overcrowded holding pens. Abuse of detainees at the main detention camp had stopped, partly as a result of the intervention of the army's top lawyer in the south, Nicholas Mercer, but there was still no oversight of the dozen smaller facilities found on British bases across the city, where detainees could be held for days before being transferred to the main camp. In late April 2003 military police investigators reported that two Iraqis had died in British custody before reaching the main prison, with 'five or six more deaths requiring investigation'.[6]

This had forced Rachel Quick, Northwood's top legal adviser, into a visit. Together with Mercer, she visited one British base, where they found fifty-year-old Faisal Sadoon held in 'appalling conditions' in a metal shipping container with the temperature over forty degrees. He was accused of involvement in the deaths of two British soldiers during the invasion in the town of Az-Zubayr. Mercer felt Sadoon's treatment only underscored the need for applying the European Convention on Human Rights for detention facilities across southern Iraq. However, Quick's response was to press for Sadoon's release, seemingly more concerned with avoiding the legal issues for the British rather than getting involved in individual prisoner's cases.[7]

If he could not get the British to change, Mercer could at least make some progress re-establishing Basra's justice system. He had recently hired the idealistic young Iraqi Haider Samad as his translator, and together they had refurbished a building opposite the Ba'athist headquarters to serve as the new law courts, which Mercer called the Palais de Justice, a nod to the Napoleonic origin of Iraqi law. But the prospect of an Iraqi judge trying a looter was still months away.[8]

As the lawyers and politicians debated how to handle – or hide – the grim consequences of the occupation, individual soldiers were largely left alone to deal with the realities on the ground. One of the most difficult areas in Basra to police was a place known as Camp Breadbasket, a large warehouse complex on the edge of town used to store humanitarian aid – baby milk, clothes and sacks of grain. The camp's sprawling boundaries made it easy for Iraqis to break in and steal supplies. Gangs of up to thirty men routinely appeared at daybreak to scale the walls. The looters made off with their acquisitions using pickup trucks and horses and carts.

The Royal Fusiliers guarding the camp had grown increasingly frustrated. Every time they handed looters over to the military police, they were promptly released. On 14 May 2003 the officer commanding the Fusiliers at Breadbasket, Major Dan Taylor, decided to teach the Iraqis a lesson: he would round up the looters and make them clean up the camp. This was illegal under the terms of the Geneva Conventions, which forbid the use of prisoners in work gangs or collective punishment, but Taylor only had a cursory knowledge of international law. The operation began at dawn. Taylor's company split into four- or six-man teams armed with rifles and the small metal poles used to support camouflage netting. One of the teams disturbed a group of looters, who fled, forcing the soldiers to give chase in Land Rovers. They eventually cornered twenty of them at the far edge of the camp. They were ordered to pick up their scattered loot and bring it back to the centre of the camp.

Ra'aid Ali was one of the Iraqis detained. He had actually been employed by the British to work in the camp but was accidentally caught up in the sweep. Back in the main warehouse, he was photographed for future identification purposes with the other looters, who were then split up into groups to tidy up the camp. Ali's protestations of innocence led to him being separated and secured in a concrete bunker, where he was punched so hard in the face that his nose was broken. Ali considered himself lucky. Through a small window he watched as Iraqis were stripped of their clothes and then beaten with aerials from Land Rovers. Others were made to run with milk cartons on their heads as soldiers laughed and made fun of them.[9]

Away from Ali's view, however, far worse was about to take place. Corporal Daniel Kenyon commanded one of the Iraqi work parties in a secluded section of Camp Breadbasket. At thirty-three years of age, Kenyon was a 'lifer', a man committed to seeing out the remaining seven years of his contract. He might make sergeant if he didn't get into too many scrapes. The younger men in his team, Lance Corporals Darren Larkin and Mark Cooley and Fusilier Gary Bartlam, looked up to him. With the exception of Bartlam they had all served with distinction during the fighting. During one battle fought near the town of Az-Zubayr Larkin had pulled two colleagues from the back of a burning Land Rover, saving their lives. Bartlam, eighteen years of age, had only just signed on and had been disappointed to miss the fighting.

The problems started when Bartlam took out his camera. Cooley, an aggressive twenty-five-year-old who Kenyon sometimes struggled to control, immediately rose to the occasion, and posed with his fist raised as if striking one of the looters. The others soon joined in, as a number of the photographs attest. They show the squaddies lording it over their prisoners, who lie prostrate on the floor in a pool of water or possibly urine.

Larkin, dressed in shorts and flip-flops, leapt onto the back of another prisoner and rode him triumphantly. Next he made two of the Iraqis take off all their clothes. Cooley then forced one of the men onto all fours, with the other behind him, and made them simulate sodomy. He also thrust one prisoner's mouth against the base of another man's penis. Bartlam snapped away. Finally Cooley dragged one man over to a piece of netting and used a forklift truck to hoist him up as the other soldiers stood by laughing. The Iraqis were released a few hours later.

The incident might have ended there as Kenyon and the others left Iraq a week later. Instead, at lunchtime on 28 May 2003 Bartlam walked into the Max Spielman photographic shop in Tamworth, Staffordshire and handed over his film to twenty-two-year-old Emma Blackie. He wanted them developed within the hour. Blackie was about to take a break so she left them with a colleague. When she got back from lunch her colleague immediately told her that there was something very wrong.

Blackie looked at the pictures and called the police. When Bartlam came back to the shop she told him there was a delay in the development of the photographs. Two police officers arrived shortly afterwards and arrested him.

Bartlam was later jailed for eighteen months, and the other soldiers were given sentences ranging from 140 days to two years and dismissed from the army. The photographs created a scandal when they were released during their trial. The MOD claimed the incident was the result of a few 'bad apples', and was not prepared to inquire into the fragile morale of young men exposed to a war in which few could find real meaning.[10]

At the time Major Parker was not aware of the crude events unfolding at Camp Breadbasket, but the pressure in Basra was growing since the Americans had scotched his Basra River Force. Whitehall wanted to hand control of Basra to the locals, but that was not as easy as it sounded in Westminster planning sessions. Under Saddam a small circle of Ba'athist officials had run Basra; there was no history of democratically elected councils, public representatives or local town-hall-style meetings. The British military's first effort to create a city council had come unstuck when the Iraqi sheikh who was to lead it was revealed to have been on Saddam's payroll. For the second attempt, Brigadier Bradshaw tried to play safe by appointing civil servants. They would not be writing laws, merely overseeing the city until public elections could be organised.

Yet Bradshaw's council was instantly rejected by a local religious organisation, Fadhillah, which had ties to the popular movement created by Ayatollah Muhammad Sadeq al-Sadr. Fadhillah was not about to let the British pick the government of Basra, and was prepared to use whatever means necessary to get its members into power. They rallied 3,000 protesters on the day of the announcement, declaring that Bradshaw was trying to install a 'puppet regime'.

The protests reinforced for Parker the lack of political guidance from the FCO and the failure to create a civilian-led reconstruction effort. So concerned was Parker that he had arranged to have a private talk with Tony Blair on the prime minister's first visit to Iraq. Blair was in defiant mood. No weapons of mass destruction had been found in the country, and the media had grown bolder in their criticism of the justification for war. Blair intended his visit to turn the media's attention back to 'our boys' and highlight the victorious invasion. It was just the sort of dramatic emotional moment on which Blair thrived.[11]

He was a little put out therefore to encounter a young army major who delivered a stark assessment of Britain's achievement so far in southern Iraq. Parker had already been irked by the mood of Blair's team: the prime minister tanned and relaxed, laughing at a Campbell joke. Once Blair sat down, Parker launched straight in.

'Where are the suits?' Parker demanded.

'What do you mean?' said Blair.

'Where are the people who know how to run a city – the electricity experts, the bankers, the water treatment managers? We've done a bloody good job, but we can only go so far. Where are they?'

With the hot weather approaching, Parker predicted riots if they didn't start providing basic services.

Although late in the day, Blair had been moving forward on this front. The UN resolution had finally been passed the week before, declaring the Anglo-American occupation legal. There was still no UN offer to play a role in rebuilding the country, but Blair hoped the resolution's passage marked a turning point. Of equal importance, Clare Short had finally resigned on 12 May. Perhaps realising her failure to resign before the war had caused irreparable damage to her reputation, she had gone on 'virtual strike'. During the invasion only one DFID representative had been deployed with the military, and since then Short had refused to deploy any more to build a post-war government as she viewed the situation as '[chaotic] and hopeless' despite the offer of Treasury funds and an explicit request.[12] Only after Short's departure did the first batch of twenty-two DFID volunteers begin to arrive in Iraq.[13]

Blair assured Parker the suits were on their way and, thanking him for his efforts, carefully extricated himself. For the media Parker had assembled 400 soldiers from the brigade, arrayed in a loose semicircle bookmarked by two Challenger tanks against a suitably dusty background. Blair clambered up onto a tank to make his address.

'You fought the battle and you won the battle and you fought it with great courage and valour, but it didn't stop there,' Blair told them. 'You then went on to make something of the country that you had liberated, and I think that's a lesson for armed forces the world over.'

As Campbell noted in his diaries, 'It was OK without being brilliant.'[14]

Blair was given a brisk but hardly rapturous round of applause. After five minutes of glad-handing, he was reunited with Campbell. 'So how did I do?' he asked his spin doctor.[15]

Campbell's mood had been soured by a call he had taken from his team at the Cabinet Office. BBC Radio 4's *Today* programme had run a piece that morning from defence correspondent Andrew Gilligan with a quote from an unnamed source claiming the government had 'sexed-up' a September 2002 dossier on British intelligence. The story threw Campbell into a paroxysm of rage, recalled one onlooker at Basra. Blair was forced to soothe his agitated communications director, little realising the trouble that was brewing.[16]

Chapter 6

DEATH IN THE MARSHES

CARRYING A SHOVEL, ALI AL-BAHADILI gathered his family on top of the earthen embankment. It was early in the morning on 5 May 2003. The hardened mud beneath his feet had the consistency of concrete but slowly and deliberately he began chipping away. The morning air already had a still breathlessness, and Bahadili was soon mopping up sweat with his headdress. Behind the embankment flowed a two-metre-wide canal of bluish-green water. On one bank stood a row of uninspiring concrete houses, Bahadili's village. On the other stretched a desert of cracked mud stained white in places where salt deposits had leached to the surface. Majar al-Kabir, the nearest town, lay over the horizon.[1]

Bahadili could remember when that same view contained open waterways and dense thickets of *qasab* reeds, when his village stood on a bed of reed islands, the arched domes of their *mudhif* homes, built of the same reeds, guiding home the shallow-bottomed canoes that served as transport. Bahadili was a Marsh Arab from the Albu Muhammad tribe. For generations his family had survived on *shabut* fish, which lived in the alluvial mud of the lakes, wildfowl and the milk from their herd of water buffalo. By sixteen, Bahadili had shot his first man in one territorial dispute. Later, he lost two sons – he had six – to another feud. If life was brutish and short, it was his inheritance.

When Saddam began draining the marshes in the early 1990s, the dictator promised to develop the region, but Bahadili knew that was just a pretence. During the 1992 Shia uprising rebels had used the reed-covered waterways to strike at Saddam's army, and the dictator wanted revenge. Within a year of the last dam being built the marshes around his village went from fecund greenery to stinking ponds and dying reeds. Then Saddam's bulldozers moved in and knocked down the teetering remains of his village. Some fled to Majar al-Kabir, little more than a small market town a few years before, or Amarah, the nearby provincial capital. Most left for the heaving concrete slums of the Hayaniya district in Basra or the Sadr City slum of eastern Baghdad.[2] In a desultory effort to resettle a few of the displaced tribesmen, the government gave Bahadili a concrete house on the edge of one of the canals that had destroyed his way of life.

That morning in 2003, as he chipped away at the canal embankment to unleash the waters of the canal, Bahadili was digging to reclaim the past and making a defiant statement of his tribe's independence. When he broke through, water gushed into the desert, and by mid-afternoon a small lake had formed. Word soon spread that a Marsh Arab revival was under way, and other embankments across the central marshes were breached.

Saddam Hussein's destruction of the marshes and the subsequent plight of the Marsh Arabs had struck a particular cultural chord in the West. The Garden of Eden was believed to be located between the Tigris and Euphrates rivers, along with the more historically verifiable relics of the Sumer civilisation. The first British colonial officers to arrive in Basra with the expeditionary force in 1916 were not disappointed by the lush scenery they found upriver. Gertrude Bell, the explorer and heiress, had managed to penetrate the exclusive male club of Middle Eastern colonial officers to land a job with Percy Cox's fledgling administration in Basra. As she wrote gushingly home, Iraq meant 'Romance. Wherever you look for it you will find it. The great twin rivers, gloriously named, the huge Babylonian plains, now desert which was once a garden of the world; the story stretching back into the dark recess of time – they shout romance.'[3]

Bell was one of the first British explorers to travel into the Albu Muhammad heartlands. Over Christmas 1916 she and Harry St John Philby, a querulous revenue officer and father of the future spy Kim Philby, journeyed by launch up the Tigris to Qalit Salih, a tribal market town on the edge of the marshes not far from Majar al-Kabir. They occupied a cinderblock house that had once belonged to a Turkish administrator and had been used subsequently by British diplomats for hunting trips.[4]

For several days they delved into the marshes, past strings of island villages that a later traveller described as 'like a fleet of lit boats at anchor in a calm sea'.[5] Reed beds chimed with the songs of reed warblers and marbled jays. Back at the house each evening they feasted on meals prepared by Bell's Iraqi boy and contemplated the glory of the British empire as the sun set in a guttering orange orb over the reeds. In the middle of the Great War, the acquisition of Mesopotamia, with its grand associations, offered would-be imperialists some solace.

The British colonialists also had a scheme for draining the marshes, originating with Cox's number two in Basra, Captain Arnold Talbot Wilson. A driven man and an instinctive reactionary, Wilson liked to sleep on the ground and once voyaged home to England shovelling coal sixteen hours a day after volunteering to work as a ship's stoker.[6] Wilson dreamed of turning the marshes into a great grain basket for the empire – provided the locals could be brought into line. Failing that, 'the Government of India would administer it [Iraq], and gradually bring under cultivation its vast unpopulated desert plains, peopling them with martial races from the Punjab'.[7]

Wilson's plans did not survive the uprising he provoked by refusing to countenance Iraqi autonomy. The Albu Muhammad tribe were among his fiercest opponents. But the plans to turn the marshes into agricultural land resurfaced thirty years later with Frank Haigh, an engineer advising the pro-British monarchy in the 1950s. Haigh's scheme consisted of a series of canals, embankments and sluices on the lower Tigris and Euphrates. Its main feature was a huge drainage canal, the so-called third river, running between the existing Tigris and Euphrates.[8]

Part of the scheme's appeal to the Iraqi and British governments was that it might finally bring ungovernable Marsh Arab tribes like the Albu Muhammad under control. The tribes knew what was up. As the third river was dug over the next forty years – part of the work carried out by the British engineering firm Murdoch MacDonald, a forerunner of the Mott MacDonald group, the UK government's favoured contractor in southern Iraq in 2003 – the tribes routinely harried construction workers. By the 1980s sporadic flare-ups against marsh drainage had become a low-level insurgency among the Albu Muhammad, exacerbated by the war with Iran.

Their leader was Sheikh Abdul Kerim Mahud al-Muhammadawi, otherwise known as Abu Hatem, self-styled Prince of the Marshes. He would have been instantly recognisable to colonial officers like Bell and

Philby: charismatic and grandiose but petulant and cunning. Saddam Hussein finally caught him in 1989. From his prison cell Abu Hatem learned what happened next: the killing of hundreds of thousands of Shias following the 1992 uprising and the completion of the third river, duly named Saddam's river.

Abu Hatem escaped from prison in 1997. He returned home to find a lunar landscape and stark choices. Saddam had done what neither the British empire nor the Iraqi kingdom had managed: he had drained the marshes. The Albu Muhammad were scattered, and the marshes, with their fecund water and rich deposits of cultural memory, were gone. In the UK Baroness Emma Nicholson, a Liberal Democrat MP, expressed dismay, striking a Bell-like note as she declared, 'They have a unique way of life that is historic and we owe them every effort to bring the marshes back.'[9]

In Amarah Abu Hatem vowed revenge on Saddam. Within a few days of the Coalition invasion, he began his own revolt against the dictator. His power might have waned as his tribesmen left the land, but he could still field 500 tribesmen to storm the Ba'ath Party headquarters in Amarah, followed by Majar al-Kabir and Qalit Salih. When 1 Para arrived in Maysan three weeks later, they were surprised to find no looting and Iraqi police manning checkpoints across the province.

Maysan was the only province in Iraq to liberate itself; and while the Marsh Arabs recognised that Britain and America were the impetus for finally ridding the country of Saddam, they had no intention of relinquishing power. 1 Para's commander, Lieutenant Colonel Tom Beckett, got the impression from meeting Abu Hatem that the tribal leader would tolerate the British as long as they brought money.[10] Abu Hatem christened the British base outside Amarah Camp Abu Naji, a nickname given to Britain's imperial troops during the First World War, translating roughly as 'strong one'.[11]

The British reception was cooler in Marsh Arab towns like Majar al-Kabir. Captain Jonny Crook was in the first Land Rover patrol to try and enter the town in mid-April. When Crook's patrol turned off Route 6, the highway linking Basra to Baghdad, and drove towards Majar al-Kabir they barely got a mile down the road before a pickup truck flagged them down.

'Please respect our wishes and do not enter the city,' said the Iraqi driver in immaculate English.[12]

The soldiers of 1 Para were only faintly familiar with the region's history and how it had bred a culture of suspicion against outsiders. Having taken charge of the province, they were not going to let a bunch of tribesmen with guns pick and choose where they could go. The journalist Patrick Bishop captured the Paras' uniquely aggressive culture when he wrote, 'The Parachute Regiment was one of the youngest in the British army. But in its short life it had developed a strong identity and a powerful sense of its own capabilities and worth. Its purpose was to cause the maximum damage to the enemy with minimal or no support. Its spirit was summed up in its motto, *Utrinque Paratus* – Ready for Anything.'[13]

1 Para began regular patrols through the town, including house searches that used dogs to look for weapons. A detachment from the Royal Military Police tried to train the local police. Clearly, the British were unwelcome: Para Lieutenant Ross Kennedy, 8 Platoon commander, was greeted most days on his entry into Majar al-Kabir by people pulling fingers across their throats. Dislike was mutual: the British soldiers felt unappreciated for overthrowing Saddam. The town quickly developed a reputation among the British for being packed with fickle and ungrateful 'ragheads' who deserved what they got.

The question of who was really in charge of the town remained unanswered. On 22 June Kennedy arrived in Majar al-Kabir determined to assert British control. Fresh out of Sandhurst and nervous about his first command, Kennedy stationed his troops at Majar al-Kabir's police station, with a number prominently placed on the roof. The station sat on a slight rise on the town's western edge, with a view over its cramped streets and sagging power cables. Some of the soldiers had their tops off, their white bodies glinting in the sun.

Kennedy heard, before he saw, the crowd chanting, 'Death to America.' Several hundred Iraqis were coming up the street, chucking stones. Clearly the locals had not got the message that the Paras were not, in fact, American. Then the sound of shots rang out from the roof. Kennedy ran upstairs to discover that one of his lance corporals, Mark Weadon, had been firing live rounds into the air.

'What the fuck do you think you're doing?' Kennedy asked. He told his men to switch to plastic bullets, as they would have done confronting a crowd in Northern Ireland.[14]

The stone-throwing increased and two baton rounds were fired, one of which bounced off the ground, hitting a protester in the face. Kennedy called up the base, and a quick reaction force of Land Rovers and Warriors, twenty-eight-tonne armoured personnel carriers, quickly set off. Just in time Abu Hatem arrived at the police station in his Land Cruiser, an AK-47 on the seat beside him. The crowd quickly calmed. He assured the protesters he would raise their grievances with the British and stop them conducting weapon searches. By the time the rescue party showed up the crowd had largely dispersed.[15]

The next day Kennedy's commanding officer, Major Chris Kemp, held a meeting with the town's tribal chiefs in which he agreed to stop the invasive house searches and to develop a 'joint approach' to weapon confiscation. But Kemp was not about to stop patrolling. On 24 June he ordered all twenty-four members of 8 Platoon back into town to 'show a presence'.[16] Kennedy's request for additional rubber bullets was denied. The bullets had already been boxed, ready to return to Britain when 1 Para's tour ended in a week's time. That left him with just thirteen rounds before his men would have to switch to live rounds. The platoon set off for Majar al-Kabir at 9.19 a.m., arriving half an hour later.

The same morning Sergeant Simon Hamilton-Jewell of the Royal Military Police, or Red Caps as they were called after the colour of their headgear, was preparing to pick up a British commander in the town of Al-Uzayr, an hour away. Before setting off he informed his commanding officer, Lieutenant Richard Philips, that he planned to stop at the police station in Majar al-Kabir en route to discuss ongoing refurbishments with some Iraqi contractors.

Hamilton-Jewell was a forty-one-year-old from Chessington, Surrey. Most of his unit had already returned home, making Hamilton-Jewell the most senior NCO among the remaining twenty-five military policemen who would see out the tour. Under his command that morning were five other Red Caps, including Lance Corporal Thomas Keys, who had distinguished himself as an eighteen-year-old recruit in Sierra Leone, and Corporal Simon Miller, newly promoted and looking forward to riding his 600cc Suzuki motorbike in a few days' time.

Hamilton-Jewell had not been informed about the earlier riot involving Kennedy or that the Paras were returning that morning to patrol the town. Not expecting trouble, Hamilton-Jewell and his men had left on their mission with only fifty rounds of ammunition each and no reliable means of communication. The army's standard issue Clansman radio was notoriously ineffective, and the battle group was under a standing order to give every patrol a satellite phone. But when Hamilton-Jewell asked headquarters for a satellite phone he was told there were none available.

Lieutenant Kennedy was also unaware of Hamilton-Jewell's meeting at the police station. By the time the Red Caps arrived there, Kennedy had already parked half a mile away at Abu Hatem's headquarters and was leading a patrol into town with some tribal militiamen. Kennedy knew about Kemp's deal with the locals and that the presence of British soldiers on the streets might be provocative. He hoped that the presence of Abu Hatem's men would act as some sort of cover, but the tribal leader was not as influential as he had led the British to believe.

As Kennedy led his joint Iraqi and British force into town, an SUV sped up to the patrol and a shaven-headed militia leader who Kennedy recognised stuck his head out of the window.

'What are you doing? The British said no more searches,' said the Iraqi.

'We're not here to search,' Kennedy replied. 'We want to patrol with the militia.'

'If you continue you will come under serious attack. Not safe.'[17]

After a short discussion, Kennedy agreed to take a shortened route and turned his patrol around. However, the decision was unpopular. Back at the Abu Hatem headquarters, Kennedy's senior non-commissioned officer, a gritty Scot called Sergeant John 'Jock' Robertson, puffed up his chest when he heard that Kennedy had backed down.

'We're not getting told what to do by those fuckers, Ross. You've had your shot. You stay here . . . We've got to show presence – show we're not scared.'[18]

Ready for a fight, Robertson took off with his squad in a couple of Pinzgauers, a type of soft-skinned truck. They immediately ran into trouble. As they entered Majar al-Kabir's market, a crowd of young men moved to block their way. Robertson ordered the convoy to proceed. The crowd parted, slowly chanting, '*La la, America. La la, America*.' A voice from the local mosque's loudspeaker was shouting instructions. An interpreter with Robertson translated: 'They are coming for your weapons. Arm yourselves.'[19]

Robertson, taking the bait, ordered his men out of their vehicle to form a cordon to walk the Pinzgauers through the market. 'Push them back. We can't afford to go static,' he yelled.[20]

The crowd began to pelt the soldiers with stones.

'Permission to fire baton rounds,' one of his men, Corporal John Dolman requested. Nicknamed the Dolmanator, a few days before he had shot and killed an unarmed Iraqi at a checkpoint after the man had struggled with him. It had been the company's first kill, roundly toasted that evening. Like any front-line unit, they felt killing the enemy was part of the job.

Robertson gave permission and Dolman quickly identified a ringleader and felled him from twenty feet. The youth collapsed and the crowd backed off. Dolman fired two more rounds, dropping two more protestors. This time the crowd surged forward. Robertson fired seven rounds in quick succession, and a space opened up for the Pinzgauers to push forward.

'Keep the momentum going,' yelled Robertson. Ahead a vehicle was blocking the route with its driver standing casually at the door. Robertson charged up and brought his rifle up to the man's face.

'Get your fucking vehicle out of here or I'll kill you.' The man hastily got back into the vehicle and moved.[21]

At the rear of the convoy Dolman was getting pelted with stones and was out of plastic bullets. He fired a burst of live rounds over the crowd's heads. That prompted an Iraqi gunman to appear at a first-floor window and fire at the British soldiers with an AK-47. Robertson and Dolman each returned fire with single shots, striking the gunman in the chest. The confrontation continued to escalate. Robertson killed another Iraqi as he darted between vehicles, before the British soldiers mounted up and sped away from the crowd.

Less than a mile away, and still unaware of each other's presence, both Kennedy and Hamilton-Jewell heard the sound of gunfire. Hamilton-Jewell was just adjourning his meeting with the contractors. He prepared to go outside. Half a mile down the road, at the militia building, Kennedy hurried his men into their vehicles. He knew Robertson had gone into the town centre, but with Robertson's radio communication not working, he had no idea of his sergeant's status. By taking to the streets, Kennedy hoped he might spot him. A mob was starting to form outside the militia building as Kennedy's men drove away. They had made it less than 200 metres down the street when they were hit with gunfire and a rocket-propelled grenade slammed into the building opposite. Kennedy's vehicle screeched to a halt, and everyone jumped out looking for shelter – no one wanted to be caught in a soft-skinned truck or Land Rover if an RPG struck. The shoulder-fired anti-tank weapon could punch through an inch of steel.

As his men looked around urgently for cover, Kennedy spotted the police station nearby, where Hamilton-Jewell's men had been meeting with the contractors, but he saw no sign of a British presence. As Kennedy later told an inquest into the tragic events that followed, if he had known the Red Caps were at the station he would have gone to help.[22] With AK-47 rounds flying through the air, and the mob closing in, Kennedy ordered his men back into their vehicles, but the driver of the lead truck shouted that he couldn't get the engine started. Kennedy's men pushed as the driver put the truck into first and thumped the accelerator. The truck sped forward as the Paras scrambled into the rear. Two gunmen appeared in a side alley; they were felled by a volley of gunfire as the truck drove away, leaving the baying mob behind.

With Robertson and his men retreating from the main market, and Kennedy's men having just escaped, the mob of angry Iraqis was looking for someone to vent their anger on. What had begun as a rock-throwing scuffle had escalated into a full-blown revolt, with every man in Majar al-Kabir on the street with a gun. Part of the crowd broke away and headed towards the police station.

It is not clear the extent to which Hamilton-Jewell knew about the commotion in the town centre, but he was standing at the station's entrance when the mob approached. The Iraqi police guarding the station took one look at the crowd and fled, sprinting back into the small compound and scrambling out windows on the other side. If there was a chance for Hamilton-Jewell and his men to escape then this was it, with the crowd still a few hundred metres away. The local police chief was the last to leave.

'Please, Mr Hamilton-Jewell, leave with us,' he implored.[23]

The Red Cap sergeant refused. Maybe he did not understand the circumstances or the sheer force amassing against him and his men. Or perhaps he believed that as a military police officer he represented law and order, as well as the courage of the British army. Either way, he stood firm.

Several miles away Sergeant Robertson's men had retreated from the town centre but were being harassed by gunmen. Fearing that Kennedy was still at the militia headquarters and under attack, Robertson was trying to circle back around to the building, not realising that Kennedy had left and was in turn looking for him. Robertson had not made much progress, having been repulsed by the same angry crowd that was marching on the police station, so he sought cover behind a river embankment, where he was finally able to reach headquarters on his satellite phone.

'It's Sergeant Robertson. I'm in Majar al-Kabir! I've been in contact for twenty minutes!' he yelled.[24]

A platoon was ordered in by helicopter, with one section to rescue Robertson and another to aid Kennedy, who had just called in from the other side of town. A ground force of tanks and Land Rovers would also support the rescue. Lieutenant Philips, the Red Cap commander, mentioned that Hamilton-Jewell's men might be in the town's police station, but no one was certain and the information was not passed on.

Robertson was now in real trouble. As he and his men crouched behind the embankment two RPG rounds struck his vehicles on the road. One was burning. With the mob closing in, the only option was to seek cover in the nearest house. He burst through the door to find a low dark room and a few women in *abayas* climbing over themselves to get away. A man tried to escape at the rear, but he was hauled back in, and the entire family roughly searched. Through the interpreter, Robertson told the man to go and check to see if they had been followed.

'If he's not back in five minutes, I'll kill his family,' he said.[25]

The man returned a few minutes later to report gunmen had surrounded the burning British vehicles. Just then Robertson heard the sound of a Chinook helicopter arriving. He scrambled onto the roof with his satellite phone and succeeded in getting through to the operations room. He explained that the helicopter needed to circle in on their location to avoid being shot at by the crowd. Robertson also prepared to fire a mini flare to signal to the pilot their location. This would also signpost their position to the mob, so he and his men would only have a few seconds to get out of the building and to the chopper before they were surrounded.

As he fired the flare and sent a plume of red smoke arcing into the air, Robertson watched in horror as the Chinook, disregarding his instructions, flew straight over the mob. As it hung in the air, looking to land, the helicopter's undercarriage was riddled with bullets. Standing on the Chinook's ramp, Private Phil Johnson felt a round rip through his inner thigh. Beside him, another round had ripped off half the face of one soldier;

others, including his sergeant, were sagging against their seats or writhing on the floor in agony. An RPG passed between the rotor blades, and the pilot was forced to swerve the chopper to the right to avoid a direct hit, knocking Johnson off his feet. In total there were seven injuries, two of them head wounds, and the pilot took the decision to pull out.

From his rooftop Robertson watched his only chance of escape disappear. He made another call to headquarters: 'Where the fuck [are the tanks and Land Rovers]?' he shouted.[26]

Back inside, Robertson tried to boost everyone's courage. 'Remember, lads, you're fucking paratroopers. It's backs to the wall fighting now. Whatever happens we're going to go down fighting.'[27]

They burst out of the house just in time to avoid being surrounded by the mob. Dodging and weaving down the street in pairs, they reached the scrubland on the outskirts of the town. Movement was hazardous – the ground was crisscrossed with earthworks and irrigation ditches which they had to hurdle – but ducking and weaving they appeared to be losing their pursuers. They also could clearly see the tanks and Land Rovers of the ground rescue force pulling into town, and these were now drawing the angry crowds.[28]

Working their way around to the rescue party, using the earthworks to shield them from sight, Robertson realised that unless their rescuers advanced deeper into town the force would not be able to provide supporting fire when he and his men broke cover. Unfortunately, Robertson had just mistimed a leap over a ditch and landed up to his neck in fetid warm water, putting an end to his satellite phone and his only means of communication. There was only one way to get the message across – someone was going to have to leg it. He turned to Private Freddy Ellis. 'Get your kit off, clean fatigues, and run over to where you see those [Land Rovers].'[29]

It sounded like a suicide mission, but Ellis, who once picked out his molars with a compass for a bet, was not overly fazed. There were a good 400 metres to cover. He stripped off his webbing and took a deep breath. Head down, zigzagging left and right, Ellis galloped over no-man's-land as Iraqi and British combatants turned to watch. He arrived breathless but intact; the rescue party advanced and Robertson's men were saved. The news was greeted with relief at headquarters. Little did they realise that half a mile away Hamilton-Jewell and his men were about to pay the price for Robertson's earlier recklessness.

At the police station a crowd of Iraqis had surged towards the entrance, forcing Hamilton-Jewell and his men back inside the main building. Corporal Miller's blood was later found smeared on an inside wall of the doorway, suggesting he may have been shot early in the retreat. The Red Caps barricaded themselves into a storeroom on the ground floor. They fired no shots. Hamilton-Jewell would have known that with limited ammo any resistance would be short-lived.

As the Red Caps desperately tried to hold back the crowd, news of their predicament had at last reached other British units. A local doctor, Firas Fasal, had heard that British soldiers were being held at the police station. After trying in vain to secure their release, he went to warn Major Kemp, the commander in charge of the rescue efforts. Kemp, unsure who the servicemen might be, preferred to focus on rescuing the soldiers he knew about. Fasal offered to try again to parlay with the mob and set off back to the station. The last reported sighting of the Red Caps was by a village elder who had also sought to free the men. Pushing his way through the crowd, he saw one British soldier lying injured on the floor, possibly Simon Miller, and another crouching in a foetal position emitting noises of despair.[30] The elder was pushed back. Shortly afterwards gunmen forced their way into the storeroom and the torture began. Hamilton-Jewell and his men were kicked and beaten. Feet and rifle butts targeted their groins. At one stage one of the Red Caps took out a picture of his family. A bloodstained photograph was later found on the floor. Then the men were shot, one at a time.

By the time Fasal reached the station all the Red Caps were dead. The doctor hurried back to Kemp, who made the quick deduction: with Robertson's men now safe and Kennedy's company in radio contact, the dead soldiers could only be the missing Red Caps. Fasal returned again, this time to bring out the bodies. They were mostly too disfigured from bullet wounds and the marks of beating to be recognised.

ON THE MORNING THAT TOM KEYS DIED his parents Reg and Sally were visiting a garden centre near their five-acre smallholding on the hills above Bala. An ambulance driver in Solihull for nineteen years, the Keys had recently moved to Wales to fulfill their dream of country life after Tom, the youngest of their three boys, left home. Keys was putting out bird feed in one of his barns when his wife called from the house. He hurried over.

Sky News was reporting that six Royal Military Police had died on patrol in Maysan Province. The moment he heard, Keys had a sinking feeling. There were only seventy-five Red Caps in his son's section. He knew that fifty had already returned home at the end of the tour. There were seven women among those remaining who would not have left the base. Tom's chances of survival were grim.

Ten minutes later there was a knock on the door. Keys answered. There were two army officers outside. He knew what was coming, but it still felt like being hit in the stomach by a sack of cement when they told him. Keys realised the officers must have been waiting at the bottom of the hill for them to return.

That evening Sally told him, 'This isn't even the worse of it. That comes when it sinks in.'[31]

The six deaths in Majar al-Kabir made front-page news, the heaviest single combat loss for British forces since the 1991 Gulf War. Criticism immediately focused on the Red Caps' lack of a satellite phone and ammunition. Tony Blair was shocked by the deaths, said his adviser David Manning. General Walker and the other defence chiefs were quick to assure the prime minister that the event was a one-off.

Tom Keys was buried on a windswept Welsh hillside a few days short of his twenty-first birthday. Birthday cards had arrived all week. Sally Keys was so distraught, she couldn't bring herself to attend her son's funeral.

Chapter 7

MINISTRY OF CULTURE

MILES PENNETT FELT THE HERCULES C-130 begin its descent into Iraq. In the dark windowless hold he could just make out the bulky silhouettes of the other soldiers. Departing from Brize Norton eight hours earlier on a rainy Oxfordshire morning, Miles had joked around with his unit and ignored the other soldiers as they crammed into the hold. A few minutes from landing he felt almost tenderly towards his fellow passengers.

The Hercules skimmed over the tarmac and roared to a halt. The desert air rushed in. This was not Pennett's first war zone. At Strathclyde University he had joined the Territorials, Britain's part-time army comprising 35,000 men and women who spend their weekends marching through the countryside in combat gear and who account for a quarter of the country's fighting strength. In 1999 Pennett dropped out of his architecture course – 'too long' – and signed on full time with the Queen's Own Yeomanry, serving in Kosovo and then Bosnia. His parents were disappointed, but certain aspects of the military clearly suited Miles.[1]

He had delicate features and a head of curly brown hair. Quick-witted but lacking focus, Pennett was attracted to the dashing life of a cavalry officer. Unfortunately, soldiering in the Balkans was more akin to a job as a night porter at a rather dismal holiday camp. In 2001 he left the army and joined the TV soap opera *Emmerdale* as a runner. The once-staid chronicle of Yorkshire farmers had ditched farm life for racy plot lines in the local village, introducing cast members via the TV series *Soapstars*. Pennett was in his element and worked his way up to assistant director, training with the Territorials at weekends.

His compulsory mobilisation papers for Iraq arrived in April 2003.

A flight engineer ushered Pennett and the other passengers down the loading ramp. Across the tarmac a single floodlight illuminated the side of the airport terminal, emphasising the blackness of the desert night. He wanted a moment to take it all in, to contemplate the inky dark that connected him with the mysterious city just beyond the airport's fences, to reflect on the drama of his own unfolding life, but the heat was too oppressive. He followed the others into the terminal building, a vaulted concrete block with cheap marble on the floor and flickering fluorescent strip lights.

Pennett's job was to be liaison officer with the Danish battalion patrolling the Saudi border, but after a restless night in the tent city alongside the airport terminal – nicknamed Brookside – Pennett discovered that his job had been given to someone else. Instead he was joining the civil military affairs team, which acted as the interface between the military and the local population. In Pennett's experience in the Balkans these teams were often dumping ground for civvies like himself. A few hours later he was on a minibus to downtown Basra with Mark Clark, a rangy lawyer from Edinburgh, also part of the Queen's Own Yeomanry, and a couple of other stragglers. A beleaguered Janet Rogan met them at the gates to the former offices of Basra's electicity commission, nicknamed the Old Electricity Building. Rogan was the FCO's top diplomat in the south and had to all intents and purposes been running the civilian side of Basra's new government. Since Bremer's creation of the Coalition Provisional Authority in May, the scramble had continued in Whitehall to work out what Britain's role in the south would be. The Americans were asking for a UK-led regional administration, but the FCO did not have the staff or the budget, and DFID representatives were only just starting to arrive in theatre.

The Danes had briefly taken charge in Basra, supplying a former Danish ambassador to Syria, Ole Woehler Olsen, to run the administration and give the appearance of international rule.[2] He was largely a figurehead anyway, as little governing could be done with the tiny staff and a barely functioning local council. In his sixties, Olsen had a morose Danish sense of humour and a petite Algerian wife, Mona, who accompanied him everywhere. Avoiding the American mistake of locking themselves away in a palace, Olsen had chosen the Old Electricity Building in central Basra to be 'close to the people'. The downside was that the four-storey building was cramped and lacked a reliable water and electricity supply. The toilets periodically backed up, leading to sewage seeping through the first-floor walls. Mona Olsen did her best, decking their room with rugs from the souk and painting the walls a gentle cream, but there was little disguising that the place was dump.[3]

Furthermore, the building's position was beginning to prove a liability. Since Bremer's announcement about disbanding the army and de-Ba'athification, there had been regular demonstrations outside the gates by former soldiers and disgruntled government employees. Some protests had escalated to stone-throwing. Soldiers from the Queen's Lancashire Regiment stationed in a building opposite were frequently called on to keep the peace. Taking a holiday in early July, Ambassador Olsen had the temerity to question the security arrangements in southern Iraq before the Danish media. The US administration in Baghdad promptly suggested he extend his leave indefinitely, leaving Rogan to pick up the pieces. Somewhat reluctantly, the Brits were back in charge of administering southern Iraq.

Rogan was the type of female diplomat rarely found in the male-dominated upper reaches of the FCO, but someone whose industry had held together a number of foreign embassies. She had worked in the British chambers of commerce in Beijing and Sarajevo, before returning to London, where she proved an unflappable desk officer. With the new British commander, General Peter Wall, she had worked out a plan to fill the gaping holes in the CPA's administration in Basra. Ushering Pennett and Clark and the others into a first-floor room, Rogan sat them down in a semicircle and informed them they were the new government.[4]

A dog-eared piece of paper was handed around, listing the various ministries to be allocated. Some already had names next to them. Looking around, Pennett identified Andrew Alderson, former director at a merchant bank who had taken an extended leave of absence to go skiing prior to Iraq, who was now finance minister; Charles Monk, a lean former SAS trooper, currently a master at a prestigious public school and the new education minister; and Rupert Hill, a gregarious internet entrepreneur, who had taken the trade and industry brief. Clark confidently wrote down his name against youth and sport. 'I'll get them playing football,' he declared.[5]

Pennett's eyes glazed over as he stared at the empty portfolios for housing and electricity. After a moment's hesitation, he put his name down against culture, at a stroke making him the first culture minister for southern Iraq, a region that contained the archaeological treasures of Ur and Sumer and a rich literary history about which Pennett knew very little.

'But what do I do?' he asked as Rogan showed the new government their offices on the second floor. Pennett would share a tiny room with Clark and Monk, while Alderson bagged his own office.

'Your first job is to scope out what's going on in your ministries so we can come up with a plan,' said Rogan.[6]

Over the next week, commuting in from the airport every day, Pennett pieced together Basra's cultural map, disappointed to find that the theatre had been gutted by fire and a Shia gang had taken over the offices of Iraq's Olympic Committee. Charles Monk was discovering similar problems at the education department, where a militia leader, Ahmed al-Maliki, had appointed himself director general of education.

Maliki carried himself like a Mafia boss, a couple of menacing goons at each shoulder, and Monk suspected that his agenda was not in sympathy with the long-term goal of restoring Basra's education system. Maliki had quickly embraced Bremer's de-Ba'athification order, sacking hundreds of teachers – he was all too happy to purge Sunnis from the school system and replace them with his own Shia followers. Maliki did not stop there. Applying the rhetoric of conservative Shia ideology, he was soon removing women from positions of authority, many of whom had no affiliation with the Ba'ath Party. Next he moved to split co-educational primary schools, so that girls and boys were no longer taught in the same classroom. Then he questioned the education of any girl over the age of twelve.

'He displayed little interest in class sizes, teaching resources, or staff and pupil welfare,' Monk said.[7]

In fact, Maliki was a leading follower of Ayatollah Muhammad Sadeq al-Sadr and had organised an armed uprising in the city following Sadr's assassination in 1999. For a few heady days Basra had been in the rebels' hands before the tanks moved in, and Maliki had fled to Iran. Returning after the invasion, Maliki had been instrumental in reactivating Sadr's movement in Basra. The murdered ayatollah's son, Moqtada al-Sadr, had recently formed a militia called the Jaish al-Mahdi – the Army of the Mahdi, referring to the twelfth Shia imam, who had disappeared in the ninth century but was prophesied to return to restore an Islamic state. His call to arms was proving popular in Basra, as was Maliki's ability to put Sadr's followers on the education ministry's payroll.

Given that Maliki had put himself in charge, Monk was uncertain how to go about removing him or rein in the ever-expanding teacher payroll. This problem was compounding the deeper issue facing the administration: it was rapidly running out of money. The protests outside the Old Electricity Building were becoming ever more frequent and volatile, and most of them were about pay. Finance supremo Alderson had no budget to speak of from the British, and the $30 million the military had salvaged from Basra's central bank was almost spent. A few emails to Baghdad elicited that the Americans might have money, but it was unclear how to get it.

Taking a flight to Baghdad a few days later, Alderson visited the CPA's finance department. He had changed into his one set of civilian clothes – chinos and a polo shirt – so he would look the part and not get spoken down to for being a lowly major. He needn't have worried. He found the Republican Palace teaming with CPA apparatchiks, many of them young American Republican Party card-carriers in their first posting out of college. Ambassador Bremer had recently unveiled a thirty-page strategy paper for the restoration of government and instigated CEO-style morning briefings.[8] Preparing for the meetings took considerable time and helped create the air of an American bubble in the palace, fuelled by high hopes and PowerPoint presentations.

When Alderson asked for £30 million, the CPA's finance director was unfazed. Bremer was in the process of securing $19 billion from the US Congress. The next day a pair of Chinook helicopters arrived at Basra palace carrying pallets of cash. Over the next day Alderson handed out £8.6 million to 110,000 workers from the back of a shipping container. Alderson did not allow himself too much time to savour his accomplishment. He could see that the entire economy of the south needed to be rebuilt, from agriculture to telecoms to electricity. When he actually calculated the enormity of the undertaking, he was gobsmacked.

By then, Whitehall had at last worked out what Britain's role in post-war Iraq would be. After Bremer's shaky start, it was clear to Blair's foreign policy adviser David Manning that the UK needed to have a heavy hitter in Baghdad. Manning recommended his old boss Jeremy Greenstock, the departing UN ambassador, for the job of special envoy, and although he wouldn't be Bremer's number two – Bremer had baulked at the suggestion – he could exert a powerful influence over American decision-making with a direct line to Blair. Greenstock was due to arrive in Baghdad in September.

In Basra the Danish ambassador's departure had created an opening which had been filled by Hillary Synnott, a shrewd former high commissioner to Pakistan. His only competition for the job of 'king of southern Iraq' had come from the former governor of Bermuda. With the Italians agreeing to run Dhi Qar, a troublesome south-central province dominated by Marsh Arab tribes, and the Dutch taking Muthanna, a large south-western province mostly consisting of desert on the border with Saudi Arabia, Synnott's job would be to represent them and the British-controlled provinces of Maysan and Basra to the central government in Baghdad. Bremer, displaying an appetite for centralising power, vetoed the idea of a semi-autonomous British region, and the kingship was demoted to 'Co-ordinator CPA-South', to Synnott's mild disappointment.[9]

'There was a good deal of satisfaction at having concluded so much business. We were going to have the A-team on this,' recalled Manning.[10] Blair was also riding high. On 18 July he was awarded a US Congressional Gold Medal and invited to address the US Congress, only the fourth British prime minister to do so. The language Blair used to explain Britain and America's role in Iraq had changed markedly since pre-war days. Weapons of mass destruction were no longer mentioned. Instead the invasion was about bringing universal values of democracy, human rights and liberty to the 'darkest corners of the earth', sentiments that appealed to his audience's belief in American exceptionalism. Blair earned nineteen standing ovations for his speech.

But in the UK events were about to bring his rhetoric crashing down to earth. That same morning David Kelly, a UN weapons inspector, was found dead in woodland near his Oxfordshire home. He appeared to have committed suicide the day before by ingesting twenty-nine painkillers and slitting his wrists.[11] Since BBC reporter Andrew Gilligan's story about Campbell's manipulation of pre-war intelligence, Blair's spin doctor had aggressively targeted the BBC to protect the government's credibility. Having identified Kelly as Gilligan's source, Campbell and Hoon pushed for the weapons inspector's identity to be leaked to the press. In the end,

Kelly's name was withheld but enough details were given to leave little doubt as to who he was. They were confident Kelly would refute Gilligan's story, not realising that Kelly's subsequent appearance before the Foreign Affairs Select Committee would plunge the weapons inspector into a deep depression.[12]

Blair was on a flight bound for Tokyo and an Asian tour when he received the news.

'Shocked, he was deeply shocked,' said Manning.

At the press conference held with the Japanese prime minister a few hours later Blair looked like he had aged several years. He announced an independent inquiry into Kelly's death, telling the assembled reporters, 'I am profoundly sad for David Kelly and his family. He was a fine public servant. He did immense service for his country and I am sure he would have done so in the future.'

Before Blair could get away, the *Mail on Sunday*'s Jonathan Oliver shouted out, 'Do you have blood on your hands, Prime Minister?'

The prime minister looked pained, chastened even, but did not answer. As Blair later admitted in his autobiography, 'Probably my own integrity never recovered from it . . . Before it, we were in error; after, we were "liars".'[13]

The frenzy surrounding Blair diverted media attention from the deteriorating situation in Basra. Protests had resumed, this time over fuel and electricity shortages. With temperatures routinely over forty degrees Celsius, summer demand for power had spiked. On 9 August the city erupted into riots. British troops guarding a petrol station at a highway interchange were attacked from all sides, and a military mail van overturned and torched nearby. Then crowds appeared outside the British headquarters.

Miles Pennett heard the roar of the mob outside his office. His tenure as southern Iraq's culture minister had proved short. Rogan had reassigned him to a more 'impactful' position with the small strategic communications team. Pennett's brief was to manage the message for the local media. He was crafting his press release for the latest 'demonstration of freedom' when the deputy division commander Bruce Brealey burst in.

'Grab your body armour and weapon. It's protesters, and this time they're serious,' he yelled before disappearing.[14]

'I'd better man the phones,' said one of Miles's colleagues, but Pennett wasn't going to hide at his desk if they were under attack. He dashed after Brealey. In the forecourt the CPA south team had gathered before the flimsy metal gates being buffeted by the crowd outside. A few rocks flew over, which Charles Monk neatly dodged.

'The Lancashires next door are on patrol somewhere,' said Brealey, 'so we've got to go out there.'[15]

'What do they want?' asked Monk.

'Electricity, I think,' said Pennett.

With little other choice, they hastily formed a riot line: Pennett shoulder to shoulder with the ministers of trade and industry, health, and law and order. Brealey opened the gates to a wall of furious Iraqis. At the sight of their government, the crowd fell back, but only a few feet, and the riot line was soon locked in a desperate scrum with the protesters, pushing them back behind the gate. For a moment it looked like the crowd might break through, but locked defiantly together the fledgling government of southern Iraq stood firm. The protesters hesitated at the entrance, their point apparently made, and backed away.

Over the preceding weeks Janet Rogan's urgent reports had failed to elicit a response from London, but now the riots prompted a visit from David Richmond, Britain's representative to Iraq prior to Jeremy Greenstock's arrival. In his early fifties, Richmond was by nature warm and effusive, but the strains of the job had given him a constantly worried look. He was all too aware of the potential for rioting to get worse. Bremer was pushing for the arrest of the Shia cleric Moqtada al-Sadr after an Iraqi court had charged him with the murder of a pro-American religious leader in April. Richmond believed the rule of law needed to be upheld, but he shared with London the view that Sadr's arrest might lead to an escalation of violence in Basra. There were reports that Sadr's militia was involved in fomenting the fuel riots.

After another day of rioting on 10 August, Richmond was greeted that evening at Basra airport by Major General Graeme Lamb, the new British commander. A former director of special forces, Lamb was not a man to mince his words. The civilian effort was 'fucking pathetic', he told Richmond. 'If Whitehall doesn't pull its finger out of its arse soon, we are in danger of losing control of the south,' Lamb continued. He was already devising a military reconstruction plan to be implemented by his engineers.[16]

Richmond had met enough irate generals in his time to know how to placate Lamb. The military always liked to see itself coming to the rescue, he thought. The next morning Richmond went to the Old Electricity Building for talks with Rogan and the newly appointed governor of Basra on how to end the rioting. On the way over, his convoy was pelted with stones by an angry crowd, and he arrived at the building flustered. Janet Rogan led him to a windowless meeting room. There was dried blood on the corridor outside where a Nepalese private security guard, shot in the chest in the city, had died the previous day.

At the meeting, Janet Rogan presented a protocol for future protesters, which recommended flag-waving and singing and requested every group provide twenty-four hours' notice of a demonstration. The strictures may have had some effect – Iraqis still half-suspected the British might adopt the brutal methods of suppression of Saddam's regime. Far more effective in quelling the trouble was the military's effort to truck in thousands of gallons of fuel. Richmond returned to Baghdad fearing the lull would prove temporary. Bremer was still pushing to arrest Sadr, and although British opposition had quashed the operation, Richmond suspected trouble with the Shia cleric was brewing. Richmond tried to get guidance from London, but Blair was consumed by troubles over Kelly's death. Alastair Campbell had resigned, and the public inquiry had recently begun. To get some breathing space, Blair had jetted off to Barbados for a late-summer holiday in Cliff Richard's mansion.

Meanwhile in Basra Shia militants continued to grow in confidence. Charles Monk had received yet another inflated payroll request from the self-proclaimed education director and religious theocrat Ahmed al Maliki. Maliki hadn't even bothered to come, but rather sent his assistant, a heavyset man who had lounged arrogantly in a chair in Monk's office. Monk felt like a bank machine. Weakly he informed his visitor that he was not going to approve any more teachers' salaries without Baghdad's authorisation.

'I will deny that this has ever been said,' the man responded. 'But if these teachers aren't paid there might be angry protests on the streets.'[17]

Monk leapt to his feet. 'Are you threatening me?' he shouted and jabbed a finger at the man's chest.

Monk was considerably older than the other man, but had retained his physique from SAS days. He was not going to be shaken down by Maliki or anyone else. For a moment the two men eyeballed each other before the Iraqi slowly backed off and left the room.

Monk might have won the first confrontation between the British occupation and the Sadrists, but the question was how far the British were prepared to go in the defence of liberal democracy.

Chapter 8

OUR MAN IN BAGHDAD

BLAIR WAS NOT ENJOYING THE INFLUENCE on Washington that he had hoped the partnership would bring. Britain's failure to push the White House to plan for the post-war aftermath had been compounded by American decisions to disband the Iraqi army and cleanse ministries of Ba'athists without adequately consulting London. On the phone every week with Bush, Blair appeared to have a close and intimate bond with the president, but if the relationship was to deliver anything more than platitudes, Britain needed to start demonstrating its influence on the ground.

That was one reason for dispatching to Baghdad the diplomatic heavy hitter and former UN ambassador, Jeremy Greenstock. At sixty years of age, Greenstock was at the end of his career, but his intellectual powers were undimmed. He was thin and angular, with a protruding nose, delicate lips and cold brown eyes. At the UN he had proven a formidable negotiator, able to flick deftly between charm and aggression. Greenstock was the best in the business, and he knew it. But ever since he had helped convince the attorney general Peter Goldsmith that Britain did not need a second UN resolution explicitly authorising war, Iraq had gnawed at him. Privately, he regarded the invasion as illegitimate and questioned the strategic thinking behind British involvement. Yet he had not raised his doubts with Blair. Greenstock initially refused the job in Iraq when Alastair Campbell called. The war was behind him, and he wanted to focus on his retirement job at an international think tank. Then Blair intervened and he changed his mind; perhaps he should help with the rebuiding, he thought. Greenstock asked to be Bremer's deputy, but after the American declined he warmed to the idea of being a roving troubleshooter. By standing outside the Coalition Provisional Authority's chain of command he could offer more effective criticism. He had known Bremer since the 1970s, when Greenstock had served as political first secretary at the embassy in Washington, although there was little rapport when the two met briefly in Washington in the summer of 2003 to discuss the political process in Iraq.

Greenstock sympathised with the pressure the American was under, but he did not agree with Bremer's strategy for moving forward. In early September, just before Greenstock's arrival in Baghdad, Bremer published an article in the *Washington Post* outlining his seven-point plan to hand over power to the Iraqis.

The plan had been drawn up by Bremer's so-called 'brainiacs' – a group of policy wonks with little experience of Iraq.[1] The first steps involved creating a panel of Iraqis to write a constitution; elections would then be held to form a sovereign government to which the Americans would surrender control. There was a conspicuous absence of any dates for either an Iraqi election or the American departure. Bremer argued that time was needed in a country currently lacking electoral laws and political parties.[2]

The plan drew the immediate ire of Iraqi leaders like Grand Ayatollah Ali al-Sistani, the spiritual head of the country's Shia majority, who did not want the constitution-writing process overseen by the Americans. Sistani had already issued a fatwa demanding elections, and few Iraqis dared oppose the cleric.

Greenstock also disagreed with the protracted political process. Set a date for elections, he thought. Momentum would be lost by dragging out interminably the formation of an Iraqi government, and provoke an already fractious population to believe the Americans did not intend to leave. The fragile security situation had recently been underscored by the bombing of the UN headquarters in Baghdad in August 2003, which had killed the UN's envoy Sergio de Mello. The British alternative, favoured by the FCO's political director John Sawers, was to set an early date for elections and in the meantime declare the group of Iraqi exiles whom Bremer had brought together to advise him, known as the Iraqi Governing Council, a provisional government.

The differences between the two men was brought to a head on Greenstock's second day in the job when Colin Powell, the US secretary of state, visited Baghdad. In the US the African-American general had been a popular and trusted figure, whose advocacy of the invasion to counter Saddam's weapons of mass destruction had damaged his reputation. According to his deputy Richard Armitage, Powell was growing increasingly frustrated with his bit-part role in foreign policy and felt he was being wheeled out as the 'good' American whenever US plans were in trouble. He was also less than enamoured with Bremer, who acted like his direct line to President Bush put him above Powell and any need for consensus-building.[3]

Greenstock and Bremer met Powell in the US administrator's office, the two locally based men sitting opposite the secretary of state. When Bremer had finished explaining his plan to Powell, Greenstock chipped in, 'Jerry [Bremer's nickname], we're behind you and these seven steps. I think we've got a structure here. It's very important. But we also have to keep our minds open to what we will do if one of the steps doesn't work. We need a plan B.'

Greenstock had made this point to Bremer earlier that summer, but before he could raise the British position of an early handover and a clear election timetable, Bremer snapped at him, 'I told you before, Jeremy, and I'm telling you again, the president has signed off on the seven steps. There will be no deviation. We are going to drive this through. Do I have your support or don't I?'[4]

Greenstock bristled at the rebuke. Colin Powell skilfully changed the subject but the secretary of state was clearly disturbed by Bremer's inability to listen to criticism. Armitage was later attributed Powell's disengagement from Iraq to Bremer's attitude. The knock-on effect of the meeting was to lessen British influence in Washington, where Powell had been an important ally.

Returning to his office in Saddam's palace Greenstock nursed his bruised ego and fired a disgruntled memo off to London. Since August there had been a startling reorganisation in Whitehall. With David Manning now ambassador in Washington, Nigel Sheinwald had taken over his role as Blair's foreign policy adviser. The two men could not have been more different. Where Manning was urbane, measured and intellectual, the fifty-year-old Sheinwald was overbearing and sometimes contemptuous of Whitehall's bureaucracy. 'Getting his own way was always top of the agenda,' noted one FCO official who had crossed swords with him.[5]

A former head of the FCO's media division, Sheinwald understood the need for clear, crisp strategy that could translate into headlines – in fact, many in Whitehall noted that Blair had found a new enforcer to compensate for Alastair Campbell's departure. Under his supervision was a recently formed Iraqi steering group, which would bring together department heads from the FCO, MOD and DFID. In addition, a new Iraq directorate, under the command of Dominic Asquith, would oversee FCO policy.

One member of the steering group, present at the first meeting, recalled Sheinwald's tempestuous opening.

'Where the fuck is Iraq's electricity?' he shouted. No one knew, and British officials spent the next week nervously researching the answer.

'What is a civil servant in Whitehall meant to know about that?' one official wondered.[6]

Sheinwald played another important role – as the link man between Blair's inner circle and the administration in Washington. Blair had held his first video conference with Bush in July, and they had become a weekly affair, the two men eyeing each other up over the grainy feed, senior aides fluttering in the background like shadows.[7] Iraq took up most of their time although there was rarely any progress on the two subjects Blair pushed: the need for a strategic communications team in Baghdad to present the positive aspects of the occupation, and the rapid deployment of Iraqi police to curb the continuing lawlessness.

Condoleezza Rice, the US national security adviser, had emerged as a vital conduit for British views. Manning had developed a close relationship with her by providing an accommodating sounding board for her ideas. Sheinwald's role was to project British opinions more forcefully, including Greenstock's views on the unnecessarily complicated seven-step plan to Iraqi sovereignty. But on the important decision of whether to accelerate the handover of power to the Iraqis, Rice remained 'undecided'.

The week after Bremer's run-in with Greenstock, the US adminstrator flew to Washington to give testimony to Congress as lawmakers considered his request for $19 billion for Iraq's reconstruction. In a series of meetings at the White House Bremer learned that opposition was mounting to his plans. The Pentagon was advocating setting up a provisional government and handing over power as early as April 2004. Rice had also picked up on the resistance of Iraqi leaders to writing a constitution with the Americans in power. Bremer managed to fight his critics off; he considered a constitution vital to stop Iraq sliding back into dictatorship. But he wondered where the quick handover idea kept coming from.[8]

He was about to get one answer. Second in command at the State Department, Richard Armitage was a pugnacious former marine who liked to keep his ear to the ground. When Bremer's right-hand man, thirty-eight-year old Scott Carpenter, visited the State Department on that same trip, Armitage laid into him, accusing Bremer of sidelining State and keeping them in the dark. 'The only reporting the entire administration is getting on Iraq is Jeremy Greenstock's, and it doesn't sound too good,' he declared.[9]

Bremer was furious when Carpenter recounted the incident on the flight back. No organisation can thrive with separate lines of reporting, Bremer stormed. He could not have Greenstock undermining him at every turn as if he knew better. At the next morning conference in Baghdad with Greenstock the body language between the two men was clear. Whenever Greenstock spoke, Bremer seemed to wince slightly, noted Carpenter, who began scheduling meetings without inviting the British representative. Given that their two offices were directly opposite a large atrium where the joint UK–US governance team sat, the tension made for awkward moments.[10]

The Anglo-American relationship that Blair had gone to war to strengthen was coming under serious pressure. In fact, it was increasingly difficult to find areas where British and American views matched. The training of the Iraqi police was a debacle. To rebuild Iraq's security forces, Bremer had initially hired Bernie Kerick, a former New York City police commissioner, who had made his name in the aftermath of 9/11. Kerick spent a few weeks driving between photo shoots before disappearing, leaving the pieces to be picked up by his deputy Doug Brand, a former assistant chief constable from Yorkshire. Brand felt the priority should be building up the capacity of the Ministry of Interior. There was no point in the Coalition churning out police officers if the Iraqis lacked the capacity to maintain and monitor the results – a strategy that put Brand at odds with the US focus on numbers.[11]

Brigadier Jonathan Riley, another Yorkshireman, was grappling with similar problems trying to rebuild Iraq's armed forces. Riley had some sympathy with Bremer's decision to disband the army: the systematic looting of barracks meant the army had nowhere to live anyway. Far more disastrous was the American effort to create a new one. Riley initially joined a team of just four US officers and three reservists, including two car salesmen. The only document they had to work from was a PowerPoint slide showing the pre-war deployment of Iraqi forces. Paul Eaton, the earnest two-star American general running the operation, had been set the task of building nine divisions over three years and was in an understandable state of despair given his resources.[12]

Like Doug Brand, Riley wanted to build sustainable structures at the ministry, but the US secretary of defense Donald Rumsfeld, visiting Baghdad in September, had other ideas. Rumsfeld's handling of the war was attracting increasing criticism in Washington, but there was still a swagger to the leading architect of the invasion. Rumsfeld told a large crowd at Camp Victory, the US military base in western Baghdad, that he wanted Eaton to create the nine divisions three times as fast.

'Yes sir,' said the hapless Eaton.

When Rumsfeld was informed of the probable costs – $2 billion in the first year – he considered the sum too high. 'Now that's gold plated,' he responded sarcastically.

Riley broke the awkward silence. He knew a bully when he saw one, and being a Brit gave him a certain licence to speak his mind. The cost of operating a single British or American armoured division in the field came to the same amount, and Rumsfeld was being offered an entire army. Rumsfeld sat back, thought for a moment, then slapped his thigh. 'OK, you got it,' he said.

'Of course, it took most of that year for the money to get through the system,' recalled Riley.

As for rebuilding the economy and creating jobs, that task had fallen to Tom Foley, a former venture capitalist at Citicorp turned private-sector consultant. Foley looked at Iraq's moribund state enterprises, which employed 30 per cent of the country's workforce, and reached the startling conclusion that they should either be sold off to the private sector or closed down. What Iraq needed, he believed, was exposure to the harsh realities of the free market.

Eric le Blan, a suave French businessman hired independently by Iraq's finance ministry to do much the same job as Foley, had come to a very different opinion: Iraq needed jobs and gentle integration into the world system. Some of Iraq's industries had potential as private companies, it was true, which was why le Blan arranged for Basil Rahim, the director of the British investment company Merchant Bridge, a £100 million fund for whom le Blan also worked, to discuss with Foley possible public–private partnerships with a few Iraqi firms. Rahim came from an old Baghdad family, and his sister had just been appointed Iraq's ambassador to the US.

The meeting started amiably enough, but when Rahim had the temerity to question Foley's more sweeping plans for privatisation, the American flew into a rage that culminated in him taking Rahim in a headlock and literally shoving him out of the room. Afterwards le Blan joked with Rahim about Foley's extraordinary defence of his radical free market philosophy.[13] Foley's plans were shelved when Greenstock threatened to withdraw British support for the Coalition Provisional Authority, a rare occasion when the British veto was actually deployed against American plans.[14]

The strains were also starting to tell on Charles Heatley, the young British spin doctor. He had kept his job with the advent of the CPA and over the summer had been one of the Coalition's main spokesmen, dealing primarily with the Arabic media. Most days he would sit in at Bremer's morning meetings, watching the ambassador flash through decisions. At the end he would have a quick meeting with Bremer to finalise communications strategy, before rushing to the podium in the Baghdad Convention Centre in the Green Zone to be grilled by the press. The minute gap between decision and declaration was exhilarating; it sometimes felt like he had made the calls himself. He found he had a politician's ear for numbers, which he fed seamlessly into his answers: a thousand kilometres of electric piping laid here, a dozen schools refurbished there.

The press, however, had grown tired of Heatley's charms. The more numbers he threw at them, the more hostile their questions became. Part of him understood what drove the media – bad news made headlines. It wasn't so long ago that he had been sitting by the pool in the Hamra Hotel poking fun at the failings of the Americans with the reporters, but as the months passed and the security situation prevented travel outside the Green Zone, Heatley increasingly believed in the positive spin he was giving the news. He soon found himself hating the media.[15]

In early October Bremer returned to Baghdad determined to salvage his seven-step plan by offering to recognise his council of Iraqi advisers as a provisional government if they in turn would convene a constitutional convention. But with Grand Ayatollah Sistani's fatwa calling for elections first, few Iraqi politicians were keen to embrace Bremer's idea. Scott Carpenter finally resolved the deadlock by suggesting an interim constitution that would pave the way for a provisional government and elections – thus satisfying Bremer's desire for some form of constitution to be in place before an American handover of power. After the elections another – permanent – constitution would be written, and further elections held.

It wasn't exactly elegant. Greenstock dubbed it the 'two chickens, two eggs' solution and was starting to doubt whether Bremer would be able to push it through. From conversations with Sheinwald he had picked up on an equally sceptical mood in Washington. With Bush's re-election campaign just over a year away there were the first signs of edginess. Bush had recently announced a revamped Iraq team. Rumsfeld, whose grasp on the chaos he had created was tenuous, was removed from managing Iraq, and Rice put in charge. She in turn appointed Robert Blackwill to bring Bremer under greater supervision. Blackwill was a senior adviser to the National Security Council and had a reputation for being a blunt cudgel.[16]

Greenstock wondered if Bremer was being set up to carry the can. He had little sympathy for the US administrator but feared that the White House, its eye already on Bush's re-election, was preparing to ditch elections altogether in favor of a speedy handover to the Iraqis. Greenstock was sufficiently concerned by the prospect to fly with Jack Straw to Washington on 13 November to brief Powell and Rice on the British position: a too-hasty handover without elections would undermine the whole mission. Bremer was also summoned back by Rice to explain his latest thinking. Both Bremer and Greenstock had the distinct impression they were briefing against each other, although, for once, their positions were not too far apart. At a National Security Council meeting Bremer once again fended off the Pentagon's calls for a rapid handover. A July end date for the CPA was finally agreed with a March deadline for writing an interim constitution. But on the crucial issue of how to choose the provisional government that would take charge, Bremer opted for American-style electoral colleges rather than a simple poll.

Back in Baghdad, Bremer announced the accelerated plans for a handover. After the imperial disdain with which Bremer had treated them, the Iraqi Governing Council was alarmed at the prospect of his early exit.

They quickly signed on the dotted line. Any sense of progress did not last for long, however. Grand Ayatollah Sistani denounced the plan and insisted on elections first. Within days the Shia members of the group were back-pedalling.

Bremer's return to the drawing board might have been a moment of quiet satisfaction for Greenstock if the security situation had not been steadily worsening. Greenstock had also detected Iraq fatigue in London. Previously, when he wrote a gloomy assessment on the state of the country's economy or army training, he was sure to have Sheinwald on the phone within half an hour, possibly even the prime minister, anxious to get a detailed breakdown. Blair was tuning out. It was not simply that a man whose persona was built around optimism was struggling with endless bad news. An alarming sense of powerlessness had gripped everyone in his team, confronted by the American political machine and the dark processes of Iraq. Britain was getting sucked into a quagmire.[17]

In London before Christmas 2003, Greenstock telephoned Sheinwald from his London home to vent his frustrations.

'There will be opportunities for terrorist cells to grow in Iraq if we lose control,' he told Sheinwald. 'Al-Qa'eda will build up there and in Afghanistan. And one day this will explode in the streets of London.'

Sheinwald tried to joke his way out. 'Now I know you've been in Iraq too long,' he said.[18] But the thought lingered for Greenstock.

Chapter 9

NEOCOLONIALISTS

By October, seven months into the occupation, the British administration in the south had at last accepted the trappings of imperial power and moved out of the deteriorating Old Electricity Building and into Saddam's former home in Basra. Hilary Synnott felt a little awkward about the British occupation moving into a palace that symbolised for Iraqis corrupt and excessive rule while the rest of Basra still struggled for basic services. As though to compensate, Synnott issued a memo for all employees to call the palace Al-Sarraji after the park that had once stood there.[1] Few bothered.

Raised in the dying days of the empire, Synnott, like many of his generation, had been taught that Britain's imperial history was tainted by bigotry and racism. More recent politically correct views were even less approving. The BBC had hosted an online discussion based around the view that 'the Empire came to greatness by killing lots of people less sharply armed than themselves and stealing their countries'.[2]

Nowhere was the need to overcome the legacy of empire felt more keenly than at the Department of International Development, the government ministry Blair had created in 1997 and charged with eliminating world poverty. Since its inception, DFID had developed an earnest enlightened philosophy towards the developing world. Instead of funding large-scale programmes that required Western firms to build and operate them, and often assumed a paternalistic attitude to the host government, DFID would develop indigenous capacity to choose and run their own development projects.[3] The department mandated that 90 pence in every pound of aid money would bypass Western firms and be given directly to recipients in the poorest countries.

The result was a government department lauded by other aid agencies and the UN but unfortunately ill suited to the demands of southern Iraq. Despite the departure of Clare Short, Iraq still provoked unease in the department, where officials saw it diverting time and resources away from poorer countries. Many were suspicious of President Bush's motives in Iraq and felt aiding the country would be helping to advance his political agenda. Then there was the fact that Basra had no functioning government or civil society to deal with, so aid agencies had to go through British and American intermediaries. DFID officials found themselves completely at odds with many of the grand ambitions of the British authorities in Basra, which went well beyond their mandate. Part-time soldier turned finance minister Andrew Alderson wanted to completely overhaul southern Iraq's creaking socialist economy, and had requested DFID funding for thirty-seven additional staff to do so. Meanwhile, the military was asking for £77 million for an emergency infrastructure programme to restore power and water supplies that would see Royal Engineers doing much of the building work.

The DFID's permanent undersecretary Suma Chakrabarti thought it unacceptable that Britain in the twenty-first century should be discussing how best to rule the natives.[4] He had joined DFID at its inception and had played a key role in shaping the department's philosophy over the previous six years. In many ways Chakrabarti, a second-generation Briton, embodied the UK's post-colonial transformation, but he was also pragmatic enough to realise that unless the DFID started doing more, the department might become a scapegoat for the occupation's failings.[5]

In early October Chakrabarti travelled to Basra with the new DFID minister Hilary Benn to assess the military's proposals. He found Major General Graeme Lamb in a foul mood. On 15 September soldiers from the Queen's Lancashire Regiment had arrested Baha Mousa, a twenty-six-year-old hotel receptionist, and six others after weapons were found in the hotel safe where they worked. Hooded and cuffed, the Iraqis were taken back to the Lancashires' base in central Basra. Despite Nicholas Mercer's efforts to, the theatre-wide ban on hooding detainees that his headquarters had issued had not been repeated by subsequent commands. Over the next thirty-six hours Mousa and his colleagues were subjected to vicious abuse by up to eight British soldiers. Corporal Donald Payne, a thirty-three-year-old from Preston, took delight in beating the Iraqis, describing their cries of pain as 'the choir'.[6] Mousa, who had complained of suffocating under his hood, was singled out and beaten separately in another room. His ordeal only ended with his death. At a post-mortem he was found to have ninety-three separate injuries.[7]

Graeme Lamb could not excuse such abuse, but he did point to the six months of mistakes and mismanagement. He was looking for answers and was only partially mollified by Chakrabarti's offer of £12 million in assistance from the DFID when they met at Basra airport.[8] The remaining £65 million that Lamb wanted for the overhaul of Basra's infrastructure would have to be found from the Americans once the projects were under way. After the meeting Lamb jokingly told his chief of staff he could 'take the pins out of the DFID doll',[9] a reference to the combative relationship between the military and Chakrabarti's department.

Chakrabarti also had his reservations. As he later told the Iraq Inquiry, 'If you just do them [reconstruction projects] through UK military designing, doing the actual work and project-managing it, you are not really involving the Iraqis in any sense whatsoever. You may feel they should be grateful to you because you have done this for them. Frankly, my experience is that people aren't grateful if you have done something for them; they are only grateful if you help them do something for themselves.'[10] But at the time he kept his thoughts to himself.

While the reconstruction effort stuttered along in Basra, the British were also confronted with the challenge of governing the rest of southern Iraq: tens of thousands of square miles of desert and marshland, dotted with hundreds of small towns and villages. American forces in central Iraq had quickly found themselves competing for influence with a determined and deadly insurgency that drew its strength from the Sunni tribes. In the south Shia militias like the Jaish al-Mahdi, which followed the anti-occupation cleric Moqtada al-Sadr, were also growing in strength in provincial towns. So far they had lacked the violent edge of Sunni groups around Baghdad, but the British could not afford to give them too much space to develop.

The task of creating friendly provincial councils and guiding the reconstruction effort in these Shia backwaters fell to a handful of civilian administrators who would work alongside the military as de facto governors in each province. The FCO had struggled to find the right candidates for the positions; not since the heyday of empire had the British civil service produced officers who combined a sophisticated understanding of tribal dynamics with the ability to live for months in isolated and hazardous conditions. Forced to look beyond its ranks, the FCO was relieved to find a handful of candidates with the unusual résumés needed.

One of them belonged to Mark Etherington, a former paratrooper captain who had left the army and ridden across Africa on a motorbike, returning to work with the European Community Monitoring Mission in Bosnia. At forty, Etherington still carried himself like a soldier. He had a Roman countenance, with a square jaw and heavy nose and sandy hair just turning grey. Etherington had grown up in Kuwait in the 1960s, where some of the last traditions of empire were observed: gin and tonic at the club and the national anthem every morning at school. He had an instinctive sense that British values were worth sharing. Abrupt and impatient at times, Etherington was also thoughtful and introspective, which explained why he had left the army for the challenges of nation-building. He would become governor in Wasit province, closer to Baghdad than Basra, with Ukrainian troops providing military support.

A second résumé, no less impressive, belonged to Rory Stewart, who had left the FCO two years before to walk across the Middle East to India. Stewart was even more a child of empire than Etherington. His father had been a senior British diplomat in Malaysia when the colony was handed over. Though the family had ultimately retired to their estates in Scotland, the hankering for life overseas had never left Stewart. He attended Eton like many a colonial administrator before him, demonstrating exceptional academic ability. While he was at Oxford he was selected to tutor the young princes William and Harry. He tried Sandhurst but was too much of an individualist for military life. The FCO at least offered the romantic allure of the gentleman traveller. His 3,500-mile odyssey across the Middle East had given him a deep insight into the pride and poverty of the Islamic world. When Stewart was interviewed for a job in Iraq, he was thirty years old, still scrawny and tough from his hiking, with an excitable and mobile face that rippled and flickered with his quick thoughts. He would be given the job of deputy governor in Maysan province, the troubled tribal area where the six British Red Caps had been murdered a few months before.

Another potential governor was a civil servant with a no less unusual background. John Bourne was a fluent Arabic speaker who had worked in Kuwait for a number of years. Another old Etonian, he was also a qualified vet whose family owned a country estate. A few years before, he had switched from the FCO to the Department for Environment, Food and Rural Affairs, working at the Cabinet Office on European agricultural policy. In his late forties he still longed for the open spaces of the Middle East. He would take charge of Dhi Qar province, supported by the Italian military. A final candidate was Emma Sky, a self-declared peacenik and former British Council worker in Manchester, who had volunteered for Iraq and was already working in the northern city of Kirkuk. She would become governor of the mixed Kurdish and Arab province, working alongside American troops.

For those in the FCO with a keen sense of history, turning to such an eclectic although undoubtedly talented group came with a pedigree. In 1915, when the British were contemplating how to run a potential Middle Eastern dominion carved out of the remains of the Ottoman empire, they had turned to an errant group of archaeologists, journalists and rogue diplomats to populate the newly christened Arab Bureau in Cairo. Among those attracted to Arabia were T. E. Lawrence and Gertrude Bell. After the invasion of Mesopotamia they were joined in the fledgling administration by talented Indian colonial officers like Harry St John Philby, the Arabist Harold Dickson and Bertram Thomas, a sailor's son who became the first Westerner to cross Saudi Arabia's Empty Quarter.

Back then Britain had little idea what to do with its newly won territory but while the politicians debated at home there was a country to run. The likes of Philby, Dickson and Thomas served as district officers in isolated outposts across southern Iraq. Thomas's descriptions of his work in Suq al-Shuyukh, in the marshes just north of Nasiriyah, bears comparison with the experiences of the class of 2003: the isolation, the desperate conditions and the wily locals seeking to take advantage of his ignorance.[11]

When Stewart arrived as deputy governor in Amarah, the capital of Maysan province, in September, he had already read Thomas's memoirs and taken to heart the advice it contained on what Arabs respect in a government: 'power, the will to use that power, and a genuine concern for the welfare of the people'.[12] Climic House in downtown Amarah did not have quite the same loneliness as Thomas's spare lodgings. Amarah was a bustling town of 250,000 people on the northern fringes of Iraq's marshes. Located on the highway between Basra and Baghdad, and thirty miles from the Iranian border, Amarah had traditionally served as a trading and smuggling hub. Its market usually heaved with vendors selling cheap Chinese wristwatches and radios, sacks of ground turmeric and cumin, and haunches of lamb and goat.

The CPA headquarters was a short distance down the street, a drab two-storey building which had recently been equipped with a cafeteria serving burgers and burritos, and picnic tables on an open patio overlooking the confluence of the Tigris and its smaller al-Kahla tributary. Portacabins provided rudimentary accommodation. The name 'Cimic' was derived from civil–military affairs, referring to the military's small team of specialists who worked with the Iraqis on reconstruction. The governor's unmistakable pink offices stood just opposite, the two buildings forming the heart of Maysan's administration.

Like Thomas before him, Stewart had a problem with the local sheikh. Within a few days of his arrival, Abu Hatem, the Prince of Marshes, came calling. Since the murder of the Red Caps some in the military had speculated that their ally might have played a role in their deaths. The wily sheikh had left for Baghdad a few days after the murderous event. He had met Bremer, who took a shine to him and appointed him to his advisory panel. Hearing about the CPA's plans to appoint provincial councils, Abu Hatem returned to Maysan shortly after Stewart to make sure he got the top job.

Ninety years earlier Thomas and his fellow district officers had regarded the tribal leaders as local aristocrats and, much like their counterparts in the UK, the natural leaders of society. For those colonial officers with aristocratic credentials, a feudal system of government was one of the attractions of Arabia after the unfortunate way in which democracy had diluted their power and standing back home.[13] Such a system was of course not an option for Stewart, despite his own landed pedigree. His mission was to build a broadly representative provincial council to oversee reconstruction projects and local security forces. The challenges Stewart faced were twofold: the province's professional and educated class was tiny and restricted to Amarah. Outside the main city, tribal rule still dominated. Somehow Stewart had to forge a consensus between the town and the country that would not immediately fall apart, or allow tribal leaders like Abu Hatem to cow the city elite. In addition, religious groups like Moqtada al-Sadr's needed to be engaged and prevented from undermining the council.

When Stewart met Abu Hatem in October there was an immediate mutual antipathy. Stewart viewed Abu Hatem as little more than a wayward thug with no interest in democracy. Abu Hatem saw Stewart as a precocious upstart who would not last more than a few months. The only thing they agreed on was their shared dislike.[14]

If that was an inauspicious beginning in Maysan, one hundred miles to the north Mark Etherington was also running into difficulties with the locals. He had been appointed governor in Wasit, a central province the northern fringes of which touched the suburbs of Baghdad. The largest town was Kut, with a population of around 300,000. The British military had endured a humiliating surrender to Turkish forces there during the First World War, but the town had since returned to dusty obscurity, a truck stop between Basra and Baghdad. Like Stewart, Etherington favoured building councils that balanced the feudal ambitions of the sheikhs with the professional classes, but he drew the line at working with Sadr's followers.[15]

Since the British had resisted American efforts to arrest Sadr, the cleric had made a bold effort to seize power in the south, besieging Sistani's headquarters in Najaf and shooting up a police station belonging to a rival political party in Karbala. Sadr was organising faster than the British: in mid-October he announced his own provisional government for Iraq and declared any government established under British or American auspices illegitimate. Many suspected Sadr's militia was behind the growing number of threats received by those who worked with the Coalition in central and southern Iraq. Etherington considered Sadr's followers to be a dangerous rabble but for the former paratrooper to triumph he needed to start demonstrating the benefits of siding with the occupation.

On 4 November Etherington and the other governors travelled to Baghdad to hear the latest American plan. At the Madrid Donors' Conference for Iraq the previous month the United States had pledged $19 billion for Iraqi reconstruction.[16] Before a packed auditorium a young man in chinos from the CPA's Office of Policy Planning and Analysis presented a PowerPoint show that flashed through a bewildering list of objectives: privatising state-owned enterprises, doubling energy production, reforming university curriculums and holding elections.[17] Etherington was astonished by how out of touch the Americans were. What he and the other governors and military officials needed were simple goals to address failing infrastructure and chronic unemployment.

Even the American military felt left out of the loop. 'Did you just say that you have briefed this plan to the highest levels in Washington without consulting any one of us?' US commander General Ray Odierno asked incredulously. The lumbering general was working closely with Emma Sky, the governor in Kirkuk.

'You're being consulted now,' replied Bremer icily. The meeting adjourned in stunned silence, as if everyone in the room was privately digesting the knowledge that the plan had little to offer them.[18]

John Bourne, the British governor of Dhi Qar province, drew a simple conclusion from the meeting: ignore Baghdad whenever possible. Bourne was also critical of the British administration in Basra, which seemed a microcosm of everything that was wrong with the FCO – insular, hierarchical and penniless. Since arriving in Nasiriyah, a hundred miles to the north-west of Basra, Bourne had taken to touring the country meeting tribal leaders. He had quickly realised that reconstruction projects like painting schools – favoured by the Italian forces supporting his mission – would do little to win over the public.

The lasting structures Nasiriyah needed were not physical buildings but a government that would gel the different tribes and political parties. At that he had proved remarkably adept, setting up his provincial council within a month of arrival. Where others thought the Iraqis grandstanding and grasping, Bourne understood that his discussion with the tribal sheikhs was really a form of play, a meander between fact and fiction and the elaborate dictates of courtesy, with neither side taking the other too seriously.

In many ways Bourne resembled the most ebullient district officer of that earlier generation of colonialists. One of the last Britons to rule the province of Dhi Qar was Harold Dickson after the First World War. Not a man destined to rise far in the FCO, Dickson was nonetheless the unassuming glue that held the region together. On leave in France after the war, he had boldly proposed marriage to an English bank clerk he had met behind the till named Violet. The two decamped to Nasiriyah a few months after their wedding in Bombay with a set of crockery from Harrods, a case of fine wine and a dozen starched white shirts. They had soon started entertaining the local sheikhs, giving Dickson deep familiarity with the lives of his subjects and the opportunity to speak for hours in the elaborate patterns of formal Arabic.[19]

Bourne adopted similar open-door policy at CPA headquarters in downtown Nasiriyah, following tribal customs. The result at times bordered on anarchy as Arab sheikhs wandered around the small compound, in and out of offices, accosting Bourne and other members of his staff with pleas for jobs, electricity and protection against sheep-stealing. At least it was transparent government, Bourne noted wryly to his staff in the dining hall at the end of another wearying day.

He was also ready to explore what democracy in the region might mean. Bourne's deputy Adrian Weale had started holding local elections for town councils, some of the first in Iraq's history. Weale, a former intelligence officer turned Second World War historian, had arrived in Nasiriyah a few weeks before Bourne, and was immediately accosted by a small delegation from Al-Rifai, a town of 125,000 in northern Dhi Qar. The delegation was incensed by the corruption and venality of the council the US military had appointed on their sweep through the area during the war. They demanded elections.

Weale had no idea how to hold them, but he promised to look into the matter. His wife was a borough councillor for Kensington and Chelsea. He emailed her and asked if she could send over some voting guidelines. The main obstacle Weale faced was lack of census data – no census had been conducted since 1974 for fear of exposing the population growth of the Shia compared to the Sunnis. The only reliable data for each town was the UN-supplied ration card, provided to each head of household during the sanctions. Weale formulated a plan to give one vote per ration card and promptly held the election in the local school at the end of August. It was crude but the townsfolk accepted the result. Within a few weeks other towns began to approach Weale. On his arrival Bourne gave him free rein to expand his programme, and by the time Weale left at the end of his tour in November he had held a second election in Suq al-Shuyukh, Bertram Thomas's old stamping ground in eastern Dhi Qar.[20]

The initiative might have ended there for lack of staff, had not Bourne's team been joined by a young US State Department official called Tobin Bradley. He was a rarity among American officials in Iraq – a career diplomat who spoke conversational Arabic. In the middle of his third posting to Brussels Bradley had felt an unabashed call to service and accepted a job in Iraq. He had a deep-rooted idealism combined with a passion for order that evoked the kind of attitude made famous by Graham Greene's *The Quiet American*. Bradley found the chaos in the CPA compound intolerable. The first thing he did was set up a front desk to manage access.

Bradley saw the potential of Weale's election scheme. Properly organised and systematised it could be rolled out to every town in the province, a powerful statement of the West's intent and just the sort of grass-roots democracy-building the country needed. The next election was held in Al-Dawaya, where for the first time they met opposition from a local sheikh. On the day of the election the Iraqi potentate stood outside the school hosting the election with a dozen gunmen. Bradley calmly approached. 'You are an important and respected leader of this town. If there is any violence today, it will be on your head,' he said. The sheikh looked alarmed and fled. Bradley was briefly surprised at his own forcefulness and success, before looking behind him to find a couple of Italian tanks approaching, which he had failed to hear above the beating of his heart. The elections went ahead without further interference.

Bradley immediately began working out how he could improve future elections and increase participation. He worked with the burgeoning student and women's groups to get the word out and devised a 'city council in a box' programme to support elected city councils. For £1,200 each council received office supplies, basic furniture, a computer and printer, a refrigerator and sometimes a car for council use – the sort of perks Bradley hoped would encourage locals to stand for office. He also changed voting rules to allow two votes per household – one for a man and one for a woman – to encourage female voters.[21]

Each night he excitedly shared his thoughts with Bourne. The two men shared a tent, although Bourne was a workaholic who spent most nights at the computers located in the CPA recreation room.

Despite his success, Bourne's dismissive attitude to the hierarchy in Baghdad and Basra would catch up with him. By December Bremer had belatedly recognised the need for a 'jobs drive' to counter the worsening insurgency elsewhere in the country. Every province would be given $2 million to create jobs and report progress back to Bremer. It was a gimmick, but Bourne had duly made an announcement on local television announcing that jobs would be auctioned off at the local football stadium.

He had not thought through the plan: on the day thousands showed up to find only a few hundred positions available. A riot broke out and was filmed by the Arabic network Al Jazeera.

Bremer blamed Bourne. 'This undermines everything we're working for,' he fumed. He immediately called Greenstock and insisted Bourne be sacked for gross incompetence.[22] Greenstock defended Bourne, but British officials in Basra were flagging other issues.[23] On a visit to the Dhi Qar CPA headquarters incoming commander Major General Andrew Stewart had been shocked by the disorder – even with Bradley's desk system. Despite Bourne's political achievements, Dhi Qar province lagged in other measurable criteria, such as projects completed and money spent.

Bourne's failure to build support within the CPA had also made him vulnerable to the shifting politics of the Coalition effort in Iraq. To the FCO's delight, the Italian prime minister Silvio Berlusconi was pushing to have his own governor in the south. A suicide bomb attack on the Italian compound in Nasiriyah had killed seventeen Italian policemen in November, and Berlusconi wanted to reaffirm his country's commitment to Iraq. The obvious position to take was Bourne's. He was sacked shortly after the New Year, returning disillusioned to his job in London at the Department for Environment, Food and Rural Affairs.

His replacement raised eyebrows. Barbara Contini was a forty-four-year-old Italian who favoured low-cut blouses and interacted with her male staff by grabbing their arms and whispering '*Amore*' in their ears, fine in Italy perhaps, but offensive in conservative southern Iraq. Concerned about her effectiveness, the British authorities in Basra kept Bradley on as Contini's political adviser and asked Rory Stewart, deputy governor in neighbouring Maysan province, if he would move across to Dhi Qar.

Stewart was ready for a new challenge. Over the last few months his star had been steadily rising as he proved himself adept at everything that Bourne was not. A natural administrator who managed to combine charming idealism with a knack for working the channels in Baghdad and Basra, he had secured $110 million from Baghdad for a school-system reform and devised a lottery system for allocating jobs that managed to keep Iraqi expectations in check. However, the arrival in Maysan of an American governer, Molly Phee, the previous autumn had reduced his influence, and he was happy to repeat his success elsewhere.

Shortly before Stewart's departure for Dhi Qar he received a lesson in the perils of neocolonialism. Despite Stewart's opposition, Sheikh Abu Hatem had succeeded in manoeuvring his brother Riyadh into the governor's job. Stewart still felt Abu Hatem was bad news, and Phee had mixed feelings about the tribal leader. Abu Hatem was a powerful counterweight to the growth of the Sadrists, but his sense of self-interest was overwhelming. Unsurprisingly, Abu Hatem's brother had staffed his office with tribal affiliates from the Albu Muhammad clan, and had hijacked Stewart's job scheme and treated it like his petty cash box. His tribal rivals the Beni Lam, urged on by the Sadrists, decided to revolt. In mid-January, thousands gathered on the street between the governor's office and the CPA compound, chanting, 'Death to the governor.'

The crowds melted away by midday, but the next day they returned. Riyadh now contacted Stewart, asking for British help in clearing the street. Stewart, tired of his bruising encounters with Abu Hatem, told him to sort the matter out himself.

'This crowd has a right to freedom of association and freedom of speech,' he told Riyadh over the phone.

'This is Iraq, not Britain,' replied Riyadh.[24]

Outside, the protesters had begun to lob bricks at the governor's building, followed by grenades. Riyadh's bodyguards responded by shooting into the crowd, killing two and wounding twenty. Stewart, now alarmed, changed his tune and asked British Lieutenant Colonel Bill Pointing, commander of the Light Infantry, to protect the governor's building. At the same time he called up Riyadh, asked him to go home and at the very least to stop shooting.

Shortly after midday a couple of sections from A Company of Pointing's Light Infantry emerged from Cimic House in a protective phalanx, riot shields raised, and pushed down the street, provoking a barrage of bricks but succeeding in scattering the protesters. As soon as the phalanx began to withdraw, however, the protesters returned. After a succession of charges down the street, the protesters began lobbing bricks again. Stewart watched from the rooftop of Cimic House as the Light Infantry stood resolutely under a barrage of projectiles, culminating in a firebomb that exploded harmlessly between the crowd and the soldiers. More bombs were lobbed at them by an Iraqi youth in a blue T-shirt until he was finally shot through the head by a British sniper.

Stewart found himself in the awkward position of having ordered British soldiers to defend the rule of a corrupt tribal sheikh installed by his brother through a manipulation of the democratic process. Sporadic violence flared throughout the afternoon and only ended when the Light Infantry withdrew from the streets altogether. The crowd dispersed, although Stewart soon received a call from Riyadh to say that looters had broken into the governor's building, which had been left unguarded. When he asked A Company commander Lieutenant Colonel Jonny Bowron what was going on, he was told the military had the situation 'under control' and that further British intervention risked inflaming the situation.

For Stewart the incident brought into stark relief the issues faced by Britain. The ability to use force was a prerequisite of imperial rule, as Bertram Thomas taught. In 1918, on his first week as an administrator in Iraq, Thomas had accompanied two gunboats up the Euphrates to coerce a rebellious sheikh to accept British rule. They bombarded the sheikh's tented village and 'the bloody head of one of [his] innumerable progeny was recovered from the topmost branch of a distant palm tree'.[25] But wielding power required belief in the mission and, unlike the Americans, Stewart feared Britain lacked that conviction. The next morning Stewart walked over to the governor's office to find the windows smashed, doors kicked in and all the furniture stolen except for the governor's massive desk, too large for the looters to take. Riyadh was staring forlornly out through an empty window frame.

'Why did your soldiers not protect this building from the crowd? You sent home my security forces, dissolved the police line and took responsibility for the building. How did you then let the crowd get in and steal everything? Would you let the mob go stampeding into your own office and loot your computer equipment?'

Stewart had no answer.

Chapter 10

UPRISING

After six months as Britain's special representative to Iraq, Jeremy Greenstock left at the end of March 2004. His relationship with Paul Bremer had settled into curt formality. Bremer's aides minimised contact between the two men by arranging meetings without him, and Greenstock continued to send bleak memos home warning that the CPA's failure to rebuild the security forces or stimulate the economy had brought the country to crisis point. Implicit in his criticisms were Greenstock's frustrations at being sidelined.

It was clear that the Anglo-American relationship in Baghdad was in need of repair, a task that now fell to Greenstock's deputy, David Richmond. Bremer still felt Richmond had acted weakly in not supporting the capture of Sadr the previous summer, but over the following months, as his dislike for Greenstock grew, he had come to appreciate Richmond's insights and found his British manner unthreatening.[1] They did not exactly become friends, but at least Bremer jokingly referred to him as the man who had saved Moqtada al-Sadr.

Richmond took the jibe with good grace. As Sadr had grown in power, the Briton had come to share Bremer's view that a tougher line was needed against the Shia cleric. Bremer had finally broken the deadlock with the other Shia groups by bringing in a UN team to assess whether elections were possible before the US handed power over to a provisional Iraqi government, a key demand of Grand Ayatollah Ali al-Sistani. If there was not enough time before the July handover date, then Sistani wanted the UN to supervise the creation of a provisional government. But despite the compromise, Sadr showed no interest in participating in the political process, and his Jaish al-Mahdi threatened to derail any future Iraqi government.[2]

Over the previous few weeks, Sadr's newspaper *Al-Hawzat* had published a series of highly critical articles about the Americans and the Iraqi politicians who worked with them. One story likened Bremer to Saddam. Another accused American soldiers of abusing Iraqi prisoners – an accusation that would later prove correct. Bremer was furious and decided to close down the newspaper.

The task of carrying out the order fell to the CPA's communications team, which had first flagged the inflammatory newspaper articles. Since Charles Heatley's departure in January, the running of media operations had increasingly fallen to a Republican Party appointee, Dan Senor. In his early thirties, lanky and effete, his dismissive manner irritated many journalists and captured the with-us-or-against-us mentality of the Bush administration. Despite Blair's push for a stronger strategic communications team to increase the good-news stories coming out of Iraq, there was no sign the message was getting across.[3] Little thought was given to how Iraqis would perceive the decision by the Americans to censor a newspaper for criticising them.

Sadr was just waiting to be provoked. The same day the CPA press team and a company of soldiers closed down the newspaper offices, a furious mob marched on the palace. Violence seemed likely until they were met at the gate by a flustered Senor and Heatley's replacement Gareth Bayley, a fluent Arabist, who succeeded in defusing the situation.[4] Richmond, by now a veteran of protests, mentioned the incident briefly in dispatches back to London. He considered the matter put to rest, but in fact a storm was gathering.

The confrontation with Sadr was soon overshadowed by events in Fallujah. On 31 March gunmen ambushed four American contractors working for the security firm Blackwater as they drove through the city. Their SUVs were set ablaze by RPG fire, and the charred bodies of the men paraded through the town, the images broadcast live on Al Jazeera.

In the White House Condoleezza Rice immediately saw the attack as a defining moment for the occupation, with America's ability to deliver law and order on the line. She agreed with Bremer's assessment that immediate action was required to bring the perpetrators to justice. Iraqi leaders in Fallujah would be offered a chance to hand over the killers. If they failed to do so, the US Marines would be ordered in. Military planning for the assault on the city began at once. Bremer did not inform Richmond about the decision, and the senior British officer in Baghdad, Major General Andrew Figgures, was left to fill in the gaps at the commander's daily briefing. The prospect of holding an entire city to ransom created a deep sense of unease in London. There was no question of UK forces taking part in the offensive, but no effort was made to stop the Americans.[6]

Events continued to gather pace in the south. On the evening of 3 April, at Bremer's urging, the US military arrested Muhammad al-Yacoubi, the editor of *Al-Hawzat* and Sadr's top lieutenant in Najaf. Sadr's supporters had protested every day since the newspaper's closure, and Bremer was keen to push back. The next morning the Sadrist uprising began when a crowd descended on the CPA's headquarters in Najaf, which was lightly guarded by Salvadorean troops. The Spanish regional commander tried to defuse the situation by issuing a statement blaming the Americans for the arrest, but the insurgents didn't distinguish between the occupying nationalities. The compound was soon under attack. A Salvadorean soldier was killed along with a dozen Iraqis. Fearful of inflaming the mob further, the Spanish commander refused to deploy troops in support.

The angry scenes were repeated in Amarah where Cimic House was besieged. In Kut Mark Etherington's attempt to mollify the crowd ended up in a slanging match between him and a local cleric. Nasiriyah was the exception. Anticipating trouble after Yacoubi's arrest, the feisty Barbara Contini had left the compound the night before to strike a deal with the local Sadrist leader Aus al-Khafaji. In exchange for peace, Contini promised Italian troops would not enter the city and instead withdraw from their base on the banks of the river opposite. When her American political officer, Tobin Bradley, found out, he warned her that the offer flagrantly contradicted Bremer's orders to take on the Sadrists, not appease them.[7]

The next morning Bradley awoke to the sound of gunfire, and scrambled out of bed to discover the Jaish al-Mahdi had seized the two bridges over the Euphrates linking the main Italian base outside the town and the CPA compound without the Italian military resisting. Contini's deal had not only been disregarded, but the Sadrists were taking advantage of her perceived weakness. With Rory Stewart on holiday in Scotland, it fell to

Bradley to write the memo to Baghdad apprising them of the situation. The Italians' stance incensed Bremer. The marines were beginning their assault on Fallujah, and the last thing he wanted to hear about was deals with the enemy. 'It's in the British sector,' Bremer shouted at one of his aides. 'Get the British to sort it out.'[8]

On 5 April David Richmond arrived at his office opposite Bremer's in the palace to hear the news of heavy fighting in Fallujah. The British representative was staggered by the scale of the violence but he barely had time to read the overnight memos or assess his own feelings when the Coalition received its next blow. Sadrist militias had seized the governor's offices in downtown Basra – the heart of the local administration that the British were trying to build up. By mid-morning there were hundreds of black-shirted men lining the rooftop, including Ahmed al-Maliki, the erstwhile director general of education,[9] and the head of Sadr's political office, Abdul Sattar al-Bahadili. The turbaned cleric was seen striding across the rooftop brandishing a scimitar.

Major General Andrew Stewart, the new British commander in Basra, could scarcely contain his annoyance with the Americans' aggressive approach in shutting down Sadr's newspaper and invading Fallujah. When Bremer called and instructed him to 'take back the building', Stewart disregarded the orders. Uncomfortable though it was to have militiamen preening before the television cameras, he felt it was largely symbolic, better to be ignored. He did not have enough resources to take on the militia and didn't want to run the risk of a major battle in the middle of the city. His brigade commander, Nick Carter, would try and talk the militia off the roof so that Stewart could tackle the crisis in the Italian sector, which the Americans were also pressuring him to resolve.

Stewart's forty-minute helicopter ride to the Italian base just outside Nasiriyah took him over the western fringes of the marshes. The handful of villages along the margin looked peaceful from above. In the past Stewart had made sure the trip coincided with lunchtime, when the Italians were at their best. They appeared to have spent much of their aid budget on building an exquisite dining hall in the base, with fresh pasta made on site. By the end of each visit the giant white moustaches of the Italian commander, Brigadier Gian Marco Chiarini, were invariably stained red with tomato sauce. This time the meeting was perfunctory. Chiarini promised that he would take action and drive out the Sadrists.[10]

By the time Stewart returned to Basra several hours later, Nick Carter had got the militia to withdraw from the governor's offices with a little forceful diplomacy. Surrounding the building with tanks, he had negotiated in person with Bahadili from the steps outside, shouting up to the cleric via a translator. Stewart anticipated Italian troops would meet with similar success when they moved back into Nasiriyah the next morning.

At dawn the following day, in the CPA compound in downtown Nasiriyah, Tobin Bradley woke to the rumble of explosions. He slept in his office on the second floor of the CPA building with just a thin layer of concrete over his head. Scrambling downstairs on the orders of the American private security team who guarded the diplomats, he made his way to Contini's room. It was the only one on the ground floor that did not have a wall backing on to the street. As the sounds of explosions got louder, he saw Jeremy Nathan, another American diplomat, on his hands and knees, banging desperately at Contini's door. It swung open, revealing Contini dressed only in a nightie.

'Come here, my darlings,' she said. Sitting back down on her bed she opened her arms and pulled the two men towards her. They sat there locked in Contini's surprisingly strong grip until the sound of explosions eased.

True to his word, Chiarini had ordered Italian troops back into the city, but not to retake the bridges. The Italian major defending the CPA headquarters had discovered the previous day that he was out of ammunition, so a resupply run was needed. Chiarini thought he could do so without breaking the deal Contini had struck with Khafaji to stay out of the city, but militiamen had opened up on the convoy as soon as it approached the city. The Italians had got through to the CPA compound and dropped off the ammunition but had no desire to provoke a larger battle. They now pulled out through another storm of fire, after which the city fell quiet, once again in the hands of the Sadrists.

Ninety miles to the north-east, the situation in Kut was also rapidly deteriorating. The previous day the Ukrainians had followed the example of the Italians and Spanish and withdrawn to their base, a fenced aerodrome two miles outside town on the other side of the river. Fearing a Sadrist coup, Mark Etherington had driven over to see the Ukrainian commander Brigadier Sergey Ostrovskiy in Camp Delta and cajole him into action. Instead he found the Ukrainian preparing to meet the Sadrist leader for talks. Etherington was appalled. Now was the time to show strength, but clearly the Coalition's junior partners either disagreed with American strategy, or did not support it enough to risk their lives defending it. Just when it all mattered, the Coalition was falling apart.

Driving back to CPA headquarters, Etherington brazened his way through a Jaish al-Mahdi checkpoint. That night he lay awake for hours seething with frustration. Every moment the militia was in control of the streets his authority ebbed away. The only thing worse would be to abandon his headquarters in Kut. He had just created a provincial council that he felt balanced the aspirations of the province's small professional class with tribal and religious groups, and could, given time, stand up to Sadr's followers. Etherington resolved to stick it out, come what may.

In Baghdad Richmond was reaching an opposite view. There was little he could do to stop the Americans escalating the conflict, and the dangers facing British personnel, for whom he was directly responsible, made him feel queasy. If one of his CPA teams in the south was captured or killed, it would call into question the whole UK effort in Iraq. As he tried to manage the stream of alarming reports from the south, Richmond counted a dozen requests from the prime minister's foreign policy adviser Nigel Sheinwald for detailed analysis on everything from power supplies to Sunni reconciliation. He wished Blair had shown this level of interest before the situation reached crisis point.

Needing to clear his head, Richmond went for a walk in the garden at the back of the Republican Palace. It was hot and humid outside. Richmond took a deep breath before he realised he was not alone. The UN's special representative Lakhdar Brahimi was pacing under a grove of palm trees. Richmond had scarcely seen him since the fighting began, and he realised for the first time the strain the old Algerian diplomat must be under. Having concluded that elections were not possible before July, Brahimi was overseeing the creation of a provisional government. He was in effect bailing the Americans out of a political crisis just as they seemed determined to provoke an even greater military one. The sight of someone more distressed than him gave Richmond a sudden feeling of calm.

'Hello, Lakhdar. How are you doing?' he said, approaching Brahimi.

Brahimi stopped, locked him with a watery gaze and launched straight in. He had spent the last few hours watching Al Jazeera's coverage of US air strikes on Fallujah. Dead women and children were shown being dragged from ruined houses. As an Arab, as a Muslim, he could no longer countenance America's actions. 'I don't think I can stay any longer,' he told Richmond.

Richmond understood the dangers at once. The UN – Brahimi's presence – was the only thing keeping the political process going.

'I feel the same way,' he told Brahimi, 'but if you leave we'll have chaos.'

'What can I do?' asked Brahimi with a despairing look.

'Let me speak to the Americans,' said Richmond. 'I'll try and make them stop.'

He left the garden unsure whether his words would have any effect but hoping that Brahimi's threat was a political bombshell that would force a rethink. Back in his office he wrote the most strongly worded memo of his career. It did not take long for Sheinwald to call. Blair's foreign policy adviser had been feeling as powerless as Richmond to affect the Americans. He believed Britain's role was to stand beside the Americans, but even he could see that this time they needed to be stopped.

'Who else knows about Brahimi?' asked Sheinwald. 'And can you keep a lid on it? I'm calling Rice.'[11]

Unlike Manning, Sheinwald had struggled to establish a rapport with his American counterpart. Rice appreciated Manning's courtly gestures, so English and genteel, and recognised the effort it cost Sheinwald to behave the same way. She listened to Sheinwald as he strained to sound measured, describing Richmond's encounter.

The resignation of the UN's most senior diplomat on the ground would make the situation 'internationally untenable', Sheinwald warned.

Rice agreed with the assessment. She had supported the decision to go into Fallujah but blamed Bremer for unnecessarily provoking Moqtada al-Sadr by closing his newspaper. She promised Sheinwald she would raise British concerns at a National Security Council meeting planned for the next day, but warned it was going to be hard to shift Bush.[12] The next morning Robert Blackwill, Rice's man in Baghdad, informed Bremer of Brahimi's possible resignation. Barely containing his frustration, Bremer met the Algerian diplomat and urged him not to resign. The US administrator ran through Sadr's crimes over the previous year, but Brahimi was only concerned about ending the US assault on Fallujah. He again threatened to quit.[13]

'The international community in Iraq must stay united,' was Bremer's response.[14]

That same morning, in Kut, Etherington awoke to find the Ukrainians, under pressure from the Americans, out in force on the streets. Pleasantly surprised, he was just preparing to return to the Ukrainian base on the other side of the Tigris to speak to the newly invigorated Ostrovskiy when the floor rocked to the sound of an explosion. Clambering onto the hotel roof, he saw black smoke billowing from a building across the river and Ukrainian troops moving into position. A machine gunner in a Ukrainian armoured personnel carrier began firing at the building. The militia responded with a RPG round that exploded over the river, followed by another that struck the riverbank beside the compound wall less than 200 metres away from where Etherington was standing. Gunfire now erupted across the city. The Ukrainians disengaged and sped back to their headquarters.[15]

With the city unguarded and the sound of fighting close to the compound, there was an air of panic at the CPA headquarters as Etherington gathered together the forty or so members of the team. There was no question of leaving now, he told them. Those helping the defence of the compound needed to be in the main villa or up on the walls. The rest needed to find shelter in a second building, a former hotel used for CPA accommodation. Etherington had a dozen men from his close protection team and a Ukrainian garrison force of twenty to protect a perimeter almost a half a mile long. It wasn't enough and he knew it.

Neil Strachan, the political officer and a Territorial Army reservist, was in charge of reporting to David Richmond's office in Baghdad. He and Etherington had agreed on a line to Baghdad: to withdraw from Kut would fatally damage the CPA, and the cost of reclaiming the headquarters would be higher than holding. It was good fighting talk, but back in his office, covered in dust with dark patches of sweat on his shirt, Etherington felt his bravado fading, and he slumped into a chair. Jaish al-Mahdi fighters had reached the rooftops behind the compound and he could hear his bodyguards from Control Risks Group opening fire. Through his office door Etherington spotted a Ukrainian sniper knocking out the window on the staircase near his office to begin shooting. The loud report echoed around the hallway. A mortar round landed nearby with an ear-splitting thud. There seemed to be little prospect of the Ukrainian garrison coming to the rescue, and Baghdad was a hundred miles away across the desert. Had he condemned them all to pointless deaths?

His bout of indecision lasted fifteen minutes. Then he was back on his feet and requesting air support from Baghdad over a satellite phone. If this was to be the end, then he would go down fighting. He had his suspicions that the Ukrainians were considering a unilateral withdrawal from the compound, so he asked for British or American troops to be choppered in, preferably special forces, along with supplies and ammunition. Non-essential staff could then be airlifted out.

Etherington ran outside to assess the state of the defences. The Ukrainian detachment and CRG team had checked the advance of the militia, firing short bursts from behind sandbags atop the compound wall and roof of the main building. Etherington's American deputy Timm Timmons was marshalling the troops and Neil Strachan was recording the grid references of the rebel mortar teams firing at the compound. Over at the hotel, however, morale had plummeted. The fear of death, which Etherington had felt himself, had festered as the shadows lengthened. In the hotel hallway an angry member of the American contracting company KBR confronted Etherington.

'Why didn't we evacuate earlier?' said the man, quivering with emotion. 'If we have to spend a night here we're all going to die,' said another.[16]

Etherington did his best to reassure them that reinforcements had been requested and that non-essential staff would be airlifted out. 'No one is going to die,' he insisted. It was not exactly Corporal Jones's inspirational speech from *Zulu*, and it was clear that CPA Kut was disintegrating. Etherington discovered later that the KBR man had been sending a lurid stream of emails to Baghdad prophesying that they were being overrun and questioning the Briton's leadership.

As darkness fell, the Ukrainian commander in the compound confirmed Etherington's fears: his contingent was going to withdraw at first light. Etherington immediately began calculating how he could hold the site without them. Official written correspondence with CPA headquarters in Baghdad had ceased early on in the fighting. He received an email from the private account of an officer in the CPA's regional headquarters – the email account was called deployeddaddy@ald.com – ordering him to withdraw. He ignored it. With little sense of the greater battle waging across the country, Etherington was surprised to learn the US military had no reinforcements to spare other than an overflight of jets. They offered to drop a bomb, but without a good target Etherington was not going to risk the lives of innocent Iraqis.

Increasingly worried about Etherington's insistence on defending the compound, Richmond sent the former paratrooper an unequivocal message to withdraw with the rest of his team. 'A civilian's job is not to hold ground,' his message concluded. Richmond went to bed just before midnight confident Etherington would comply.

But Etherington was dealing with a new crisis. Shortly after midnight he received a series of terrified phone calls from a British private security firm called Hart whose compound in downtown Kut was being besieged by the Jaish al-Mahdi. One of their team was already dead, and the remaining men were on the roof, where they were rapidly running out of ammunition. 'We need help,' came the desperate plea down the phone. But Etherington had no troops to spare. From what he could make out from the CPA compound rooftop, there were still dozens of fighters surrounding them. He kept the Hart situation from the rest of his staff. No one needed reminding of how close they were to a similar fate.

However, he reported the plight of the security contractors to Baghdad, and this convinced Victoria Whitford, a member of Richmond's staff, that if something wasn't done to help Etherington his position was about to be overrun. A diplomat in the Gertrude Bell model, Whitford raced over to the caravan park behind the palace and banged on Richmond's door. The British representative at once grasped the seriousness of the situation and hurried through the palace corridors to the US military liaison office. They found the staff overwhelmed by battle reports from across the country. A dozen American soldiers had lost their lives that day in Fallujah.

'I don't have assets to spare,' one senior US commander told them curtly, but Richmond wasn't going to be put off. He had never spoken to an American official so frankly, but he had to see the terrible blow to the Coalition if their entire governorate team was wiped out. 'We need American support now,' said Richmond, his voice rising.[17] After consulting his headquarters, the officer agreed to send a Spectre gunship: a Hercules with a 105mm cannon slung beneath it capable of firing a dozen rounds a minute. Richmond paused as if to ask for more, but then hastened over to the British office with Whitford. There was little they could do now but mount a vigil. The loss of the team in Kut would spell the end of British involvement in Iraq, he told Whitford and his staff.

In his darkened office Etherington sat alone and awake amid the sleeping bags and gentle snores. A few faces were lit by the glow of cigarettes. Direct attacks on the compound had eased after midnight, but there was still sporadic gunfire. He had pushed through his own fear and entered a void where he felt strangely accepting of his fate. He had read his history and gone to visit the forlorn cemetery beside the Tigris that marked the British military's humiliating defeat eighty-seven years before. Turkish forces had checked General Charles Townshend's army as it charged for Baghdad, forcing the British and Indian force to fall back on Kut and take up defensive positions.[18] A relief expedition hastily readied made little progress, hampered by dug-in Turkish positions and spring floods that turned the area to the south of Kut into marshland. Over the next five months 23,000 men died trying to rescue them. It was hard to fathom such figures, and contemplating them made Etherington realise his demise would barely register as a footnote in the history books. In the end Townshend, hero of an earlier siege of the Himalayan fort of Chitral, capitulated. Nine thousand starving men were marched across the desert to Turkey. All but a few thousand died.

Just after 3 a.m. Etherington heard the roar of an aircraft and shouts from the roof. He climbed the stairs in time to see the dark outline of the American Spectre gunship overhead, darker than the night. He watched as it fired at a machine-gun position on the opposite bank, sending up a glowing mushroom cloud, and felt a surge of exhilaration. Perhaps that would hold off their attackers a bit longer. Just before dawn he awoke from a brief sleep to find the Ukrainians lining up in the compound yard. They were getting ready to leave. Etherington was still hoping for reinforcements or at least a dedicated rescue team for the civilians, but time was running out. No one could remain once the Ukrainians left. Reluctantly he ordered the civilians to evacuate. Computers were thrown into the river, confidential files burned, and a million dollars recovered from the CPA safe.

At first there was a degree of organisation as contractors shuffled out of the hotel and villa with their bags and laptop cases, but with the Ukrainians in no mood to wait around and only a few vehicles to ferry everyone, there was soon a desperate scramble to load up the vehicles. When the

Ukrainians, unannounced, began pulling out, Etherington leapt in front of the vehicles to enable a roll-call to be taken, but they simply drove around him. Etherington took a final look at the villa, the scene of his hopes for Kut's future, before leaving with his private security team, numbed by the shock of capitulation. He imagined the militiamen who would shortly be flocking into the compound to celebrate the defeat of their British overlords once again.

As soon as the convoy reached the Ukrainian base, Etherington began preparing for the journey to Baghdad with his deputy Timmons. The two-hour drive along a desert road known for ambushes was dangerous, but his private security team agreed to take the risk. Fifteen minutes later his two-car convoy barrelled into an ambush as they entered the town of Numaniya. Sadr's militiamen had built a makeshift chicane across the main road.

The CRG driver didn't stop. Etherington heard an AK-47 being fired at them from close range and saw an Iraqi step out onto the road with an RPG. He either missed or the device failed to fire. The security team leader swerved off the road down a side street. With the militia in pursuit, the cars sped through the narrow streets before finally making it back to the open road. Rattled, Etherington called a halt just outside the town at a new Iraqi barracks he had opened only a few weeks before. American contractors were still there. On a borrowed laptop he fired off an email to Richmond, telling him Kut had fallen.

Richmond was already dealing with his next crisis. In Nasiriyah Barbara Contini had also ordered an evacuation from the CPA compound, which was now surrounded by Sadr's militia, who could attack at any moment. In fact Contini's order was slightly more nuanced. Since the raw emotion of the first day of fighting, she and her political adviser Tobin Bradley had grown increasingly distant. She suspected her American staffer of briefing against her. Baghdad knew about her deal with Sadr's militia, and political pressure on Rome was mounting for the Italians to take decisive military action. The Italian commander Chiarini had blown up Sadr's local office the day before, a largely futile gesture as the militia had already moved into the governor's buildings. It earned a stern rebuke from Contini. In the middle of the fighting she found time to give a highly charged interview with *Corriere della Sera* in which she proclaimed there would be no Fallujah in Nasiriyah.

Contini struck on an evacuation plan that seemed to target those whose loyalty she questioned. Bradley was stunned when Contini asked him and a dozen others to leave, ostensibly to lobby the Americans not to invade the city, while she and a small team remained. He was not going to argue. With the other Contini rejects, he drove to an army barracks just over the river, where they planned to pick up their escort. The Italians had occupied the barracks until the first day of fighting, when they had pulled back to their main base a few miles outside town. As the CPA convoy rounded the chicanes and entered the base, the scene resembled a spaghetti western, sand devils whipping around the deserted buildings. There were no Italians waiting for them.

Within a few minutes Bradley was spooked. Where the hell was their escort? They were sitting ducks like this. He called up Contini but could not get through. Finally, after half an hour, a couple of Italian armoured personnel carriers nonchalantly showed up. The indignity did not end there. When they got to the main Italian compound, the guards outside the gates would only admit the Italians in the group, claiming they had no knowledge of the others. Bradley and the others were left to fend for themselves, eventually making it to safety at a nearby US base after waving down an American military vehicle.

In the White House Rice had begun her day with a series of telephone calls with Bremer ahead of the National Security Council meeting. With Brahimi still threatening to resign, the American administrator was now dealing with another problem: half of the Iraqis who formed his advisory council were threatening to quit over Fallujah. Television footage showed thousands of refugees fleeing the city, with the Iraqi death toll estimated to be in the hundreds, many of them civilians. Bremer had met the advisory council members that morning and told them to stand firm. They had reluctantly agreed to issue a statement supporting the Coalition.

'Shouldn't we back off a little, Jerry?' Rice asked. 'We're reaching breaking point here. That's what the British are telling me.'[19]

Irritated that British reporting was once again undermining his position, Bremer reasserted that any sign of weakness would undermine the entire occupation. Half an hour later he presented his opinion directly to George Bush during a video conference. The president initially agreed with Bremer that 'the American people want to know we're going after the bad guys'[20] but by the evening, with a further dozen US fatalities in Fallujah that day, a second National Security Council meeting was convened with Bush clearly beginning to harbour doubts. However, he stopped short of calling off the offensive. At the NSC meeting Bremer had a dig at the British failure to retain control in the south, a comment that duly found its way back to London. Lieutenant General Rob Fry, deputy chief of the general staff for operations at the MOD, called Andrew Stewart to warn him that American displeasure at the British approach was growing. Stewart responded that there was little he could do with current troop levels and no support from Coalition partners. Ten thousand British troops could barely contain Basra and Maysan, let alone the handful of major uprisings in cities across the region. Stewart had upgraded his military 'request' to the Italians to an order to enter the city but did not know if they would listen. Most of Stewart's attention was now focused on Amarah, where mobs had been besieging the CPA's headquarters at Cimic House since the first day of the uprising. Abu Hatem, who the British had empowered, was refusing to side with them against the Sadrists.

'We're managing the situation as best we can,' Stewart reassured Fry.

The next morning, 8 April, Mark Etherington arrived in the Green Zone and marched straight over to the British office. He had heard from Timmons that an American brigade was on its way to Kut, and the sooner he got back to the city the better. Richmond, dishevelled and clearly exhausted, was relieved to see him and to hear that there had been no casualties. After recounting the events of the past few days, Etherington asked to return to Kut immediately.

Richmond was reluctant but eventually agreed to let Etherington assist American forces to retake the city, but only as long as US troops remained in Kut. He wasn't going to trust British lives to the Ukrainians again. Etherington left the meeting frustrated, not fully appreciating the strain Richmond was under. That morning the Iraqi Governing Council had started to disintegrate as feared. Interior Minister Nouri al-Badran had resigned, followed quickly by Human Rights Minister Abdul Basit al-Turki. Abu Hatem had 'suspended' his membership, and the veteran Sunni politician Adnan Pachachi was on the verge of jumping. Richmond was scurrying between Iraqi leaders trying to shore up support before Bremer finally reached the conclusion that the offensive in Fallujah had to stop.

The next day, 9 April, the offensive was halted with all but a small area of downtown Fallujah captured. The Americans had lost thirty-nine men, the heaviest losses in an operation since the invasion. An estimated 1,500 insurgents were dead, and over 120,000 Iraqis displaced, with western Baghdad now overflowing with angry refugees. Fallujah itself was a ghost town of bombed-out buildings and shuttered shops. It was not clear whether the break in fighting was a ceasefire or merely a pause to enable Bremer to shore up political support in Baghdad. Either way, the incoming British deputy commander in Baghdad, Lieutenant General John McColl, was determined to prevent further operations in Fallujah, which he viewed as a form of collective punishment.

McColl's predecessor Andrew Figgures had warmed to the American military machine, but McColl came from a stiffer tradition. He preferred to keep his American generals at a distance, giving him time to react. The US military had requested the use of British special forces for any renewed offensive in Fallujah. McColl turned them down. It wasn't quite a national veto but the implication was there. Sporadic fighting continued until the end of the month, when an unhappy compromise was struck between US forces and Iraqi fighters in Fallujah. Security for the city was handed over to a hastily formed Iraq unit under the command of a former Ba'athist general. It was a dangerous compromise that effectively 'gave the city to the *mujahideen*', one US officer noted at the time.[21]

The situation in southern Iraq was only marginally better. Moqtada al-Sadr was holed up in the Imam Ali Shrine in Najaf and unrepentant. The US military had recaptured Kut within a week, and Etherington had returned, although he had all but given up hope of re-establishing CPA rule. Amarah remained tense and dangerous. In Nasiriyah Andrew Stewart had upped the ante with the Italians, threatening to send a British brigade to the city if they did not take action. The Italian military promptly entered the city, enabling Bradley to return.

That did not stop Bremer from doling out retribution where he could. He blamed the British for the collapse of the south and tabled a formal request to the British embassy in Washington for Andrew Stewart's removal from command. Stewart kept his job after sending a written defence of his actions to Baghdad, but it was a reminder of how low Anglo-American relations had sunk during the crisis. They would get even worse. One evening in April Irfan Siddiq, a twenty-four-year-old diplomat from Leeds who had played an important role shaping Iraq's interim constitution, entered the British office in Baghdad looking ashen-faced. He had just been taking part in discussions with the Americans as to who should be Iraq's first prime minister.

'I've got some bad news,' Siddiq told Richmond. 'I've just been told that the British are excluded from the talks.'

'What? That's ridiculous!' exclaimed Richmond.[22]

The reason emerged via Bremer's adviser Scott Carpenter a few hours later. Negative reporting by British officials during the uprising had not gone unnoticed in Washington, to the frustration of American officials on the ground. During one National Security Council meeting in April Condoleezza Rice's special envoy to Iraq Robert Blackwill had sought to reassure the White House that the political process remained on track, despite the uprising.

'You're lying,' Deputy Secretary of State Richard Armitage declared, holding up a sheaf of papers. 'The British are reporting that [UN envoy] Brahimi is on the verge of walking out and the political process is about to collapse.'[23]

It was the same problem that Bremer had faced with Greenstock the previous year: British reporting undermining the over-optimistic views of American officials. Blackwill was not the type to ignore such behaviour. The British clearly could not be trusted. From now on they would be briefed the following day on developments, and Britain's effective involvement in shaping Iraq's political future was over. The same afternoon Siddiq was barred from the talks an irritated Sheinwald talked Rice into reining in her special envoy. Blackwill relented a week later, but by then the next disaster was brewing.

Chapter 11

DANNY BOY

THE PHOTOGRAPHS OF US SOLDIER Lynndie England grinning as she paraded an Iraqi prisoner around on a dog leash destroyed whatever vestige of moral legitimacy the Coalition had left after the April uprisings. David Richmond had known about abuse in Abu Ghraib, the country's largest prison, since Christmas, but the full horror only struck home after seeing the images that became public in late April. A few rotten apples in the US military, ran the official line from Downing Street. Privately, Richmond and others suggested the incident was indicative of America's overly aggressive attitude. However, British officials were about to get a rude shock three days later when the *Daily Mirror* published pictures of alleged abuse of detainees by British troops. One picture showed a British soldier urinating on an Iraqi. The pictures were eventually revealed as fake, but not before they had circulated throughout the Arab world.[1]

As the MOD tried desperately to limit the damage, the British reputation in the south was about to take a further blow with the recent arrival in Amarah of the Princess of Wales's Royal Regiment to replace the Light Infantry. Since the rioting outside Cimic House the tension had eased, if only slightly. By force of will, Molly Phee, the American governor of Maysan, had succeeded in holding the fledgling council together. A forty-one-year-old Irish American, Phee had rich auburn hair and soft, winsome features that belied her no nonsense approach to the province. When a Sadrist mullah had asked her to put on an *abaya* during a meeting in Cimic House, Phee rounded on him. 'If you're saying I'm not virtuous, you're bringing shame on my family, and you wouldn't want that,' she said. Faced by the prospect of a vengeful Phee clan, the mullah wisely backed down.[2]

Yet even Phee had met her match in Abu Hatem, the Marsh Arab leader. He was still furious with the Coalition for choosing a member of a rival militia group, the Badr Brigade, as police chief, and had temporarily disappeared. Phee hoped a united council would force him into line. She did not think for a moment the aging tribal leader would strike a deal with the rabble-rousing Sadrists against the British. But that was precisely what the wily Sheikh was about to do.

The trouble started on the Princess of Wales's Royal Regiments' first day in Amarah. At thirty-eight, Sergeant Dan Mills was affectionately known as granddad by his men. Like many of the NCOs of his generation he had never seen real action. Preparing to deploy to Iraq he had eagerly listened to news of the Light Infantry's battles with the Jaish al-Mahdi, but feared that the Sadrists would make peace before he got there.

The handover had been cursory as the homesick Light Infantry hurried to leave. Mills had no real idea where he was going when he and his men set off in their Snatch Land Rovers, a lightly armoured vehicle only capable of deflecting small-arms fire.[3] One of the downsides of constant troop rotation was that the departing soldiers took with them all their experience, knowledge and Iraqi contacts, leaving newly arriving troops to repeat the same mistakes. Heading south, Mills lingered at a road junction and set up a temporary vehicle checkpoint, before moving on to a police station a little further down the street. Amarah's main bridge over the Tigris crossed a little further to the south, and they pulled up under its shade near a three-storey building festooned with pictures of a turbaned cleric. Mills did not realise he was now standing directly in front of the headquarters of Moqtada al-Sadr and that to those inside the building the squad of British soldiers was highly provocative. As Mills's men moved past the building to the police station, an unshaven Iraqi officer in flip-flops ran over to Corporal Darren 'Daz' Williamson, the second in command, and told him that the British were not welcome in the neighbourhood.

As Mills and his men headed back to their vehicles there was a shout of 'Gunman top window!' Before they had time to mount up, a grenade sailed over the wall of the compound. They dived for cover as the grenade exploded and the road was raked with gunfire. Williamson screamed, 'Fuck. I'm hit, I'm hit.'[4] He struggled to his feet, blood running down his left leg. Mills ducked out into the gunfire and helped his mate into the back of one of the Land Rovers, where the medic got to work on Williamson's leg.

As they radioed the operations room at Cimic House for backup, bullets ticked against the side of the vehicle. The rest of the platoon was badly exposed. Mills leapt out of the back of the Land Rover, spotted an alleyway behind which his men could return fire. As the wounded Williamson was driven to safety, the soldiers sprinted for cover. Mills looked back around the wall in time to spot a fighter creeping down the street towards them. He waited until the man's body filled his sights before firing three times. The third bullet hit him in the head and he crumpled to the ground.

'Yes,' Mills screamed. 'Have some of that, you fucker.'[5]

Militiamen were appearing around the side of a mosque and on the rooftops overlooking the street. An RPG slammed into a nearby wall. A second struck the empty Land Rover, which began billowing smoke. Fuck, thought Mills. There goes our ticket out of here.

The Princess of Wales's Royal Regiment's commanding officer, Matt Maer, heard about the attack a few moments later. He was visiting Molly Phee at Cimic House to announce that the military would no longer be guarding the building. The recent uprising had exposed the vulnerability of the base.

'Let me know when you leave,' retorted Phee, annoyed at Maer's unilateral declaration, 'because I'll be leaving the same day you do.'

Maer tried backtracking. His colleagues described him as emotional, not a trait usually encouraged at Sandhurst, but Maer knew how to use his passion to galvanise his men.[6] Placating Phee was another matter. He was interrupted with a note from the operations room with news of Mills's firefight. He left without telling Phee where he was going. A patrol of Land Rovers from the Argyll and Sutherlands was on its way to Mills, but Maer saw no point staying in Cimic House, where the company commander Major Justin Featherstone was managing the response, or wasting time heading back to Camp Abu Naji.[7]

Driving south along the river, Maer quickly made out the bridge and the smoking remains of a Land Rover beneath it. Maer's aim was to use the bridge for cover, dismount and approach on foot, but even before they reached it, RPGs struck the convoy and a blast rocked Maer's Land Rover. After the split second it took to see that no one was hurt, Maer shouted 'Floor it' at his driver and they sped under the bridge as machine-gun fire raked the road and another RPG landed a few feet away. They stopped half a mile past the kill zone, safe but nowhere near Mill's men.[8]

Back at the bridge, the Argyll Land Rover patrol had already linked up with Mills. With militia swarming over the rooftops attempting to surround them, the combined patrol fell back into a defensive position further down the alley. The quick reaction force of armoured Warriors from Camp Abu Naji was under attack on the outskirts of the city, and it was not clear if they could hold out long enough for their rescuers to break through. Mills knew it was time to escape and that the only way to do that was to pile into the Argyll's Land Rovers. Moving in pairs, the soldiers sprinted up the alley towards the vehicles.

It looked like they were going to escape with only a single injury. The operations room tried to pass the good news on to Maer, but the radio equipment, shaky all afternoon, now failed altogether. As far as Maer knew, Mills was still in dire trouble, and he decided to charge back into the ambush site.[9] Driving under the bridge again was not an option, so they dismounted a hundred metres short. They were soon pinned down, with gunmen working their way south to cut off any retreat. Maer and his staff sergeant made a dash for a nearby house and tumbled over the compound wall as the rest of his men spread out over the road junction, ducking behind whatever scraps of cover they could find. With bullets clipping the tarmac around their positions, Maer's men returned fire whenever the militiamen poked their head around buildings or ran into the open.

They had just settled into a rhythm, co-coordinating their fire, when a militiaman fired an RPG, and a fireball engulfed their positions. The dust settled to reveal a nightmare scene: blood was splattered on the wall and one of the soldiers, Lance Corporal Philips, was lying face down on the ground, his left leg convulsing. Every second was vital if he was to survive. He was bundled into the back of a Land Rover and driven through a hail of gunfire to safety at Cimic House.[10]

At the British base Molly Phee discovered what was going on when she walked past the operations room and found officers struggling to respond to the crisis. She picked up the phone, called a contact among the Sadrists and asked him to call off the attack. What the hell had Maer been thinking, she wondered, driving into a firefight without telling her? Ten minutes later her contact spoke to the Jaish al-Mahdi leader Ahmed Abu Sajad al-Gharawi, assuring him that British were not trying to storm Sadr's headquarters and wished to withdraw. Gharawi agreed to pull back his fighters.

Maer heard the fighting dying down and assumed the approaching Warriors were responsible. The noise of their grating tracks was soon echoing around the street. He sent them over to Mills at once. When the sergeant and his men emerged from the alleyway, they found the Warriors waiting for them and scrambled in. Maer was also able to cram everyone else into the remaining vehicles and drive away.

He learnt about Phee's role in his rescue when he next spoke to her. The battle group's first day in charge of the city had been a disaster. Coming so soon after the Coalition's defeats in the neighbouring provinces of Kut and Nasiriyah, the British could ill afford further humiliation at the hands of Sadr in Amarah. After a frank discussion Phee and Maer agreed that the Jaish al-Mahdi posed a deadly threat to Coalition efforts in the city and needed to be confronted if British forces were to have any credibility on the street. Maer promised to work closely with Phee to build local political support for any action. Phee, in turn, accepted that a military operation was the only way to force an insurgent leader like Gharawi to the negotiating table. She also saw how defanging Gharawi would strengthen her hand in the wrangling match with Abu Hatem.[11]

But their plan unravelled quickly. Ten days later Maer launched a snatch operation targeting members of Gharawi's network in the Khadeem estate of south-west Amarah. But again a straightforward mission provoked unintended consequences. They managed to drag nine Iraqis into the back of a Warrior and get back to base safely, but as the men carried no ID it was difficult to tell who they had got. With the Abu Ghraib scandal dominating the news, the prisoners were carefully handled back at the camp, and none revealed his identity. The Jaish al-Mahdi responded by kidnapping three Iraqi policemen and threatening to murder them. That afternoon a resupply run to Cimic House was hit by an attack as intense as the previous fighting. Private Johnson Beharry drove his Warrior through an ambush with the vehicle filled with smoke from an RPG strike and his head poking out of the cockpit so he could see to steer.

Maer then decided to storm the militia headquarters in Amarah with a brigade-size force of a thousand men. The fighting was less intense than expected. After surrounding the building with tanks, British soldiers burst through the entrance only to find the headquarters deserted. At dawn Corporal Mark Byles and his squad stormed the building. Three eight-tonne truckloads of weapons were recovered, including a quantity of Russian anti-tank missiles stacked on the shelves of the nearby town library.[12] The victory was small but felt good. Maer called on Phee just after lunch and she sang his praises. With the Sadrists suffering a blow, they could now confront Abu Hatem. A few days later she summoned the tribal leader to a meeting. At her suggestion, Major General Andrew Stewart, the commander Bremer had tried to remove from office, flew up from Basra to lead the talks. Phee wanted Abu Hatem to be under no illusions that his time lording it over the province was coming to an end. The Marsh Arab leader seemed to sense he was walking into a trap. As soon as Stewart started speaking Abu Hatem flew into a rage, accusing the British of bringing chaos and disorder to Maysan so they could justify their occupation as necessary to bring peace.

'You want Iraq like this. That way you can stay for ever!' he shouted.[13]

Stewart told him not to be ridiculous, but Abu Hatem was not done. The British had undermined his authority and were cutting deals with his enemies. They had left him with no choice but to declare war so he was joining the Sadrists to kick the British out. He left with a dramatic swirl of his robes. Stewart broke the stunned silence in the room, suggesting feebly, 'He'll fall into line; he hasn't got the support.'

The next afternoon, 14 May, Abu Hatem arrived in Majar al-Kabir. It was true that his stock had fallen over the previous year. Spending his time in Baghdad, he had neglected to maintain the steady flow of gifts and money that reminded the townsfolk of the tribal name many bore. The Sadrists had grown powerful opposing the occupation and had little interest in backing the tribal hierarchy that Abu Hatem represented. But he was not above pandering to the deep-rooted anger of the masses to get his way. That morning fighting had led to an explosion in the Imam Ali Shrine in Najaf, where the Americans continued to surround Moqtada al-Sadr. Iraqis were furious at the damage to the holy site. Emotions were high in Majar al-Kabir, and the town centre was packed. Muhammad al-Fartosi, a shopkeeper, heard Abu Hatem harangue the crowd.

'Abu Hatem told us we must fight against the occupation. We all cheered,' said Fartosi.[14] Khuder al-Sweady, who worked as a clinician in Majar al-Kabir's small hospital, also recalled the anger that day. 'Our holy shrine was being attacked by the occupying forces, and Shia were dying at the hands of the Americans,' he said.[15]

The young men in the crowd quickly scattered, among them Sweady's nineteen-year-old nephew Hamid. Few had any military training, and there was no real plan. Half an hour later a minivan ferried some of them out to the highway a couple of miles away, and the first group began digging themselves in. The ground was already furrowed with trenches from before the war. An embankment on the west side of the highway provided natural cover, with a drainage ditch behind it and a network of narrow wadis. By mid-afternoon two hundred fighters had gathered, strung out over a mile.

They were still unloading weapons into the trenches when two British Land Rovers approached from the south. Taken by surprise, the gunmen could only fire off a few rounds before the Land Rovers burst through. Another British patrol, led by Captain James Passmore, was quickly alerted and rushed to the ambush site, but did not arrive until some time later. Passmore knew the area; the intermittently manned police checkpoint at the turning to Majar al-Kabir was known to the soldiers as Danny Boy. This time the tribesmen were dug in and ready. The first fighters stayed hidden, drawing Passmore deeper into the ambush. When they finally opened fire on Passmore's patrol, the British soldiers were pinned down.

His vehicle's radio was down, so Passmore was forced to use his satellite phone to call up the switchboard operator in Whitehall, which meant he had to poke the phone antenna out of the vehicle as the bullets whizzed by.

'Is this an official call, sir?' the operator asked.

'It's an official firefight!' Passmore shouted and was duly connected to the operations room at Maer's headquarters at Camp Abu Naji.[16]

Sergeant Christopher Broome was part of the quick reaction team that day. He set off to rescue Passmore, but his two Warriors were ambushed before they got to the Danny Boy checkpoint. The ambushers were protected by the trenches, and the only way to clear them was for the troops to get out of the vehicles and chase them down on foot.

Reassured that a second rescue party was on its way from Abu Naji, Broome ordered his men out of their vehicles with bayonets fixed. Corporal Mark Byles led the charge. He moved through the trenches, stabbing and shooting with his rifle. Iraqi fighters scrambled over the dirt and ran for their lives. Byles and the other soldiers chased them through a system of narrow gullies while Broome drove his Warrior into a drainage ditch and up the other side to provide support with his machine gun. He soon lost sight of Byles but could hear a gun battle. Leaping out of his vehicle without a weapon, he chased after Byles and found him in a drainage ditch, returning fire. There were three dead Iraqis on the ground and four prisoners lying on their fronts. They looked like teenagers and were terrified. Broome quickly cuffed, blindfolded and hooded them.

As Broome's men finished clearing out the maze of gullies, Major James Coote arrived to take charge of the mopping-up operation. He ordered prisoners to be rounded up. There were over half a dozen. Lance Corporal Mark Keegan, injured in the groin during fighting to the south of the checkpoint, was propped up on the ground near one of the vehicles, one of two British wounded so far. He watched as one of the first prisoners to appear, a sandbag hiding his face, was punched in the head by a passing soldier. Others followed suit.

'Every time he kept crying out, he was knocked to the floor with his hands tied behind his back. Then they'd leave him for a bit until he started squirming, lift him up again until he started moaning again,' recalled Keegan. 'His face must have been pissing out with blood underneath that sandbag.'[17]

Other Iraqis were roughly treated as they were rounded up. One of the captured fighters, Ahmed Jabar Ahmood, claimed he was struck on the head by a rifle butt and tripped.[18] Coote was inside his Warrior getting ready to leave when he received an order from Maer, visiting Basra at the time, to bring the bodies of any dead Iraqis back to base for identification purposes. Removing enemy dead from the battlefield was not common practice, but the brigade commander, Andrew Kennett, thought the killers of the Red Caps might be among the corpses.[19]

Coote wasn't exactly thrilled by the order, but he duly emptied a Warrior, and Broome and his men began dragging the bodies out of the trenches. There were over a dozen. As Broome recalled, 'At first we didn't want to make skin contact with the bodies, so we picked them up by their clothes. However, due to the fact they were wearing loose clothing, when we picked them up by their cuffs their clothes started to fall off. We were covered in blood from head to toe . . . Killing someone, then looking at them afterwards as you place them in your Warrior, is something I could have done without, personally.'[20]

Traffic had resumed along the highway. Iraqis pressed their faces against the windows of passing cars to catch a glimpse of the grisly scene.

Arriving at Abu Naji, Broome's men unloaded the bodies. At first the Warrior's door wouldn't open, so Broome's driver Private Taylor volunteered to climb over the bodies and unblock it. He emerged several minutes later 'freaked out' and sprinted off. Once they had unloaded the bodies, a medic came over to proclaim them dead. Broome and his men stood there covered in 'blood and stuff, soaked through, sweaty, filthy and feeling like vomiting'.[21] The bodies were then photographed and checked for useful intelligence. There was not much.

Meanwhile the prisoners were processed and led to a detention facility in a former Iraqi barracks. Several were injured and moaning.[22] Hussein Fadhil Abbas was one of the prisoners. Before being blindfolded at the battlefield, he believed he had seen his friend Hamid al-Sweady among the group. He had been injured in the leg. In the dark he now called out, 'Hamid, are you there?' and received a 'Yes' but no more. He heard the sound of screaming and cries from other parts of the building. Throughout the night prisoners were led away to be interrogated. Several heard the sound of shots.[23] One claimed he was the subject of a mock execution. At one stage Abbas heard the shouts of about five soldiers in the cell.

'They came in and then it started . . . There was the sound of choking. Like this: kkkkkhh khhhhh,' recalled Abbas.[24]

At Camp Abu Naji the next morning Maer informed Ahmed Fausi, leader of the Majar al-Kabir council, about the bodies and that they would be returned before sunset so they could be buried within twenty-four hours, following Islamic custom.[25] Fausi didn't mention it, but Majar al-Kabir was already alive with rumours that the British had removed the corpses. By the time the bodies were picked up from the British and brought to Majar al-Kabir hospital, a furious crowd had gathered outside.[26] Abu Hatem and his brother, Governor Riyadh, were standing at the gate as the bodies were carried through the crowd on stretchers. As they were laid out, hospital clinician Khuder al-Sweady, who had witnessed the town's anger that morning, now grimly surveyed the aftermath. One of the first body bags to be unzipped contained his nephew Hamid. His death certificate noted that he had a bullet wound to his neck but offered no suggestion as to time of death. Khuder al-Sweady, however, is clear. As he washed his young nephew's body, he tried to straighten Hamid's neck. It fell limply to one side.

'When I tried again, it fell to the other side. What does this mean? It means the neck was broken. It was execution by hanging,' Khuder suspected.[27] He later claimed his nephew was inadvertently caught up in the fighting, and that the ambush against the British was the work of outsiders.[28]

Other bodies also appeared to be inexplicably disfigured. Some corpses had flattened skulls; others were missing eyeballs. One of the bodies was that of Heider al-Lami. His death certificate recorded that his penis had been severed.

Abu Hatem, whose rally had incited the violence, knew how to capitalise on the moment. 'Look at how the British have mutilated the dead,' he shouted to the crowd outside the hospital. 'They have stolen the eyes of the dead.'[29]

The Majar al-Kabir police chief remonstrated with him, saying that the injuries were from the fighting and there had been no abuse. Governor Riyadh turned, took out his British-issued 9mm Glock pistol and shot the police chief in the head, according to witnesses.[30] The incident did not end there but gathered pace. A few hours later representatives from an Iranian-backed militia group called Thar'allah arrived with a video camera to film the bodies. They were soon distributing the images around the souks in Amarah and Basra, and printing flyers showing blurry pictures of Western soldiers apparently beating Iraqi civilians.

On the British side no effort was made to quell the public relations disaster. The military's press team insisted all was quiet in southern Iraq. Journalist Stephen Grey recalled a bizarre scene with Maer a few weeks before, in which the PWRR commander, flanked by two press officers, insisted that life was 'rather dull' in Maysan. 'Reading some media reports, anyone would think we are in some kind of war zone out here!' he exclaimed sarcastically.[31]

But the truth could only be covered up for so long. In early June insurgent leader Moqtada al-Sadr finally bowed to pressure and announced a ceasefire, but little would change in Maysan. Iraqis who had lost sons, fathers and brothers wanted revenge and they appeared to be only waiting for the next offensive to begin.

Chapter 12

OUT OF IRAQ

FOR ALMOST THE FIRST TIME SINCE becoming prime minister, Tony Blair was visibly depressed. His aides all noticed it: the attention to policy was still there, but the zest and optimism were gone. Iraq was wearing him down. The Hutton Inquiry into David Kelly's death had reported at the end of January 2004 and cleared the government of charges that they had unduly exposed Kelly to the press or manipulated a pre-war intelligence dossier, as the BBC had alleged. Gavyn Davies, BBC chairman, and Greg Dyke, the director general, promptly resigned. It looked like a Blair victory, but the media quickly labelled the inquiry a whitewash which left unanswered the fundamental question: had the prime minister lied to the public to take the country to war?

The following month Blair was forced to agree to a second public inquiry, this time into pre-war intelligence.[1] No politician likes being called a liar, but for Blair the charge threatened to strip him of his moral certainty, his only protection against the mounting horrors in Iraq. He had taken to secretly visiting the families of injured or dead soldiers like a housemaster visiting the dormitories of his pupils. In a rare moment of candour with his arch-rival, Blair revealed to Gordon Brown, 'I can't get out of Iraq. I'll never turn this around.'[2] Tessa Jowell, minister for culture, media and sport, observed, 'Iraq turned his hair grey. It was in early 2004 that you first saw a big physical change in his appearance.'[3]

Blair's biographer Anthony Seldon believes the prime minister came close to resigning during that troubled spring. Iraq was disintegrating and his plans for an Arab–Israeli peace were scuppered by the resignation in September of the moderate Palestinian prime minister Mahmoud Abbas and Israel's discovery of a ship bound for Gaza loaded with weapons and a Hezbollah bomb-maker. Condoleezza Rice would later comment on Anglo-American Middle East strategy, 'We suffered from terrible luck.'[4] The one bright spot was the Libyan leader Colonel Muammar Gaddafi's decision to renounce weapons of mass destruction. At home Blair was facing a backbench rebellion against his plans to introduce top-up fees for universities.[5]

In March Blair gave a speech in Sedgefield justifying the war in Iraq as part of the greater 'war on terror'. 'The nature of the global threat we face in Britain and round the world is real and existential,' he declared. 'It is the task of leadership to expose it and fight it, whatever the political cost; and the true danger is not to any single politician's reputation, but to our country if we now ignore this threat or erase it from the agenda in embarrassment at the difficulties it causes.' The same month Blair told Brown he would be announcing plans to step down after the Labour Party conference in September. But Brown feared such an announcement would throw the Labour Party into turmoil ahead of local and European elections in the summer and give his rivals a chance to run for the leadership. Brown counselled against announcing the departure, a decision he came to regret.[6]

Then, in the depths of Blair's personal and political crisis, salvation loomed in the form of another international adventure. A NATO conference in Istanbul had been scheduled for the end of June 2004. Top of the agenda was Afghanistan, where the US-led military mission had been neglected because of events in Iraq. From as early as 2002, George Robertson, the bluff NATO secretary general, had been pushing for an expansion of the alliance's role in Afghanistan to rescue the organisation from its post-Cold War doldrums. At the previous NATO conference in Prague the former UK defence minister had secured NATO's commitment to take over 'select areas' of the country from the Americans. Two years later NATO's only operation outside Kabul was a German provincial reconstruction team in the northern city of Kunduz. Initially the Germans had been unable to secure enough helicopters from fellow NATO members to allow them to deploy. The war in Iraq, and the damaging rifts it had created in the international community, was one reason for the failure.

In Downing Street Blair began to see Afghanistan as an opportunity not just to tackle a resurgent Taliban and the burgeoning opium trade, but to bring the West together again and restore his own reputation as an international statesman.[7] The UK's only battle group in Afghanistan at the time, a force of around 1,000, was located in the northern city Mazar-e-Sharif and was seen as having exhausted its usefulness in one of the country's more peaceful regions. British forces could have much greater strategic impact in southern Afghanistan, where the Taliban were strongest, Blair believed. There were already encouraging signs that the Canadians – who had refused to take part in Iraq – were prepared to send a sizeable force to the south as NATO looked to expand its operations to other parts of the country.

A few weeks before the NATO conference British defence chiefs raised the idea of deploying the headquarters of the Allied Rapid Reaction Corps to the country. The headquarters, based in Rheindahlen, Germany, was one of the jewels in the crown of the British military. Set up in 1992, it was intended to provide an agile response to crises in post-Cold War Europe and was soon deployed to the Balkans. Other NATO nations now contributed almost half of its manpower, but the operation was still a British-run affair.

In March 2004 the Americans had asked the British to deploy the ARRC to Iraq to take over all nine Shia provinces of the south. The prospect of further British involvement in Iraq as the Sadrist uprising broke caused considerable alarm in London. 'No one wanted to send more troops to Iraq,' recalled one senior officer, 'but we didn't want to turn down the Americans either.'[8]

Lieutenant General Richard Dannatt, ARRC commander, recalled how the discussion circled back to the nature of the American request. 'The Americans had asked very nicely but said they would not put us under pressure,' Dannatt noted.[9]

Blair double-checked with his Cabinet colleagues: did that mean the Americans would be OK with them not sending ARRC to Baghdad? Realising a critical watershed had been reached for Britain's involvement in Iraq, Defence Minister Geoff Hoon assured the prime minister that the US military was not expecting the ARRC to deploy to Baghdad. In his opinion it would be better to keep it out of Iraq.[10]

But having raised the possibility of an operation, Richard Dannatt was one of several voices that urged the defence chiefs to send the headquarters to Afghanistan instead. He viewed Afghanistan as a more important strategic consideration for the UK, given the terrorist threat posed by Pakistan and Britain's large population of Muslims with links to the Indian subcontinent. The move would also hasten the handover of the country from American to NATO control and provide a powerful platform for any British contribution to southern Afghanistan.

Hoon was uncomfortable with the idea of committing more troops to Afghanistan while the situation in Iraq remained uncertain. The crucial argument was delivered by Lieutenant General Robert Fry. As director of operations at the MOD Fry held one of the most powerful positions in the British military, a fact not lost on Fry, who was widely regarded as one of the most accomplished of a new political breed of officer. His job was to do the strategic thinking for the defence chiefs. A Royal Marine, his career had ticked all the right boxes. He had commanded 3 Commando Brigade and worked at the Directorate of Special Operations, but his strength lay in politics, where his formidable intellect and eloquence meant that few could argue against him. At fifty-two, he was stocky and self-assured, with only a hint of his hometown Cardiff accent. As Fry saw it, the British military stood at the apex of its power, and he was in a unique position to harness it. That power rested on close strategic alignment with the Americans and the willingness of Britain to put its troops in harm's way.[11]

Fry's argument to the defence chiefs was simple. The military's job was largely done in southern Iraq, and it was time to get out. At the same time the Afghan campaign was a mess, with the Americans distracted by Iraq and the NATO plan stalled. A bold intervention in Afghanistan, including both the ARRC and additional forces on the ground, could breathe new life into the moribund campaign and present the perfect justification for exiting Iraq. Chief of the Defence Staff Michael Walker admired Fry's nimble thinking, although characteristically Walker wanted 'more planning'.[12] Within a week of sending out requests to NATO allies alerting them to a possible ARRC and deployment to Iraq, Fry sent out a fresh batch with Afghanistan pencilled in instead. Blair was happy – and determined to arrive at the Istanbul conference with something to deliver.

The conference got off to a good start when the Dutch, timid since the debacle in Srebrenica, signed up for a reconstruction mission in Bagram, taking everyone by surprise. 'Once the Dutch were in, every self-respecting military needed to be,' said one UK official following the talks.[13] In a scene oddly reminiscent of the nineteenth-century scramble for Africa, NATO members jostled for slices of territory. Germany took the lead in Mazar-i-Sharif, with Norway playing a supporting role; the Czechs and Danes took over in Feyzabad; Canada was the UK's main rival for the southern half of Afghanistan.

It was just the sort of global coming-together that Tony Blair thrived on. At the conference table in Istanbul's Hilton hotel, George Bush, sitting next to Blair, slipped the prime minister a piece of paper. On it was the news that Paul Bremer had disbanded the CPA and passed sovereignty to the new interim Iraqi prime minister, Ayad Allawi, a former exile leader who had emerged as the compromise candidate after Grand Ayatollah Sistani signalled his approval. The handover was two days earlier than advertised. Blair had been concerned that leaving early sent the wrong signal but at a late dinner with Bush the night before had given his approval.[14] The departure of Bremer from Iraq and the official transfer of power to the Iraqis was one of the most significant moments in the occupation thus far. Across the note, Bush had scrawled, 'Let freedom reign.'

Blair shared with the US president an excited smile. He was delighted with the choice of Allawi, a secular Shia who had once run Saddam Hussein's network of informants in Iraq's universities, before turning against the dictator and living in exile in the UK, where he worked as a doctor in the Surrey suburbs. He had earned his spurs as an opposition leader when in 1978 Saddam Hussein sent a henchman to his semi-detached to bury an axe in his head. Having survived the attack, he had gone on the MI6 payroll for a number of years, and later supplied the British government with some of the most flagrantly misleading intelligence before the war, namely that Saddam Hussein could deploy weapons of mass destruction within forty-five minutes. He looked like an ageing Mafia don, which helped him gain a reputation as a hard man – as Blair's aides liked to appreciatively titter, just the type of leader Iraq needed.[15] Blair made sure he was the first to call him once the NATO meeting broke up. 'We're going to back you all the way,' he enthused to Allawi.[16]

From the plush conference chairs of their five-star hotel, the leaders celebrated the first step towards a peaceful transition of political power to the Iraqis – although neither the US or UK had yet to announce a timetable for withdrawing troops. The next day, 29 June, Blair announced the

ARRC deployment, with the Afghan president Hamid Karzai standing beside him in his particular blend of ethnic clothes.

At the Ministry of Defence Fry began work on planning how a deployment in Afghanistan might affect British forces, principally those engaged in Iraq. His findings were presented at the end of the summer in a twenty-page strategy paper. The most eye-catching feature of the work was a graph whose implications were to prove deeply controversial. It depicted troop numbers steadily decreasing in Iraq before rising again in Afghanistan. The inherent risk in the scheme – that the Afghan deployment relied on a successful outcome in Iraq – was not lost on some. If the withdrawal did not progress smoothly in Iraq, the British military would be fighting a war on two fronts.

Defence Minister Geoff Hoon did not like the plan. The 1998 Strategic Defence Review had planned for only one deployment at a time, meaning the UK would be severely stretched if it had to fight two medium-sized operations at the same time. But when Geoff Hoon raised his concerns to Michael Walker, the chief of the defence staff told him to speak up in Cabinet if he had objections. He said nothing.[17] In any case, Walker dismissed the guidelines set out in the Strategic Defence Review, saying it was underfunded from the start and not a credible guide. The army had not been properly resourced for Iraq either – before Iraq Walker estimated a shortfall of £1 billion on the military's annual budget – yet once the invasion had started, the Treasury had provided emergency funding of £500 million for so-called urgent operational requirements. That had covered everything from armoured vehicles and helicopter radios to ammunition and unmanned aerial vehicles. The additional funding allowed the military to spend its budget on big-ticket items like aircraft carriers and fighter jets.[19]

It was a risky approach to managing the military budget that left many officers nervous about the implications for funding once the war in Iraq ended. The prospect of Gordon Brown becoming prime minister was greeted with further dismay. The current chancellor was seen as a typical Labour politician with little sympathy for the military or the Iraq operation. He always made sure military requests for emergency funding were answered – to appear to have done otherwise would have been political suicide – but he thought Blair was far too soft on the defence chiefs, telling him, 'You've brought this [Iraq] all on yourself by making the service chiefs believe that if they press you'd give way.'[20] Brown was conscious that his own stance now was going to shape the spending priorities of his own premiership.

Tensions between the MOD and the Treasury had come to a head in December 2003. The previous year a new chapter had been added to the Strategic Defence Review promising a modest increase in defence expenditure. The agreement coincided with a new accounting method at the MOD. Instead of traditional cash accounting, where the MOD planned for future requirements and set budgets accordingly, the ministry started using resource accounting. The method, popular in industry, is designed to take account of depreciation in the value of assets over time. The idea is to force resource managers to dispense with stock they do not need and order more efficiently, 'on demand'. Crucially, the Treasury allowed the ministry to convert any efficiency savings it made into cash. Given that the MOD had an asset base of between £70 and £90 billion, the ministry had become a giant bank account.[21]

The Treasury had initially offered an overall increase of 3.6 per cent in the MOD's budget for 2004. Without including the urgent operational requirements, the offer reflected an actual budget cut, leaving defence chiefs fuming that Brown was conducting a 'Defence Review by stealth'.[22] But by converting savings into cash, the MOD accounts men had devised a 3.6 per cent rise of £1.3 billion. Gordon Brown was horrified and believed the ministry was grossly manipulating the new accounting method. On 26 September 2003 he wrote to Blair and Hoon informing them he was guillotining the MOD budget. Having optimistically expected a larger budget, Trevor Woolley, the MOD's financial director, was now faced with a similar-sized shortfall. The defence chiefs predictably chose to protect the budgets for their individual services. The compromise position was cutting the joint budget – the largest element of which happened to be for helicopters. The decision was to have deadly consequences.[23]

The defence chiefs still had to cajole Brown the following summer into a further budget increase during the 2004 Comprehensive Spending Review. On 3 July General Walker even travelled to the prime minister's country retreat at Chequers to threaten Blair with his resignation if the services did not get more money. Coming after the high of Istanbul, Blair prevailed on his chancellor to provide an extra £800 million of funding. Just as with Iraq, the prospect of a campaign in Afghanistan ensured the Treasury purse strings were kept loose.[24]

When Fry presented his strategy paper to Michael Walker and the other defence chiefs at the end of the summer, no serious questions were raised. Their enquiries focused on NATO coalition-building and the future campaign. Gordon Brown was also starting to realise that Blair had no intention of leaving office soon. In June he had confronted him, one of those moments that Blair later described as physically intimidating. 'When are you going to F-off and give me a date?' Brown shouted. 'I want the job now!'[25] But after Istanbul Blair wasn't going anywhere. On 18 July, shortly after the Butler Report had cleared his government of charges of manipulating pre-war intelligence, Blair told the chancellor he would not be forced into declaring a date and in September declared his intention to serve a full third term if he won another election.[26]

Chapter 13

POLICE

For William Kearney there were only three types of policemen: drinkers, womanisers and thief-takers. During his twelve years in Special Branch he'd seen a few of the first two. Thief-takers – and he counted himself as one – were rarer, and most of them ended up at the Branch, the closest thing the UK has to the FBI. Working in Liverpool in the 1980s, Kearney had busted drugs rings before moving south to serve on the security detail for the royal family. Kearney considered the Branch the only place where real police work was done, and the other officers' aggression and drive matched his own.

Kearney's dad was a second-generation Irishman who had worked as a labourer in Birkenhead and instilled an old-school fear of God into him. After leaving school at fifteen, Kearney had signed up with the Royal Green Jackets and completed a few tours of Northern Ireland at the start of the Troubles. He liked the camaraderie of the soldiers, but could not stand the officers and the trappings of class and privilege that would always separate him, a rough Scouser, from the Sandhurst toffs. He had bought himself out of his army contract with a little of his dad's hard-earned cash before signing on with the police, where he met his wife, a fellow officer, two years later.

Looking around the room containing a dozen British policemen in Az-Zubayr police training academy in July 2004, he knew what type he was looking at: drinkers. They had even built a little bar for themselves at the entrance to their accommodation block from a couple of oil drums, with withered palm tree leaves as cover and a portable fridge to keep the beers cold. The policemen hardly looked up as he and the deputy senior police adviser Dennis Jackson, a soft-spoken Scots constable from the Ministry of Defence Police, entered the room. Kearney was introduced as the country director for ArmorGroup, the private security firm that would be running police training programmes alongside serving officers like Jackson.[1]

Kearney had already picked up on the tone of the British occupation since arriving in Basra a few days before. The palace had the air of the end of a party – everyone trying to get out before the clean-up starts. Since the CPA's departure the previous month there had been an exodus of staff. Andrew Alderson, the erstwhile finance minister, had made a final push at DFID to keep his team in place. Over the past few months they had spent £16.5 million on projects like laying water pipes that seemed to balance the demands of quick job creation – they were all dug by hand – with longer-term strategy.[2] Millions more had been committed to projects but not spent. Alderson feared the reconstruction effort would collapse without his team and wrote an impassioned letter to Hilary Benn, the DFID minister, urging him not to abandon Basra and the other southern provinces.[3] He received no response. In early June Alderson suffered the further ignominy of being told that, due to increased security risks, his team would start leaving at once.[4]

Morale was understandably low on the civilian side, not helped by the arrival of an American consulate in the palace complex with a large budget and little interest in following the British lead. The Americans promptly selected one of the larger villas for their building, which they then surrounded with a line of concrete barricades. The only access between the British side of the compound and the American was through a single metal door guarded by Nepalese security contractors. At least the British diplomats could pride themselves on having the better set-up. A pool was in the process of being built complete with a modest bar largely restricted to FCO and DFID personnel. It was reminiscent in spirit, if not appearance, of the fine old clubs of the British Raj, Kearney thought on his first visit. Not to be outdone, the UK special forces detachment in Basra had a bar in their accommodation, nicknamed Brookside Close, with a DIY jacuzzi on the roof. The scene was altogether more raucous than over by the pool. Female FCO officials would often come over after dark in their golf buggies. The diplomats' private security guards – a requirement since the start of the year – were billeted nearby and could be persuaded to make beer runs down to the logistics base at Shaibah, twenty minutes away. At one stage the drunken races in golf buggies around the palace grounds in the early hours of the morning led to so many crashes that a senior officer warned his civilian counterparts that he would bar military helicopters from airlifting anyone suspected of injuring themselves while drunk.[5]

Kearney knew how you handled drinkers. You got your head down, worked hard and, before you knew it, you were being turned to for advice and running the show. At forty-eight years of age he wore his hair cropped to hide the grey and worked out in the gym for two hours before breakfast, giving him a barrel chest and a bodybuilder's swagger. Off duty he might wear a flashy shirt and a diamond earring, but at work he was a tightly wound spring.

Kearney was hungry for success. Ten years before he had taken early retirement from the force and set up a private security firm, Balmoral, trading on his experience as a royal bodyguard. It was not a glamorous job but it made him part of the burgeoning world of private security firms that had sprung up in the mid-1990s. With the end of the Cold War, tinpot dictators were free to plot their own coups and counter-coups. Sandline, run by Tim Spicer, a former officer in the Scots Guards, made money from a botched attempt to defeat rebels in Papua New Guinea and gunrunning in Sierra Leone – despite a UN arms embargo. Another company set up by a former British officer was Defence Systems Limited, which rose to prominence in the 1990s by guarding oil installations in Colombia and diamond mines in the Congo.

Private security firms were about to reap the rewards of the post-9/11 world. US Secretary of Defense Donald Rumsfeld's preference for a stripped-down military gave the private sector key roles supplying and guarding his forces. In March 2003 Tim Spicer, now working for a new company called Aegis, was awarded a $240 million contract for guarding American bases, the largest post-war contract won by a British company, dwarfing even those won by American security companies like Blackwater. 'Without any assistance from the British government,' Spicer noted. He had enjoyed a cold relationship with the FCO ever since he had suggested it knew all along about his gunrunning.[6]

Another beneficiary was ArmorGroup, an American firm that had recently taken over Defence Systems Limited. They had retained the British staff, including the man who was to become Kearney's boss, David de Stacpoole, a former colonel in the Irish Guards. Educated at a Benedictine Catholic school called Worth Abbey in Sussex and scion of a family with estates in western Ireland, he knew how to work the system.[7] ArmorGroup won an early contract to guard Basra palace.

In January 2004 de Stacpoole heard about another big contract in the works: to train Basra's police force. Ever since Blair had been pushing for police as a priority, the Iraq steering group chaired by Nigel Sheinwald had been struggling to implement the prime minister's wish. The British effort had met with a measure of success in Baghdad. After the disastrous tenure of Bernie Kerick, the New York City police commissioner who had tried to run Baghdad like Queens, his deputy, a former Yorkshire assistant chief constable called Doug Brand, had gone back to basics at the Interior Ministry. He was firmly of the view that creating an Iraqi police force could only be done sustainably by the Iraqis themselves. The previous autumn, over a number of late-night meals with Nouri Badran, the interior minister, he had sketched out what an Iraqi police force might look like. The British and American military were keen for the police to have a paramilitary role rather like the Italian *carabinieri*, but Brand pushed Badran towards a force structure that was more recognisably British, with a unit for investigating serious crimes and an independent prosecution service to assess which cases to bring to trial. Brand feared that if the police became militarised it risked recreating the blunt tools of repression used by Saddam. The police needed to be the upholders of civil society and the rule of law.

Unfortunately, the Americans took a different view, preferring to flood the streets with police rather than 'waste time' tinkering with the ministry. In early 2004 the US military established a new organisation called the Civilian Police Advisor Training Team under the command of Major General Paul Eaton. Modelled on a similar programme for the Iraqi army, the police training team aimed to turn out 200,000 new police recruits in less than six months. The majority would be funnelled through a training programme in Jordan, where the US had spent a billion dollars renting facilities. An American security firm called Dyncorp would provide trainers – mostly retired officers who seemed to have been dredged up from every small-town sheriff's department in the US.

The dangers of such a numbers-driven approach were clear to some.

'We were sacrificing quality for quantity. We had no idea who we were giving guns and badges to,' said Andrew Rathmel, a British security consultant working in the Interior Ministry.[8]

The British effort in the south was not doing much better, but for different reasons. The FCO was struggling to find serving British police officers to work at the police training academy in Az-Zubayr, twenty miles to the west of Basra. The Home Office refused to help, viewing overseas adventures as outside its remit and thus avoiding the awkwardness of demanding resources from its already strapped police forces. Instead the Association of Chief Police Officers, the primary agency for developing policy, led by the chief constable of Hampshire, Paul Kernaghan, was requesting volunteers from the constabularies.

'It was a classic case of no one in government having the ability to marshal resources to meet a national crisis, which is what Iraq was,' said Doug Brand.

Hilary Synnott, the earlier head of the CPA in the south, had requested ninety-one officers at the start of his tour. He got three.[9] In April 2004 the dribble of volunteers stopped altogether after a suicide bomber struck the entrance to the Az-Zubayr academy, killing three Iraqis and injuring two British soldiers. The six police advisers at the academy were evacuated to the palace, and all twenty-two in the country were sent for immediate leave in Dubai with their families, at considerable expense. Phil Read, senior police adviser in the south since January, had resisted American efforts to send down Dyncorp trainers to Az-Zubayr, reflecting widespread distrust of the American effort in Baghdad.[10]

It was becoming increasingly clear that Britain would have to come up with a new approach, using private security firms like ArmorGroup to provide police trainers. The idea was distasteful to some civil servants, although once the £5.3 million contract was awarded the issue of which government department would run the operation was quickly resolved. James Tansley, the man who would later host the Queen's Birthday Party in Basra, was preparing a glossy brochure of the FCO's recent achievements in preparation for the department's submission to the 2004 Comprehensive Spending Review, at which government departments would jockey for money from the Treasury. A police training contract was not the natural preserve of the FCO, which had had no contract management facilities since the creation of the DFID in 1997, but the department moved quickly to claim it in the hope of justifying a bigger budget the following years. 'We featured the work prominently in the brochure,' Tansley noted.

As soon as Kearney heard about the job he put his name forward. He had trained police and done his tours in Northern Ireland and knew how to impress a former officer like de Stacpoole by playing up his streetwise credentials. De Stacpoole hired him on the spot to manage a 120-strong team of mentors, mostly ex-Royal Ulster Constabulary. The idea was that once Iraqi police recruits had been trained, Kearney's men would carry out regular check-ups on the local stations. They would be deployed initially in Basra and Maysan under the supervision of serving police officers brought in by the FCO.

In early March both men were in Basra on a reconnaissance mission. The FCO security manager, John Windham, was an old Wellingtonian and former Irish Guardsman like de Stacpoole, and the two men knew each other through their private security work. Kearney stood quietly to attention watching de Stacpoole, who resembled an overgrown choirboy, sweet-talk the seasoned security manager. Kearney had never watched the old boy network in action before, and he was impressed by the easy language of privilege. De Stacpoole was obviously pleased by how the trip had gone and the potential to expand the contract. On the way back he told Kearney there could be more jobs here, if he 'used his peasant cunning'.[11]

Kearney kept his mouth shut. He was not going to jeopardise his new job and had already got a taste for life in Iraq. He couldn't care less about the palace scene and the FCO types at the poolside bar; he had been hooked by that dusty drive through Basra, by the vividness of the sunlight and the exotic otherness of the Iraqis in the street. This was what he had been waiting for.

A few days after his arrival that summer he and his fellow trainer Mike Cole were standing in the training yard of Az-Zubayr police academy with Major Mohsen Shaqir and his thirty or so recruits of the newly formed Tactical Support Unit. Most of them looked like typical recruits, feckless and itching for their first pay cheque, but Kearney liked what he saw of Shaqir, a trim officer in his early thirties who wore a thin moustache and gave off an air of subtle condescension.

Kearney soon had them running around the yard and doing push-ups. Before ArmorGroup's arrival the police mission had focused on changing police culture in Basra. Under Saddam Hussein they had been little more than traffic cops notorious for corruption and brutality. Phil Read, the senior police adviser in Basra from January 2004, had in mind the British model of community policing. This calls for police, instead of focusing on investigating violent and organised crime, to build strong bonds with the local community as a way of fostering a more law-abiding society. John Alderson, a former chief police commissioner for Devon and Cornwall, had popularised the idea during the British industrial strife of the 1970s, and it had become popular again after the 2001 Bradford race riots. 'For police to serve their communities they need patient and nuanced engagement with their communities,' said Read, a former chief superintendent of West Yorkshire Police.[12]

With that as the ideal, Read had started training Iraqi police recruits in basic human rights. Early on, Kearney witnessed one class in which a British police officer put four pictures up on a whiteboard: Saddam Hussein, Pol Pot, Adolf Hitler and Margaret Thatcher. 'Saddam Hussein, good or bad?' the officer asked the class, speaking through an interpreter. There were a few shouts of 'Bad.' Kearney could see they were trying to figure out what the British wanted to hear.

Kearney found such exercises pointless. The only way to change Basra's police culture was to take them through practical examples and build up personal relations with officers like Shaqir. In one of his first sessions with the TSU recruits he and Mike Cole took them through a few scenarios.

'Imagine Mike here is an old woman, and I am a thief, an Ali Baba,' Kearney said, 'and I come up behind Mike and take his wallet. What would you do?'

This time one of the older recruits, a twenty-four-year-old called Muhammad, raised his hand. 'We would take this man and beat him,' declared Muhammad.[13]

'OK. Let me tell you what we do in England,' said Kearney. It was a good job he had started with the basics.

Back at the palace for lunch after one of his first mornings, Kearney was joined in the queue outside the cafeteria by a gaunt-looking Haider Samad, the Iraqi interpreter who had greeted British forces with such optimism at the start of the occupation. Kearney had picked up on a certain attitude to Iraqis around the palace, a tone he recognised from when his wife Linda, who was half-Bajan, entered a room of white officers. The racism in Basra palace was, if anything, more overt. He had already heard

Iraqis described by the military as 'corrupt, lazy and devious'. Gruelling though the sessions had so far been, he was not about to start labelling Iraqis. What he wanted to know was what was going on inside their heads.

'All right, mate,' he said to Haider.

Haider gave him a cautious smile. Since Nicholas Mercer's departure in January he had worked with his replacement at the Palais de Justice, without striking up the same rapport. Before the handover he was politely informed that the British army would have no more use for his services as the brigade headquarters was withdrawing from the palace to the airport, and Iraqis were now in charge of the justice system. He had applied for a job at the consulate but missed the cut and only had a few more days of work left.

Basra was no place to be unemployed with the British government on your résumé. Haider saw the crowds of young men lining up outside the palace gates each morning looking for work and felt their growing resentment. The fighting between Sadr's men and the British army might have stopped, but the militia remained on the streets. Anyone who worked for the foreigners was considered a traitor and infidel. A few months before two Iraqi sisters who worked in the CPA's laundry had been gunned down on their way to the palace. Haider saw in the militia the same army recruits and deserters he had shared a cell with before the invasion, the same victims swept along by events. Saddam's revenge, he nicknamed them – Basra's underclass, now dressed in black shirts and running petty extortion rings and beating up women who didn't cover their faces with a *hijab*.

Losing his job came at a cruel moment for Haider. He was planning to propose marriage. It had taken six months, but at last he had worked up the courage to contact Nora al-Sadoon, the Sunni girl he had met at the wedding party several years before. She was not yet married and still remembered him. But initial soundings of Nora's father did not bode well. He was suspicious of a penniless Shia from the slums.

Haider responded by enlisting the help of Nora's uncle, a family friend. Ali was a man singularly devoted to the good things of life – women and food – as evidenced by his egg-shaped frame, half a dozen children and a string of concubines around the city. Haider became a regular visitor to Ali's house in the Mutheneb district, enjoying his refreshing candour. When he explained to Ali that he had managed to save a thousand dollars over the past year for his wedding, Ali declared it would never be enough to persuade Nora's father before announcing his own plans to take a second wife. His job at the port in Umm Qasr always threw up business opportunities. The Jaish al-Mahdi had recently moved in to set up a smuggling business, but Ali knew how to run rings around them.[14]

Haider was leaving Ali's house one evening after asking him for job suggestions – which Ali had supplied in abundance – when a gang of four men, their heads wrapped in scarves, approached him. He was already a dozen metres from Ali's house, caught in a pool of light cast by a single street lamp. He froze as the men bore down on him. The first to reach him first raised his hand and struck him across the head, knocking him to the ground.

All four gathered over him, breathing hard.

'We are giving you a warning because you are a *sayyid*,' said one of the men in a thick Marsh Arab accent, referring to the fact that Haider was a direct descendant of the prophet Muhammad, a prestigious but not uncommon distinction in the Middle East. 'Next time there will be no warning,' said the man. Haider heard them leave, but he still lay on the ground, his eyes closed. The last time he had been beaten was at the hands of Saddam Hussein's secret police, the *mukhabarat*. He slowly got to his feet and walked home. He wanted to believe he had simply been caught crossing into the wrong neighbourhood, but deep down he knew it was because he worked for the British and he had been followed.

Back in the lunch line a few weeks later Kearney was telling Haider about his morning. He wanted to know what motivated his recruits. Haider had a simple answer. 'Don't underestimate the power of fear,' he told him.

'Who are they afraid of?' asked Kearney.

'The city is changing. The militias are everywhere now. They're building up for something.'

'Who are you working for?' Kearney asked him, intrigued.

'The British army, at least for a few more days,' said Haider.

'Right. I'll pay you double whatever they are, but I want you working for me,' said Kearney. 'We're going to knock some sense into this place.'

Haider accepted at once. Kearney was about to give Haider a slap on the back, but he saw the Iraqi flinch and opted for a handshake instead.

There's more going on beneath the surface than he's letting on, thought Kearney – which was just what he wanted to know.

On Haider's first day in the office Kearney asked him to compile a list of militias.

'Can I include the police?' Haider asked ironically.

'I want everything, Haider,' said Kearney. His policeman's nose was starting to catch the whiff of trouble.

A few weeks before, Haider had met an Iraqi doctor, Ahsen Schwekit, who had set up an NGO to investigate abuses at a number of police stations in the area, including the old Ba'athist headquarters near the Palais de Justice and another called the Jamiat. Under Doug Brand's police reforms, every province was to have a serious crimes unit to conduct FBI-style investigations. In early 2004 the Jamiat station was selected to house Basra's unit. The station, a sprawling compound in the heart of the Hayaniyah slum, soon developed a reputation for abuse and corruption. The station's white Land Cruisers with blackened windows were called by local residents the *mukhabarat*, after Saddam's secret police.[15]

Schwekit's investigation added detailed descriptions of police crimes, which ranged from the stealing of sheep to the kidnap and rape of women. Detained women were photographed naked carrying out various sexual acts and the pictures were then used to blackmail their families. In May Schwekit took his findings to the CPA's political director, Robert Wilson. He had a dozen photographs of miserable-looking women in various poses with two different men. Schwekit claimed he had obtained the pictures from an informant of his, a women's hairdresser, in whom the women in question had confided. Wilson treated Schwekit cautiously. After a second meeting in June Wilson wrote to his colleagues, 'I'd like to be a bit careful about handling this sort of report – Schwekit seems to be genuine, if a bit naïve, and he is really trying to get some sort of official recognition from the CPA and Coalition Military for him to continue his intelligence-gathering activities. But he seems to have access to potentially useful information.'[16] But nothing had been done. When Haider mentioned the Jamiat to Kearney, the latter immediately called in his newly arrived police mentor for the station, a self-possessed former RUC officer called Charlie MacCartney, who had worked for the RUC's Special Branch. At fifty-five MacCartney was still a physically imposing man, well over six feet with butcher's hands and cold blue eyes. From his nights on the prowl in West Belfast he knew when a place was rotten, and the Jamiat was that all right.[17] The station consisted of two linked compounds lined with offices rising to three storeys. The first compound, entered from the main road, housed the regular police. The station chief was Colonel Ali al-Sewan, whose corpulent frame seemed permanently wedged in front of the television in his office. In other police stations MacCartney had visited, the junior officers at least paid some deference to the boss while they got on with running the show. At the Jamiat Sewan was clearly cowed by his aggressive and cocky subordinates.

One in particular seemed to be in charge, a captain called Jaffar who ran the serious crimes unit. MacCartney recognised Jaffar as a player, but who was calling the shots? When he relayed his concerns to Kearney, the ArmorGroup boss told him the whole station was going to be put through training shortly, when they would take a good look at the personnel. In the meantime MacCartney should get close to Jaffar and wait for the Iraqi's master to reveal himself.

Chapter 14

FARTOSI

SINCE BECOMING THE LEADER OF THE Jaish al-Mahdi in Basra that spring, Ahmed al-Fartosi's goal was nothing less than the destruction of the British occupation. Already the Jaish al-Mahdi ran what amounted to shadow government in the city, meting out justice on political opponents, policing the streets and running oil smuggling rings. The first time Fartosi attacked British troops – in May 2004 – had not been a success. From that time, his men had a series of ambushes on British patrols, but they had escaped each time. Fartosi had enjoyed emptying his AK-47 at the fleeing Land Rover, but he knew that until he started killing or capturing British soldiers, his cause would lack a symbol.

Fartosi had soft, rounded features which made him look younger than his thirty-two years. He moved with the self-important swagger of a classroom bully, broad shoulders thrown back, his actions quick and deliberate. A closely cropped goatee and preference for Western dress over traditional tribal wear were a statement of Fartosi's deeper sophistication; that there was a dark philosophy behind the brutality. His clan was one of the largest in southern Iraq, numbering several hundred thousand.

Over the summer of 1954 Wilfred Thesiger, the great British explorer, had stayed for several months with members of the same clan in Maysan province. He was often asked for medical help, which meant he performed many circumcisions. It was not unusual to find men of Fartosi's father's generation who had had their foreskins removed by Thesiger. That world had disappeared when the Fartosis, along with the Bahadilis, had risen up against Saddam Hussein after the Gulf War, and the dictator had drained the marshes, scattering tribes like Fartosi's.[1] Many fled to Iran; others were forced into slums in Basra and Baghdad, where they bided their time, plotting revenge.

Fartosi had a chaotic upbringing marked by violence and political resistance. His immediate family headed the Basra branch of the clan, which had several thousand members. Fartosis could be found in most schools and on every street, with influential positions in the administration, police force and prisons. Like a Mafia don, the head of the clan exerted power over his tribesmen in return for defending their interests. Although the Basra branch had avoided the traumas which had hit their kin from the marshes, Saddam's paranoia meant Fartosi's family was soon targeted. In 1984 two of Fartosi's uncles, who ran a profitable transport company, were murdered, and his father fled to Lebanon with his young family.

Fartosi only returned to Iraq in his early twenties at the urging of his grandmother. He briefly attended the naval academy and was soon sucked into the Sadrist movement. He was drawn not by religious fervour but in protest against Saddam Hussein's rule. When Ayatollah Muhammad Sadeq al-Sadr was assassinated by the dictator in 1999, Fartosi joined the small group of plotters that led Basra's brief uprising. He narrowly escaped capture, returning to Lebanon, where he spent the years leading up to the US invasion.

Fartosi's opposition to British rule in Basra was rooted in his belief that the Western occupiers primarily wanted Iraq's oil and would subjugate its people to get it. The British had, after all, a legacy of meddling in Iraq's affairs for oil. Upon returning to Iraq in 2003 he had pledged his allegiance to Moqtada al-Sadr, the radical son of the murdered ayatollah, and joined the Jaish al-Mahdi. He found a rabble of young men with guns eager to show their fervour. In Lebanon Fartosi had spent time with the Iranian-backed militia Hezbollah. He quickly applied what he had seen, organising the fighters under his command into cells of seven or eight that could quickly assemble for an attack and then slip away. After the invasion – when any gang could claim the right to siphon off oil or rifle through the containers at the Umm Qasr docks – organisation had been vital. Fartosi's ability to calculate and the slow purposeful manner in which he twisted the organisation to his purposes saw him swiftly rise to the top.[2]

Fartosi was looking for confrontation. He had been on the roof of the governor's building during the April uprising and been dismayed when Abdul Sattar al-Bahadili, the political head of the Sadrist movement in

Basra, decided to withdraw. Fartosi's first attacks against the British, minor skirmishes, had confirmed the occupier's superior firepower, but he was convinced they were soft underneath. If he could sustain the fighting for long enough and cause enough British casualties, he was sure they would leave. Over the summer of 2004, during the truce between Sadr and the Coalition, Fartosi had been in touch with Ahmed al-Gharawi, the Jaish al-Mahdi commander in Amarah. Gharawi's battles against the Princess of Wales's Royal Regiment, in which a dozen British soldiers had been injured, had come close to all-out war, and Fartosi wanted to discuss tactics. Despite being part of the same militia, ancestral animosities divided the two commanders. The Fartosi transport business had often had to contend with extortionate demands from the Gharawi clan. 'They are little more than highway robbers,' Fartosi believed.[3] But Gharawi also came with useful connections over the border in Iran. He told Fartosi the Iranians were looking to train Iraqis to fight in Iraq for Shia supremacy. During Saddam's rule the Al-Quds Force, a branch of the Revolutionary Guard used to foment Islamic revolution ouside Iran, had trained thousands of Iraqi exiles to fight against the dictator. But with the largest of the Iranian-backed paramilitary organisations, the Badr Brigade, working with the Coalition in Baghdad, Tehran was keen to channel the anti-American sentiment of other militias to their advantage. The same camps that had trained the Badr Brigade were now offering a three-week course in bomb-making and guerrilla tactics.[4]

Yet Fartosi kept the Iranians at arm's length. The provinces of southern Iran have close affinities with the neighbouring Iraqi province: they are populated by ethnic Arabs who speak Arabic rather than Parsee and often come from the same tribes. Pragmatic leaders like Gharawi felt that history should be no barrier to doing business; if the Iranians wanted to create trouble for the Coalition they should take advantage of their mutual goal. But Fartosi was old enough to remember the Iran–Iraq war, when Iranian forces had come close to overrunning Basra, and over half a million Iraqis had died.

While Fartosi deliberated, events overtook him. On 4 August 2004 the truce between the Coalition and Sadr collapsed. A patrol of US soldiers in the holy city of Najaf provoked a gun battle outside Sadr's offices. Grand Ayatollah Sistani promptly left the city for surgery in London, which American commanders took as a signal to move into the city to finish off

Sadr, whose growing power was known to have upset the traditional clerical order in Najaf. As Al Jazeera broadcast live images of American troops fighting around the Imam Ali Shrine, the most revered site in Shia Islam, angry mobs took to the streets across southern Iraq.

The British response was to avoid confrontation and try and negotiate with the protesters. At military headquarters at Basra airport British Major General Bill Rollo was as frustrated as his predecessors had been in having to manage the fallout from another botched American operation. He flew to Amarah to mollify the sheikhs and imams. Since the departure of CPA officials like Molly Phee, the British had lost important political leverage in the province, and the reconstruction effort had ground to a halt. Rollo found himself heckled by the Iraqi dignitaries, who did not distinguish between the US and UK. The British compound in Amarah was soon under mortar fire, giving the defenders little option but to fight back.

Returning to Basra, Rollo was still hopeful of avoiding confrontation. He ordered British patrols to avoid trouble spots, such as the police station housed in the former Ba'athist headquarters in central Basra. Angry crowds had gathered since the morning. Far from dispersing them, the police seemed to be actively encouraging the anti-Western mood. In fact, none other than Fartosi himself had taken command of the station without opposition. His fighters had quickly fortified the buildings with a network of roadside bombs and firing positions. The low-key British response to the uprising had given him the space to take control of various areas of the port city. From the station his fighters ranged across Basra, harassing British bases and ambushing the occasional British patrol, but without inflicting serious damage.

Then, on the morning of 9 August, a few days into the fighting, his militia spotted British soldiers arriving at the Iraqi National Guard headquarters on the Shatt al-Arab. Fartosi quickly directed his gunmen to the building. His aim was not to seize it but ensnare the British units that would come to the rescue of their comrades. Along the road leading from the nearest British base his men blocked side streets with rubble and oil barrels for half a mile. The strategy was to steer British troops towards the police station in the former Ba'athist headquarters, where he would be waiting for them. He positioned gunmen along the entire route.

Twenty minutes later two Snatch Land Rovers came down the road, sunlight glinting off their windscreen. Sergeant Terry Bryan, 1st Regiment

Royal Horse Artillery, had had a sick feeling in his stomach all morning. The streets were usually teeming with life but today the place felt like a Wild West showdown before the shooting. It was clear the locals knew something. At Camp Cherokee, a small base in Basra's industrial district, he hadn't been surprised to hear from one of the operations officers, Captain Amber Tyson, that gangs of Iraqis, some with AK-47s, had been milling around the city centre. A patrol was besieged at an Iraqi National Guard base nearby. Tyson wanted Bryan to lead a unit to cover the escape route when the patrol made a break for it.

Bryan had scrambled his men into two Land Rovers. He had joined the military as a sixteen-year-old; after spending the next sixteen years serving Bryan could hardly remember any other life. He had just signed on for another ten years, making him a lifer, a poignant moment in any soldier's life as the lure of Civvy Street fades. Bryan had wanted to be a chef once.

Still, there was little time for reflection during a battle, and they got that soon enough. Less than a mile from the Shatt, Bryan's patrol drove into the ambush and was caught in a hail of bullets so intense he did not dare return fire. Fartosi's men had occupied the rooftops along the road, and hundred of rounds ticked off the windscreen's front and sides of the Land Rovers. The vehicles' Kevlar armour held up, and they kept on rolling as the sergeant pored over a map, looking for an escape route. A three-point turn in the road would badly expose them. The only thing to do is put our foot down and try to drive out of the killing area, Bryan thought.[5]

But after a mile they were still being shot at, and Bryan began to suspect they were driving deeper into a trap: the angry crowd that had lured them out, the breaks in the central reservations for U-turns closed off, and hundreds of armed men ready on rooftops. His heart sank. They reached a roundabout only to find all the exits blocked with oil drums and packing crates. As they drove round, Bryan's Land Rover hit a roadside bomb, which blew away its left front wheel. The other Land Rover, struck by an RPG, was now billowing flames. At that point Bryan's driver, a gritty South African called Frank Haman, spotted the police in the former Ba'ath Party headquarters to the right and a break in the central reservation. A dozen Iraqi policemen were manning the barricades outside.

Thank God, thought Bryan as they approached. Then he realised the police – supposedly British allies – were shooting at them. The situation was dire, with both vehicles on the verge of conking out. There was no alternative but to dismount. His squad barrelled out of the vehicles and into defensive positions. Bryan estimated his nine men were now surrounded by 200 militia and Iraqi police. He glanced at his fellow soldiers, mostly an assortment of young lads with less than eighteen months in the army, and felt reassured to see them looking sharp and fearless as they fired their rifles. Almost immediately one of them hit a police officer as he stood casually reloading a rocket launcher, but they were going to have to kill a lot more Iraqis before this battle was over. A grenade exploded against the building behind Bryan's head, showering him with sparks.

He got through to the operations room, where Tyson answered: 'Hello, Terry. How are you?'

Bryan was normally chatty on the intercom. This time he croaked, 'Fucking contact!' He could barely get the words out, his mouth had gone so dry.[6]

He looked round for somewhere to take cover. Given the amount of lead in the air, it was only a matter of time before someone was hit. As the section laid down covering fire, he dashed down the street looking for shelter. Taking the first alley, he was confronted by two gunmen. Bryan shot them and they collapsed in a gurgling heap. It was his first kill, but it barely registered. He felt nothing but an instinct to survive.

Bryan had spotted a semi-detached house with a driveway down one side and a garden enclosed by a wall. He called up the rest of the patrol to follow, but they were barely in the driveway when the mob began pouring over the wall. Haman was down on one knee, shooting militiamen from less than five feet, as Bryan raced to the house and kicked in the door, only to find a father and mother and three children huddled inside. Bryan got them into the cellar as his men spread out to take up defensive positions. Iraqi shooters appeared on the rooftops and began peppering the building.

Bryan, at an upper-floor window, didn't want to kill anyone else, so he fired warning shots, hoping they would back off. One guy was a stone's throw away, firing at him from behind a chest-high wall. Rounds were hitting the wall around Bryan but miraculously missing him. He looked through his sight, aimed and fired, but missed. Bryan saw the bullet hole appear in the wall and thought, Surely he's not going to be so stupid as to come back up in the same spot?[7] But, sure enough, up he popped. Bryan shot him in the head and he dropped. The operations room now piped through on his radio to inform him that a rescue force was on its way, led by Major David Bradley, commander of B Company PWRR.

But hope of rescue didn't last long. Bradley's company was also ambushed as it sped south towards the former Ba'athist headquarters. Exposed in the open turret of a Warrior, Bradley was struck by an RPG. Shrapnel blasted through his body armour and into his chest. Before he knew it, another RPG struck his rifle, and when he lifted his right hand he saw it had been sliced in half. Hit again as he was pulled inside the hatch, the left side of his body caught fire, but his soldiers, also injured, succeeded in putting this out.

As Bradley's Warrior returned to base, the remainder of his rescue force drove deeper into the ambush. Lieutenant Ian Pennell was now in charge. His soldiers reached the roundabout in view of the former Ba'athist headquarters, still under heavy fire, and it was clear that Bryan's men would be mown down if they tried to run for the Warriors. They had started to drive closer to the building when Pennell's top cover, Private Lee O'Callaghan, was shot in the chest just above the line of his body armour. He slumped down into the Warrior, forcing Pennell to return to base to try to save the mortally wounded soldier.

The rescue mission for Bryan and his men was now in the hands of two men, Corporals Terry Thompson and Andre Pepper, commanding the remaining two Warriors of the unit. With his radio cutting in and out, Thompson mistakenly believed the missing men were inside the former Ba'ath Party headquarters and used his Warrior to break through the front gates. Iraqi police and gunmen ran for cover as Thompson dismounted and sprayed the courtyard with bullets, a ferocious one-man onslaught that left an estimated twenty dead. But because Bryan's men were not inside, he was no closer to rescuing them.

In the besieged house Bryan's ammunition was running low, and the mob appeared ready to storm the building. At the rear Haman was shooting militiamen one at a time as they tried to get in through the back door. The murder of the Red Caps flashed through Bryan's mind. Better to take their own lives than end up being dragged through the streets and tortured, he decided. At that moment he saw an Iraqi aiming a gun at him, and a moment later a round cracked into the wall a few inches to his left, filling his left eye with dust. Disorientated, he staggered onto the balcony just as someone lobbed a grenade. He had time to stumble back before the blast knocked him to the floor. Shrapnel hit his arms and inside thigh. He looked over to see the section's medic, Corporal Ryan James, face down on the ground. But, as Bryan watched, James sat up and gave a thumbs up: he had just been winded.

Bryan ripped into him good-naturedly, 'Get up and fight, you lazy bastard . . . Any excuse to lie on your belly and get your head down, you medics!'

Then James showed him his other hand. In it was a grenade. He had been about to go onto the balcony to chuck it into the crowd below.

'Pass it to us,' said Bryan, ready to take the lead again.

'I can't.'

'Ryan, pass us the fucking grenade.'

'I can't. I've pulled the pin.'

Luckily, he had kept his hand over the grenade's safety lever, which had stopped it detonating.

They collapsed in laughter, a moment's respite. Bryan crawled over to James and took the grenade, careful not to let it go off. Wincing with pain, he crawled out onto the balcony and lobbed it at the Iraqis below. That quietened them down, but not for long.

Downstairs, the militia had set light to a car and rammed it into the front door, filling the house with smoke. Bryan hoped that would attract the attention of the rescue team, because they were down to their last few rounds. A few minutes later he heard the rumble of approaching Warriors. The radio in the Warrior commanded by Corporal Pepper had finally kicked in, and the two armoured vehicles had hurried out of the former Ba'athist Party headquarters.

After his suppressed terror, Bryan felt a wave of relief, but he was still concerned the rescue party would not be able to spot them. Taking a few quick breaths, he ran down the stairs and out the front door past the burning car. The militia on the street, after a moment's surprise, fell back. Some tried to get shots off, but most fled at the sound of the Warriors. Bryan's appearance attracted Pepper's attention, and his machine gun made short work of the remaining Iraqis.

Bryan and his men hurried out of the house towards the Warriors as the militia at the rear of the house gathered for another attack. Fartosi watched from a nearby rooftop as the British soldiers roared away. Later he would calculate that he had lost thirty-eight men for the death of one British soldier, but the ambush had been a propaganda success to his way of thinking. The Al Jazeera cameramen filming the burning Land Rovers would broadcast the images across the Arab world and make him look like a hero for standing up to the Westerners.

Fartosi had hoped that with both sides bloodied, the fighting would now escalate, but the next day the British returned to their stand-off tactics. Their patrols proved elusive and difficult to ambush, and the palace was too well guarded for him to assault. As his frustrations grew, Fartosi cast an envious eye north. Moqtada al-Sadr was locked in a deadly battle with American forces. His defiant last stand in Najaf's Imam Ali Shrine had galvanised the Arab world.

Meanwhile, in Amarah, Fartosi's partner Ahmed al-Gharawi appeared to have the British on the verge of defeat at Cimic House, the Coalition base in the centre of the city. Four hundred rounds had fallen on the buildings in the past forty-eight hours. Half a dozen Land Rovers were shredded and all but one of the portacabins used for accommodation were destroyed. Inside the main building, a dozen members of the civil-military affairs team that had taken over running reconstruction projects from the CPA spent their nights huddled together on office floors. The hundred or so soldiers worked in shifts, snatching a few hours' rest when they could. No one could wash properly, and there were only two toilets for a hundred men. The cookhouse served one cooked meal a day until the kitchen was hit by a mortar, which almost took off the chef's leg.

At Camp Abu Naji, the permanent British base near Amarah, Lieutenant Colonel Matt Maer knew he needed to relieve the pressure on Cimic House, six miles away. On 10 August, a week into the fighting, he sent a convoy of Warriors and Challenger tanks into the city to capture some militia. They barely reached Amarah's outskirts before they were bombarded with RPG fire, with a Challenger immobilised by a roadside bomb. Maer managed to get some of his Warriors in position, and they snatched a few locals from the streets, but this minor success was overshadowed by news from the few provincial council members who were still speaking to the British. Abu Hatem, who had stirred up trouble during the earlier uprising, was now openly siding with Ahmad al-Gharawi, the Jaish al-Mahdi leader.

Next morning at Cimic House, the commanding officer, Captain Charlie

Curry, reported that none of his Iraqi employees had shown up for work. In fact they did make a brief appearance. Curry was called to the balcony mid-morning to watch as his interpreter, the laundry men, gardener and cleaner went back and forth from the block of portacabins to the rear entrance of the base carrying whatever loot they could find before disappearing for good.[8]

That afternoon the mortaring of the compound reached an ominous intensity. Maer had tipped Curry off that Abu Hatem had changed sides once again and transformed himself into an insurgent leader. Curry made sure his sniper team, led by Sergeant Dan Mills, was on the roof. Shortly after midday prayers a large crowd gathered outside the governor's offices. Mills was watching the crowd when he spotted a small group of gunmen sneaking up the river road toward them. He bellowed out, 'Stand to! Stand to!' As his men scrambled to their posts, they also spotted groups of insurgents approaching Cimic House from other directions.

Mills reported to the operations rooms, 'RPG men ducking in and out of the alleys in front of us. Must be a dozen of them in there.'

Just at that moment an RPG shot overhead, the signal for the gunmen to open up. Mills's men responded in kind. Unlike the suicidal charges of the past few days, the militiamen were careful not to get too close. After an hour they ceased firing altogether and began to withdraw.

'What the fuck are they doing? Come on, you fucks!' shouted Mills's corporal, Chris Jones.[9] They had heard that Abu Hatem had been involved in the attack. 'Abu Shat-imself,' they joked.

This minor victory failed to steady nerves in Basra, where Fartosi was continuing to harass British patrols. The British commander, General Rollo, believed the cost of holding Cimic House outweighed the benefits of staying and promptly ordered a withdrawal. Matt Maer was not happy about it, but he passed on the order to Captain Curry in Amarah. In between breaks in the mortaring, the Cimic House defenders dug a giant pit for burning papers. The detritus of a year's occupation – PlayStations, TVs, deckchairs, Arab rugs, porn magazines – were carried into the main building, which, with the patrol boats and few remaining vehicles, was rigged to blow.

Only after they had finished did Maer call to tell Curry that he had succeeded in changing the general's mind. Curry and his men could stay and uphold the honour of the British army, provided everyone there was happy with the risk. At a hastily convened meeting Curry gathered together Mills and the others. After a unanimous show of hands in favour of staying, Curry told them, 'I don't see why we have to hand over this place to the modern equivalent of the Nazis. We'll withdraw when we're ordered to, or if we really have to. Until then we're going to sit it out.'[10] Mills led the others in a round of applause. That night was the quietest for months.

The next morning, however, the mortaring resumed and carried on through the following night, only stopping after the morning call to prayers. An eerie silence continued as the heat of the day built. Mills's men were just standing down from the rooftop when the militia and their tribal allies attacked. Three minibuses pulled up out of range on the other side of river. Twenty fighters poured out with AK-47s, RPGs and heavy machine guns. Another dozen charged up the river road towards the main entrance of Cimic House, and more arrived at the dam to the north. Behind them additional minibuses pulled up.

'Targets approaching from all sides. Repeat, targets approaching from all sides,' Mills bellowed down the intercom.

It was clear the Jaish al-Mahdi would try to take Cimic House. Every spare soldier scrambled onto the roof as the shooting started. The militiamen did not hold off this time. As they closed in, the mortaring stopped. Mills rushed about the roof, calling out targets to his men and firing when he could. The open ground to the north of the base, littered with debris, was providing ample cover for the attacking gunmen. Mills tried to disperse them with a few well-aimed mortar rounds and then heard a shout. Near the front gate a dozen fighters had appeared with a 107mm rocket, capable of blasting a hole through the compound wall.

As Mills directed fire at the new threat, a gunman sneaked across the road behind the cover of the compound wall. He reappeared scaling the iron gate at the rear entrance. The man had got one leg over when he was spotted and brought down by a burst of heavy machine-gun fire. A second fighter trying the same thing was hit in the arm. By then Mills's men had succeeded in knocking out the support frame pointing the rocket at the front gate. It tumbled to the ground and exploded on the road, sending a plume of smoke into the acid-blue sky. The snipers switched their attention to the rear gate, riddling it with bullets, before turning back to the open ground to the north. Some of the Iraqis were less than a hundred metres away, scrambling like black ants over the piles of rubble towards them. Mills grabbed a 51mm mortar and peppered the area, forcing back the assault. The volume of militia gunfire was finally beginning to lessen.

'Stay sharp, lads,' said Mills. He had barely spoken the words when he again heard the hollow *whump* of mortars. He dashed for cover. 'Where the fuck did they land?'

Popping his head up, he got his answer. They were smoke rounds. When the dense white smoke cleared a few minutes later the only Iraqi fighters to be seen were the dead and dying, strewn in piles across the wasteland, their blood already caking in the hot sun. Women in black *abayas*, waving white handkerchiefs, soon appeared, to drag the bodies slowly away. It was over.

The next morning a column of British tanks arrived at Cimic House to bring fresh soldiers to replace Curry's exhausted men. The compound resembled a disaster zone. The building was pockmarked with bullet holes and the gashes of shrapnel. The defenders emerged, unshaven and caked with sweat and dirt, like the survivors from a plane crash. The siege had racked up a number of records – the longest defensive stand since the Korean War, with 86 militia attacks on the compound since the fighting had begun, including 595 mortar rounds and 57 RPGS.[11]

Sporadic fighting continued for the next few days, but the Jaish al-Mahdi opted to lick their wounds rather than try another assault. Matt Maer took the opportunity to quietly withdraw British forces from Cimic House for good. The decision was not popular with those who had just risked their lives to defend it, but Maer saw no point in holding the building just for the sake of it. He had succeeded in forcing the Jaish al-Mahdi to retreat, but lasting peace would only come with a political settlement, and that could not be imposed by British firepower. Destructive battles like the one he had just fought were likely to lesson the UK's reputation and ability to act as a power broker.

In Basra the militia continued to harass British troops, and two more soldiers were killed in action, but at the end of August General Rollo's patience paid off at last when Grand Ayatollah Sistani returned to Najaf. The destruction wrought by weeks of fighting had turned the mood in the holy city decisively against Moqtada al-Sadr, whom many blamed for provoking the Americans. He agreed to a ceasefire and announced the disbanding of the Jaish al-Mahdi. Sistani also prevailed upon him to support elections scheduled for early the next year. A united front would deliver the ultimate prize: Shia dominance in Baghdad.

Fartosi greeted the news with dismay. The previous three weeks had shown what the Jaish al-Mahdi could do, with the British holed up in their bases and pounded every time they ventured out. He predicted they would eventually withdraw from Basra just as they had done from Amarah. After receiving orders from Moqtada al-Sadr, he agreed to stand his fighters down for now, but like other Jaish al-Mahdi commanders Fartosi started to look for ways to continue the fight. Sadr's right-hand man Qais al-Khazali had recently announced he was leaving the militia to form his own splinter group called Asa'ib Ahl al-Haq – the League of the Righteous – and fled to Iran.

Fartosi was not ready to strike out on his own or place himself entirely in Tehran's hands, but that autumn he met Gharawi's main Iranian contact, Abu Mustafa al-Sheibani. A Badr Brigade commander, Sheibani had headed the organisation's network in Baghdad during Saddam Hussein's rule, supervising the supply of weapons and money to the capital.[12] A wiry man whose face and arms were covered with scars from skirmishes with Saddam Hussein's army, Sheibani's name had surfaced during the SIS investigation into the murder of the six Red Caps at Majar al-Kabir in 2003.[13]

Raising a mob to attack British troops was not Sheibani's style; he modelled himself on the guerrilla leaders of the Palestinian Liberation Organisation or the Front de Liberation National, the Algerian resistance group that had fought the French in the 1950s. To be successful, an insurgency needed to avoid open confrontation with the powerful armies of the West. As the fighting at Cimic House had shown, the Jaish al-Mahdi was no match for British firepower. The secret lay in small cells able to strike quickly before blending back into the local population. Fartosi knew about Sheibani's history fighting Saddam Hussein, although he had a low opinion of the Badr Brigade. During the abortive 1999 uprising against Saddam Hussein they had promised troops and equipment to Fartosi's men, but neither had materialised. He considered most of them to have been corrupted by Iranian influence. Fartosi called Sheibani a 'bad' man.[14]

The meeting took place in a farmhouse outside Amarah. The men met alone, their escorts staying outside. Fartosi had brought a pistol with him just in case. Sheibani told Fartosi his men needed to wait until after the election. Then he would give him a weapon that would transform the war.[15]

Chapter 15

GET ELECTED

Iraqi Prime Minister Ayad Allawi was frustrated. More than ever the country needed a strong man, but the summer's fighting against insurgent leader Moqtada al-Sadr had exposed Allawi as little more than a figurehead. Being in charge of Iraq's first sovereign government since the invasion meant little without the money and guns to hold together Iraq's patchwork of warring sects and tribes. Handing out cash was the traditional Arab way of appeasing tribal leaders; and unless he was seen to be stamping out the insurgency he would never gain his countrymen's respect. But the Americans did not understand the Iraqi way and largely ignored him. He had not even been invited to the first meeting of the strategy group of the new American commander, General George Casey, to discuss the US's military offensive against Sadr. When Allawi complained, Casey insisted the decision to attack Sadr rested with him, the prime minister. Allawi played along; but they both knew he had little choice but to follow US directions.[1]

Like other Iraqi exile leaders who had opposed Saddam, Allawi had relied on British and American financial support since turning against the dictator in the 1970s. Although he switched entirely to the CIA following President Clinton's approval of a $100 million bill to support exile groups like Allawi's Iraqi National Accord, he had retained close links with the British. Upon returning to Iraq, he developed a closer rapport with UK officials in Baghdad, who seemed less interested in manipulating his position than the Americans. But Allawi did object to British efforts to micromanage his office. A small team had been dispatched to Baghdad on a DFID contract to train Allawi in the intricacies of governing a representative democracy. The team leader was William Morrison, the debonair managing director of Adam Smith International, an offshoot of a Thatcher-era free-market think tank that had turned to development work. Sixty-two years earlier Morrison's grandmother had been governess to Faisal II, the last of the British-installed kings of Iraq. The FCO encouraged Morrison to downplay this connection, for fear of raising alarming parallels with the colonial past.[2]

From Morrison's perspective, Allawi's office was completely disorganised. He was surrounded by a small coterie of advisers who, instead of shielding the prime minister from myriad demands and requests, competed to advance their own agendas. 'There was zero delegation of authority,' Morrison noted, a hangover from thirty years of dictatorship and the customary style of tribal leadership. It was no way to run a government, even one with the constraints faced by Allawi.

Morrison and his team set to work, creating a diary, a new website – Allawi had been using his own Yahoo account – and writing job descriptions for the premier's team. For his first meeting with Allawi, Morrison brought a whiteboard on which he had drawn a series of circles demonstrating the relationship of various parts of his potential cabinet office. Arrows pointed towards the centre, where the largest circle designated the prime minister. Allawi had been vaguely curious about what Morrison and his eager team were doing in the outer offices of the prime minister's building, but he was a man with a short attention span for organisational theories of management with his country in crisis. He took one look at all the circles before demanding, 'What is this?' and waving the young Englishman away.[3]

Allawi was thinking about next year's election and his prospects of winning. The situation had calmed somewhat in the south, with Grand Ayatollah Sistani having returned from London to Najaf, and the fighting subsided. However, the western Sunni stronghold of Fallujah was still in open defiance and had become a breeding ground for insurgents and suicide bombers. But ahead of his first visit as premier to London and Washington, what Allawi really wanted to know was whether there was any substance behind Blair and Bush's promises to support his re-election. Both leaders were convinced that Allawi's pro-Western secular outlook made him an ideal candidate to stand against the Iranian-backed religious parties that had also gained in popularity, but neither government had tackled the thorny topic of whether to actually interfere in the election and fund Allawi's campaign.

That was about to change with the arrival in Iraq of Tom Warrick, the US State Department official whose pre-war report on Iraq entitled *The Future of Iraq Project* had been lauded as the great post-war plan that never was and given him cult status among left-leaning liberals in Washington. Eager to make up lost time and as combative and opinionated as ever, Warrick was determined to get Allawi elected.[4]

Like some on the British side, Warrick had spotted a flaw in the rules for the election: by allowing the creation of multi-party lists, it encouraged the country's various political parties to band together along ethnic lines. For Shia parties like Dawa and the Supreme Council for Islamic Revolution in Iraq, the strategy for sweeping the Sunnis out of power was simple arithmetic: they would run as a united list, win the majority of seats and then divide up the spoils. Warrick considered both Shia parties puppets of Iran and loathed them. As he saw it, the only way to stop Iraq becoming an Iranian client state was to support Allawi, and that meant funding him.[5]

Warrick had already spotted how to do this. The Iraq supplemental bill before the US Congress had earmarked $40 million for 'aiding democracy in Iraq'. The money was intended for the National Democratic Institute, and the International Republican Institute, two NGOs with experience in training political parties. Instead of apportioning the money equally between the Iraqi parties, reasoned Warrick, why not use it to support America's preferred candidate.

However, the NDI and IRI objected to Warrick's suggestion that their funds be diverted to a specific candidate. The NDI's country director at the time, Les Campbell, argued that America had rarely interfered successfully in a foreign election. The CIA's history in Latin America demonstrated the erratic and often unfavourable results of tinkering. 'Any whiff of trying to influence Iraq's election would undermine the whole election process and one of the rationales for the invasion, which was to deliver democracy to the region,' Campbell believed.[6]

Warrick considered such views naïve. Having invaded the country, now was not the time to have qualms about further intervention to produce a positive outcome. Iranian domination in Baghdad would lead to a more theocratic regime, which would serve neither the US nor Iraqis. Supporting

Allawi's campaign would merely offset the funding the Iranians were giving their candidates. However, after a high-level meeting at the US State Department in September 2004, Campbell successfully fought off Warrick's proposal. American plans to fund Allawi directly were shelved.

But far from going away, the issue of directly supporting Allawi now resurfaced with a new champion, Tony Blair. British leaders had a certain pedigree when it came to engineering elections in Iraq. After a Shia revolt in 1920 all pretence of incorporating Iraq into the British empire had disappeared. Unless the costs of the administration were slashed, Colonial Secretary Winston Churchill was prepared to pull British forces out of Iraq altogether. The solution was a more subtle system of control: an Iraqi-led administration with the British pulling the strings in the background. However, the British could not simply install a new leader in the wake of the revolt. Instead, the British administrator in Baghdad, Percy Cox, had the main rival to his preferred candidate arrested after attending a garden party thrown by his wife.[7] He then held a sham referendum in which Britain's candidate, Lawrence of Arabia's friend and ally, Faisal, won a 98 per cent approval rating.

Such heavy-handedness was clearly out of the question for Blair, and there didn't appear much he could do. UK law restricted using government funds to influence foreign elections, and in any case the Iraq budget was small. Then he struck on a better idea: perhaps the British Labour Party could aid Allawi directly. He was not sure what form this help could take, but surely Millbank's election-winning machine could help craft Allawi's election plan.

When Allawi visited Blair in mid-September 2004, en route for Washington DC, the two shared a ready camaraderie, one onlooker noted.[8] 'How can I help you?' Blair asked earnestly. Allawi paused for a moment, awkward in the grand setting. He had lived almost half his life in exile in the UK, and he knew from being on M16's payroll that the British liked to do things on the cheap. Blair confirmed that direct funding was not possible, but Allawi seized on the idea of receiving strategic advice.

'Give me Charles Heatley,' he said. The former CPA press officer had impressed Allawi at press conferences with his ability to pitch his message to both Iraqi and Western audiences.

Before the meeting concluded, Heatley was summoned from the Cabinet Office. Since leaving Iraq in January 2004 the young diplomat had ridden his motorcycle across Jordan and Syria in an attempt to unwind. The media's craving for bad news had exhausted and embittered him, and the bike trip did little to help. Back at Number 10, he was as singular as before, bypassing Sheinwald's Iraq steering group and briefing the prime minister's inner circle directly. Some felt slighted. 'I warned Charles that he was burning those around him, but he didn't care,' said one colleague.[9]

If Heatley was chafing at the confines of the Cabinet Office, then Allawi was about to offer him a chance to shape Iraq's future. The last time a British official had worked so closely with an Arab leader was in 1921, when Kinahan Cornwallis, a young Foreign Office official, had accompanied Faisal on his first visit to Iraq as would-be king after the sham referendum. Heatley's imagination was fired.

His work began the next day, when he accompanied Allawi and his small entourage to Washington DC, with Allawi due to address Congress the day after that. With Bush's re-election campaign in full swing, how the press covered the event was crucial. The Bush administration wanted nothing less than a grateful Iraqi leader thanking the president for the invasion of Iraq. Bush's White House staff had written Allawi's speech, a fact that was only shared with a few.

Reading the speech, Heatley found it anodyne if not downright fawning, but he was more concerned about coaching Allawi on how to speak effectively to camera. The prime minister had never used an autocue before. Heatley spent his first evening in DC in a small suite of rooms in the Sheraton running through the speech, peppering Allawi with suggestions: 'Keep your eyes fixed ahead. If you look from side to side you look shifty. Don't fidget.'[10]

The next morning Allawi met Bush in the White House. 'What sort of support can I ask for?' Allawi asked Heatley on their way to the meeting.

'You're not in a position to ask for anything,' Heatley replied.[11]

'Then what is the point of me being here?'

Heatley had judged the tenor of the meeting correctly. According to one of Allawi's aides, Bush was warm but brusque. Nonetheless, that evening Allawi gave Bush what his re-election team wanted: a stiff but faultless performance. 'We Iraqis are grateful to you, America, for your leadership and your sacrifice for our liberation and our opportunity to start anew,' he declared. The speech ended with a ten-minute standing ovation and Allawi disappeared into a sea of eager senatorial hands.

'Allawi rocked,' enthused one American official, but the Iraqi prime minister left Washington with no money for his campaign. Tom Warrick, who was increasingly aggressive in his push for the US to back Allawi, made a final bid to swing the administration behind his cause. A week after Allawi's visit Warrick bypassed the State Department altogether and put the case directly to the National Security Council only to be turned down.[12] 'Bush said he wanted free and fair elections, and that we couldn't be seen to be biasing one side over another,' Condoleezza Rice told her adviser. 'Any suggestion that America was picking sides would have destroyed everything we were trying to do.'

But while the White House showed one face to the public, behind the scenes Bush was prepared to support Allawi if he could do so secretly and avoid jeopardising his promise of free and fair elections. In late October he signed a confidential presidential 'finding' authorising the CIA to support democratic campaigns across the region. One former intelligence officer believed this was largely a cover to allow the CIA to provide money to Allawi's campaign.[13] The tactic soon hit a snag however: under federal law presidential findings must be submitted to Congress's intelligence committee. Nancy Pelosi, the House minority leader, strongly protested and threatened to expose the whole episode in the press, effectively ending the plan.[14]

'Can I win this election?' Allawi asked Heatley on their return to Iraq. The assuredness Heatley had felt back in London dissipated upon arriving in Iraq. He knew that the country had seen bouts of violence and unrest since he had left in January but was shocked by how bad things had become. Sadr's truce in Najaf had brought calm to the south, but Fallujah was casting a pall over the country. Insurgents were using the city as a base to launch car bomb attacks and kidnappings.

In the Green Zone Heatley found the entire staff of the British embassy consumed by the deadly drama around the kidnap of British computer technician Kenneth Bigley and two American colleagues. They had been seized from their office in the Mansoor district of the capital in mid-September. A group led by Musab al-Zarqawi released a video of the men kneeling before a black flag as a masked insurgent demanded the release of female Iraqi prisoners by Coalition forces within forty-eight hours. The Americans were subsequently beheaded, but Bigley was kept alive. In two subsequent videos Bigley, dressed in an orange jumpsuit to evoke the

Guantanamo detainees, pleaded with the British premier to save his life: 'I need you to help me now, Blair, because you are the only person on God's earth who can help me.' The UK press seized upon the incident as symptomatic of Tony Blair's powerlessness, with the prospect of yet more blood on his hands.

The British embassy, housed in a former high school for the children of high-ranking Ba'athist officials, had become the equivalent of a police hotline centre. Harried diplomats manned the phones and chased down leads. At one stage the SAS had tailed a two-car convoy Bigley was travelling in, but when the two cars split at a junction they followed the wrong one. British government policy forbade negotiations with kidnappers, but a host of interlocutors was tried, including Yasser Arafat and the Libyan leader's son Saif Gaddafi. Bigley made one escape attempt in early October, after persuading one of his captors to release him. An unmanned drone filmed his desperate dash down a street in Fallujah before he was stopped at a mujahideen checkpoint and taken away.

On 7 October the news so many feared broke when his kidnappers released a video of Bigley's beheading. Foreign Secretary Jack Straw was visiting Baghdad that day to assess election preparations. He found the embassy staff fraught after their failure to save Bigley.

'We felt the country was tipping over the brink and the mood was very low,' recalled Victoria Whitford, who had transferred from the CPA to become the embassy's press officer.

SIS station chief Roger Sutcliffe was particularly incensed. With Iraq elections scheduled for January, only three months away, he took Straw aside as he mingled with staff at the embassy and told him bluntly, 'We can't have elections in this climate.'[15]

Sutcliffe was concerned that the Sunni tribes would boycott the election, and that if they did so they would be marginalised by the Shia parties, making an increase in sectarian violence likely. The SIS man noticed the FCO's political director John Sawers, who had accompanied Straw on the visit, heading over to break up their tête-à-tête. Senior civil servants do not like their ministers receiving contradictory messages, but Sutcliffe, a former military man, persisted.

'If we have elections now we're going to create a divide between the Sunni and Shia that could lead to a civil war,' he warned.'[16]

Straw did raise the issue of delaying the elections with Condoleezza

Rice – only to be rebutted. The US administration was not going to change tack right before the US presidential election. Furthermore, the political timetable in Iraq was the 'only thing stopping the country sliding into anarchy', Rice told Straw.

Yet, at the same time it was rejecting Britain's suggestions, the US had requests of its own. Fallujah had become the inevitable target for a second US assault, and the Americans wanted British troops to form part of the outer cordon around the city. The British defence chiefs greeted the proposal sceptically: the last thing any of them wanted was to be drawn into another American-created hellhole in the Sunni Triangle. The first US siege of Fallujah in April 2004 had killed thirty-nine US Marines and hundreds of Iraqis, and prompted uprisings around the country.

But once the request had been boiled down to numbers – a single battalion for not more than six weeks on the southern fringes of the Sunni tribal areas – the operation seemed eminently doable. Lieutenant General John McColl, the top British soldier in Baghdad, who had opposed the earlier assault, now recognised that action was needed to stop the country sliding into chaos. In late October a second Briton, aid worker Margaret Hassan, was kidnapped. The fifty-nine-year-old was a long-term resident of Baghdad, working for the British Council and Care International, and a popular figure in the city's slums. In a video message she pleaded with Blair to withdraw British troops from Iraq and not make her 'another Bigley'.

On the eve of the Fallujah assault the Black Watch battle group under the command of Lieutenant Colonel James Cowan was dispatched to Camp Dogwood, a desert outpost twenty-five miles to the south-west of Baghdad. The Black Watch's role was to stop the fighting in Fallujah from spilling south, and Cowan soon had his men patrolling in soft hats to show their lack of aggressive intent, as they did in Basra. Bush's election victory in early November against Democratic Party nominee Senator John Kerry heralded the start of the campaign. But as US Marines massed outside the city ahead of the invasion, it was the British who bore the brunt of Sunni anger. In the early hours of 4 November a red Opel accelerated towards a Black Watch checkpoint. Sergeant Stuart Gray from Dunfermline raised his rifle but was too late. The massive explosion killed him and Privates Paul Lowe, nineteen, and Scott McArdle, twenty-two, instantly. A further eight men were injured.

The deaths turned an already unpopular deployment into front-page news in the UK, reviving for many all the concerns about the original invasion and the sacrifice of British troops for American objectives. As the US Marines surrounded Fallujah and began to clear the city in vicious block-by-block fighting, Camp Dogwood was under regular mortar attack and tensions among the soldiers were rising. When another soldier, Private Pita Tukutukuwaqa, 27, a Fiji native, was blown up by a roadside bomb, John Kiszely, John McColl's replacement in Baghdad, flew to the camp to assess morale. He knew the pressure Cowan's men were under, both from the attacks and the extreme media scrutiny of the mission.[17]

There were no more casualties, but neither were there any illusions about the price to be paid for British support of the occupation. The fighting succeeded in ending Fallujah's role as a rebel stronghold but, as an American officer once famously said of a Vietnamese settlement, 'It became necessary to destroy the town in order to save it.' Over a fifth of Fallujah, a city of 250,000, lay in ruins. The US military suggested 1,350 insurgents were killed; some Iraqi sources claimed the death toll was closer to 6,000. Ninety-five US troops died in the fighting. Among the bodies later recovered was that of Margaret Hassan. She had been shot through the head; her legs and arms had been hacked off and her throat slit.

The fighting in Fallujah provided a stark backdrop for Allawi's election campaign. A September poll by the International Republican Institute showed support for Allawi's government in Iraq had slumped.[18] Allawi's small election team took some consolation in the fact that the offensives in Najaf and Fallujah had not made him even more unpopular. With no American campaign financing, Allawi's budget of £1.5 million was dwarfed by his rivals', the religious Shia parties. He paid the Lebanese Broadcasting Corporation to produce a slick series of adverts featuring a reworked anti-British folk song from the colonial era. There was little money for anything else.[19]

Heatley's instincts told him Allawi should confront the carnage across the country. The prime minister would need to strike a difficult note: commiserating with Iraqis for the American damage, at the same time claiming responsibility for restoring order. Allawi visited Iraqi troops in the provincial capital Ramadi, bombed during the fighting in nearby Fallujah. The visuals worked well, with Allawi looking like a man of the people as he broke bread with a few unshaven recruits, but the next trip to Najaf almost ended in disaster when his convoy was attacked by a large mob as it neared the Imam Ali Shrine in Najaf. Stones were hurled, and the police guarding Allawi's convoy opened fire. There was a brief shot of Allawi's armoured Land Cruiser peeling away from the mêlée.

'It was a disaster,' said Azzam Alwash, the deputy campaign manager. 'No one mentioned it again.'

With Allawi trailing the Shia parties by 30 per cent, his last hope rested with the campaign advice from Millbank that Blair had promised. Sent to Iraq were Margaret McDonagh, the former general secretary of the Labour Party who had helped introduce American-style war-room tactics to the 1997 election campaign, and Waheed Alli, a television executive who had created two popular shows of the 1990s: Channel 4's *Big Breakfast* and *Survivor*, the desert-island fantasy game where contestants battled for supremacy while eating worms and collecting rainwater in coconut shells.

Shrouded in secrecy, with less than two months to go before the poll, the mission got off to a poor start. Allawi and his Iraqi staff were unimpressed with McDonagh and Alli's suggestions on tactics, which ranged from analysing polling data, through working with focus groups to coordinate campaign messaging, classic New Labour ploys. 'They clearly had no idea what Iraq was like,' said Alwash, who was more concerned about the smear campaigns against Allawi in the media. Al Jazeera was calling him an American puppet.[20] 'If only they knew the truth – that the Americans are doing nothing for us,' Alwash groaned. Far less conventional was the suggestion to buy support from the Sunni tribes in Anbar province for £1 million through the SIS. If Allawi was to have any chance of victory he would need Sunni support.

But the UK money pumped into the province was not nearly enough to make a difference. When polling booths finally opened across the country on 30 January 2005 only a small fraction of Sunnis voted. Everywhere else, Iraqis defied the insurgents and voted in their millions. When the results were declared the following month the Shia religious parties, backed by Grand Ayatollah Sistani, had won a sweeping majority as Iraqis turned to their religious leaders for security. Allawi managed just 14 per cent of the vote.

UK officials braced themselves for a dramatic loss of influence over the Iraqi government. The aid to the Anbar Sunni tribes had raised suspicions among Shia politicians, who saw a throwback to the colonial days, when the British had favoured Sunni leaders like King Faisal. Heatley had left Baghdad a week before the poll, exhausted and disillusioned.[21] Another casualty of the Shia victory was the cabinet office that the DFID contractor William Morrison had laboured to construct. When Allawi's Iraqi staff left their jobs they stripped their offices. The Shia politicians jockeying for the top job showed little interest in re-creating the set-up.

Office equipment was not the only thing missing from Allawi's time in charge. Iraq's Defence Ministry subsequently revealed that almost a billion dollars had been stolen through fraudulent contracts for arms purchases in Poland and Pakistan.[22] Allawi denied any suggestion of corruption, blaming his subordinates. Meanwhile, he opted for self-imposed exile in London, leaving the scene set for Shia dominance and the dramatic rise of Iranian influence.

Chapter 16

TRANSITION

ONE OF THE LAST TIMES BRITISH FORCES occupied southern Afghanistan was in 1879. Imperial rivalry with Russia was at its height. As Russia expanded rapidly into central Asia, the British feared an invasion of their Indian empire through Afghanistan's mountain passes. Britain hastily dispatched 40,000 troops, mostly Indians, and installed a puppet ruler in Kabul. The following year the Afghans rose in revolt. Ayub Khan, son of the deposed ruler, raised an army of 12,000 tribesmen and set off to attack the British, who met them with a force of around 2,000 from their garrison in the regional capital Kandahar. The local Afghan governor had urged the British to confront Ayub Khan and had offered troops to support them, only for his men to switch sides shortly before the battle.

On a desert plain outside the village of Maiwand in Helmand the two armies met. Unused to the terrain of exposed flats and plunging ravines, the British were quickly outflanked by Khan's horsemen and forced to retreat, losing almost a thousand men in the process. This stood as the heaviest defeat of a Western power by an Asian force until the Battle of Kut, thirty-six years later.[1]

For British imperialists at the time Maiwand underscored a few unsavoury facts about southern Afghanistan that a modern-day invader would do well to consider: its Pashtu tribesmen are deadly fighters who compensate for their lack of equipment with scant regard for their own lives. Unruly and capable of changing allegiances in a moment, they are also bound to a tribal code of honour that requires them to protect their guests and repel invaders. The landscape is equally formidable. The Helmand river, emerging from the Hindu Kush mountain range in the north, supports a thin ribbon of greenery as it flows west to the border with Iran. On either side are the forbidding Rigestan and Dasht e-Margo deserts – the latter translates as 'Desert of Death'. During the summer the parched land is baking hot and subject to raging sandstorms; in the winter the temperatures drop below freezing and the passes through the surrounding mountains are choked with snow.[2]

In 2004, when the British military started planning to send troops to southern Afghanistan, the lessons of Maiwand, like many other aspects of the colonial era, seemed to belong to a distant and rather quaint past that had little to teach the present. But time passes more slowly in the tribal fastnesses of Afghanistan. The lessons of how to defeat imperial invaders had not been forgotten as events were to prove.

In November 2004 Major General Barney White-Spunner arrived in Kandahar, the principal city of the south, to make the first assessment of what type of British force, if any, should be deployed following the NATO summit in Istanbul earlier that summer. At the time the Americans had 20,000 troops scattered across the country, mostly stationed in and around Kabul. There were fewer than 1,200 US soldiers in southern Afghanistan. Nation-building efforts were limited, and the Americans were using their stretched resources to strike at insurgent leaders linked to al-Qa'eda. At Kandahar's dusty airbase, where the US maintained a small headquarters, White-Spunner was struck by the dim grasp the Americans had of the life in the desert beyond. In the neighbouring province of Helmand, rapidly emerging as the biggest opium-producing region in the world, there were just 200 soldiers.

The Eton-educated White-Spunner shared with his Victorian forebears a spirit of adventure, but with only two days in Kandahar he had little time to travel beyond the base. His lumbering frame, heavy jowls and beady blue eyes hid a sensitive, romantic side that had led him to pen a volume of military history and picturesque accounts of the English countryside. White-Spunner spent his time trying to gauge American opinions on the likely reception of British forces by the Afghans. The Americans, unsurprisingly, were entirely positive about a British mission, which would free up US forces for other tasks.

White-Spunner reported back to London that the British should send 'about a brigade', that is, upward of 3,000 men. Conveniently, that was at the lower range of the force size that the 1998 Strategic Defence View had stipulated for overseas expeditions. Although the British army could field 40,000 troops for a short operation like the invasion of Iraq, it could only support a brigade-sized force for longer-term deployments. However, given there were still 9,000 troops in Iraq, sending a brigade to Afghanistan would severely strain the army's resources.

As the planning for southern Afghanistan gathered pace, the defence chiefs knew they would have to begin withdrawing British troops from Iraq, but Whitehall had yet to decide how to begin the process. Some civil servants pondered whether the UK should wait for American forces to finish battling Sunni insurgents and restore calm to the country; others thought the Iraqis needed to establish a stable government first, which was likely to take a year or more. Ambitions to rebuild the country and establish a healthy democracy had been scaled back dramatically. Reconstruction efforts in the British-controlled provinces like Maysan had all but ended due to the violence and militia attacks. After the elections, unfriendly politicians with links to Moqtada al-Sadr had taken over the local councils which had been set up. The British military had continued to refurbish schools and lay out football pitches, and in Basra the DFID offered Iraqi administrators lessons in management and accounting, but it wasn't much. Most of the British in Iraq felt if they could just leave the country with a semblance of security in place, they would be exiting without too much embarrassment.

The task of resolving the impasse in southern Iraq fell to Major General Jonathon Riley, the new commander in Basra and the sixth since the war began. Riley adopted a simple formula: British forces would withdraw as the Iraqi army and police demonstrated their ability to manage their own security. The concept had originated at the US military's Central Command in Florida the previous year, as the Americans had begun to consider their own withdrawal strategies. It was a straightforward plan in theory. Yet, in reality, the Iraqi army had shown itself brittle and liable to desert at the slightest sign of trouble. As for the police, they were corrupt and infiltrated by the militias. During the Sadrist uprisings British troops had often found Iraqi police shooting at them, rather than with them.

Having participated in the American effort to rebuild the army in Baghdad the previous year, Riley was aware of the problems but felt that with enough attention, Iraq's security forces would improve. The question was whether the British would commit resources to the plan, not just rhetoric. When Tony Blair visited on 2 January 2005, just before the Iraqi elections, Riley told him the British effort had 'delivered nothing'. So far, the British had only trained a single ill-equipped army brigade. The Iraqi soldiers Riley needed to guard polling stations were lacking body armour, helmets and in some cases even boots.

'You tell me Iraq is the most important foreign priority of the British government, then why do I not have more resources?' he said. Blair looked startled and changed the subject.[3]

To Riley's surprise, he did get some equipment for the Iraqi security forces in time for the election. Reassured that training Iraqi forces had become Britain's primary objective, Riley's headquarters started work on a handover plan which envisioned British troops switching from daily security operations to a more supervisory role, eventually only venturing out of their bases at the request of local forces. US military planners dubbed this end state Overwatch and had developed a set of metrics to assess the progress of Iraqi forces towards it. Riley wanted to dedicate an entire battle group to training the Iraqis, with British officers embedded in Iraqi units to mentor their counterparts and provide links to the Coalition's air and surveillance assets. Knowing that the pressure for a quick withdrawal from Iraq was likely to mount, Riley made sure there was no mention of a timetable. Britain would only get out of Iraq when the Iraqis were ready.

Riley's plan was embraced in London with only one caveat. The chief of joint operations in Northwood, Glen Torpy, opposed the plan to embed British officers at company and battalion level, citing a 'dependency culture' that might develop between Iraqi and British forces. 'We're meant to be withdrawing, not getting in deeper with the Iraqi army,' Torpy informed Riley. The decision not to embed was to have serious consequences later.[4]

As Riley began to implement his plan, however, he butted heads with British civilians on the ground who had different concepts of how to best train the local police force. A former chief superintendent in Hampshire, fifty-three-year-old Colin Smith had been Basra's senior police adviser for a few months and was in charge of training and mentoring 25,000 Iraqi officers across southern Iraq. Riley had assumed that the military had responsibility over police training, but Smith thought otherwise and was not afraid to speak his mind. He liked to boast he had made Hampshire the third most effective police force in the UK, and was confident he could perform a similar transformation in southern Iraq.

Smith's plan to train the Iraqi police was based on many of the tactics that he claimed had worked in Winchester. He was critical of ArmorGroup, the security company with a £5.3 million contract to mentor the police, for not doing enough to train Iraqis in community policing, in which officers walk the beat and get to know the locals in order to prevent crime. Nor were the police learning enough about the criminal justice system, including the rights of detainees, the importance of gathering evidence and the process of working with lawyers to push cases through the courts to gain convictions. And the British military was not helping the situation by treating Iraqi police officers like army recruits. A few drills and some firearms training were not going to build an adequate force, Smith believed. Establishing the rule of law through the courts lay at the heart of the occupation's success, and Smith felt he was the man to lead that effort – not a general like Riley.

He presented a report outlining his plans to Riley in early February before a large gathering of officers and diplomats at Basra airport, including William Kearney, the ArmorGroup manager. For the occasion Smith wore two stars on his epaulettes, an indication that he considered himself to be of the same rank as the general. Riley, one of the most operationally experienced officers in the British army, looked peeved. His jutting nose and close-set eyes gave Riley a hawkish look. One glance at Smith, with his neatly-trimmed white beard and air of quiet satisfaction, and the British general let out an audible 'Humph.'

As far as Riley was concerned, the report was bunkum. Iraq had no tradition of community policing but instead followed the continental European model of an armed gendarmerie keeping the peace rather investigating crimes.

'You've got to go with the grain in Iraq,' Riley told Smith.

'That grain is rotten,' Smith reposted.[5] Watching the two men sparring, Kearney thought the debate academic. Neither raised the real issues facing the Iraqi police, such as the killing and intimidation of officers by the militia. Since the start of 2005, over 350 Iraqi officers had been killed in attacks against police stations and recruiting centres across the country. Then there was the question of how to handle the growing evidence of Iraqi police brutality, including the torture and murder of detainees. Since

Kearney had first learned of this problem the previous summer, stories of abuse had continued to emanate from Jamiat police station in Basra. Kearney suspected such violence was common in many stations and was even quietly condoned by the government as a way of silencing its enemies. He had tried to alert Smith and his predecessor, Kevin Hurley, but neither man had shown much concern.

Kearney's own efforts to tackle the problem had come to naught. Charlie MacCartney, the police mentor at the Jamiat, was struggling to get close to the likes of Captain Jaffar, the head of the serious crimes unit based at the station. Every visit the Ulsterman made was inevitably broadcast hours in advance and the place scrubbed up. He would be carefully chaperoned around the station and his questions moderated through an interpreter. MacCartney was beginning to feel as hapless as Jaffar's powerless boss at the station, Colonel Sewan, but there was nothing he could do about it. As he told Kearney, the Iraqis were bound to get up to their old tricks, and he had little desire to interfere as long as he could continue to do his job.

Kearney felt differently. In January he had heard from Major Shaqir, the first Iraqi officer he had worked with, that three middle-ranking police officers from the Jamiat had been found dead in a gutter. The men had had their kneecaps drilled before being shot in the head. The motive for the killings was unclear, but he feared they had been targeted by the militias to send a warning to those police officers who did not obey them.

'Now the Jamiat is completely in their hands,' Shaqir warned cryptically.

'Whose hands?' Kearney wanted to know, but Shaqir would not say anything else.

Haider Samad, the young Iraqi interpreter who had been so excited at the start of the occupation, was convinced that most of the police were already in league with the militia. On his route to work each morning he routinely came across makeshift barricades manned jointly by the police and local black-shirted militiamen. He recognised some of the militia as former Ba'athists as he brazened his way through or sneak down side streets.

In early April suspicions about the Jamiat became common knowledge. A joint British and Danish patrol arrested Abbas Allawi in a dawn raid. Allawi, a suspected fuel smuggler, was handed over to Captain Jaffar's unit at the Jamiat. Over the next three days he was tortured before being beaten to death and dumped outside the station. When Allawi's relatives protested to the authorities, the initial British reaction was to investigate the matter quietly. The Danes, however, were outraged, threatening to pull out of the Coalition if immediate action was not taken.

Kearney had already taken the matter into his own hands. A few days after the Allawi incident he and Charlie MacCartney conducted a snap inspection of the Jamiat. An unshaven sleepy-looking Captain Jaffar was quickly summoned. He led them to the first compound. The walls were lined with Iraqi policemen toting AK-47s. Both Kearney and MacCartney were wearing Kevlar armour and had handguns at their hips. Kearney, with his bodybuilder's physique, wore his easily, but the lumbering MacCartney looked like a sagging scarecrow.

'Where are the prisoners?' Kearney demanded. Jaffar ushered them to a passageway leading to the serious crimes unit's offices. One of the ground-floor rooms served as a cell. It had a battered-looking metal door with a small grate. The room contained a dozen prisoners in various states of undress. Kearney insisted on entering the room and looked closely at their bodies for signs of abuse. He could see none, although the prisoners were eerily silent.

He noticed Jaffar watching him with the faintest suggestion of a smile. He seemed to be enjoying the game. Determined to catch him off guard, Kearney opened doors around the courtyard. Most revealed dirty cubicles containing filing cabinets and rickety desks. He was about to give up when he opened the door of a larger room with a meat hook in the ceiling.

There were chains and plastic piping on the floor. He leapt forward and seized a piece of the pipe, which he was sure was used to beat prisoners.

'What are these for?' he demanded.

Captain Jaffar smiled. 'No problem,' he said.[6]

Kearney looked over at MacCartney. Both men were aware of the hostility in the air. 'What should we do about it?' Kearney asked.

'Nothing we can do, Billy my boy, if they want to fuck each other up,' said MacCartney.

Kearney gathered up the pipes and chains. 'Well they're not going to do it with these,' he said.

He sent a report to Smith, but other than an official complaint being lodged with the Iraqi Ministry of the Interior, which many suspected was also infiltrated with militia, no further action was taken.[7] In fact, Smith, for all his reforming intentions, was on his way out of Basra. Shortly after clashing with Riley, he had been reassigned to Baghdad. Kearney found him more aloof than ever on the troubling subject of torture. Kearney became increasingly disillusioned. He had seen evidence of torture, and yet senior officials preferred to look the other way, and even he had to admit he did not know what could be done about a problem that seemed endemic. To Kearney, the problem was bigger than just a human rights issue. If the police were systematically working as a front for the militias, the Brits were only arming their enemy by helping to train and equip them.

He was not alone in being concerned about the poisonous link between the police and the militias that Jamiat represented. Over the same period the SIS station chief, Kevin Landers, had also become aware of Captain Jaffar's activities within the serious crimes unit. However, Landers' contacts in Basra were making an even more dangerous link that Kearney had only suspected – Jaffar was in league with the man rapidly taking over the city, Jaish al-Mahdi leader Ahmed al-Fartosi.

Since the fighting the previous summer, Fartosi had become the SIS's primary target in Basra. Fartosi had kept a relatively low profile since Sadr's truce, but from the scraps of information Landers gleaned, it was clear the Jaish al-Mahdi leader was building a powerful network across the city, including within the police force. Fartosi's rise had also been aided by the January national elections, which had put Sadrists onto Basra's provincial council and into the governor's seat. Muhammad al-Waeli was a small-time businessman and the political face of a Sadrist splinter group called Fadhillah. Without its own militia, Fadhillah was dependent on the good graces of men like Fartosi, at least until Waeli was able to bend the security forces to his will. Landers wondered whether the murder of Allawi had been carried out on the governor's orders. Allawi had been a rival oil smuggler after all.[8]

By late April Landers was urging Riley to take military action against Fartosi; the militia leader was growing too powerful. Roger Sutcliffe, SIS Baghdad station chief, told the general that he could put Fartosi under surveillance and quickly prepare an operation to arrest him, but Riley didn't believe kidnapping Fartosi would be so simple and did not want to provoke the Jaish al-Mahdi with a botched raid.[9] There was little appetite for action in London either. The MOD had recently drawn up plans requiring all but 1,000 British to leave Iraq by the end of 2005 – just in time to start sending troops to southern Afghanistan in the New Year.[9] The political pressure was mounting to keep the situation quiet, even if it meant turning a blind eye to militia activity.

As Riley, Kearney and the others struggled with the innumerable obstacles to establishing security in Iraq, London was moving ahead with the deployment to southern Afghanistan. The plans had changed somewhat since White-Spunner's lightning-quick trip to Kandahar. The Canadians had pushed to take command of the overall international effort in southern Afghanistan, which now included troop contributions from the Dutch, the US, Romania and Australia, and they would be headquartered in Kandahar.[10] That had left the British military searching for another base of operations. Further investigations by White-Spunner's team suggested that neighbouring Helmand province, which produced almost half the world's supply of heroin, would be ideal given Blair's desire to tackle the illegal narcotics trade. Helmand also shared a mountainous border with Pakistan's North West Frontier Province, a Taliban and al-Qa'eda stronghold, which meant British troops could interdict incoming fighters before they could stage attacks. A less remarked-upon fact was that Helmand was home to the small village of Maiwand, where British forces had been so heavily defeated 126 years before.

But before either plan for Iraq or Afghanistan could be implemented, Whitehall went into purdah ahead of the general UK general election, scheduled for May 2005. Few outside the military realised just how much the futures of Iraq and Afghanistan hinged upon whether Tony Blair was re-elected.

Chapter 17

A PROPOSAL

HAIDER HAD AT LAST SAVED UP ENOUGH MONEY to approach Nora's father, Muhammad, with another wedding proposal. The last time they had spoken, Muhammad had been less than impressed with the penniless young man from the slums of northern Basra. But the young Iraqi had been working as a translator for the British for two years and saved a small fortune, although he had to hide his controversial job from Nora's family. Muhammad was a Sunni with a good job at the electricity plant. He might well not approve of Haider helping the Western forces that seemed intent on destroying the old order.

Haider's first job, translating for the British army, had paid a pittance, but since getting a job with ArmorGroup, the private security firm training the police, he had been earning £350 a month and able to save. He had finally amassed £3,000, the sum Muhammad had informed him was the bare minimum to winning his consent. In preparation for the proposal, Haider had used some of the money to take out a lease on a small apartment in downtown Basra. Although it was customary for the bride to move in with the husband's family, Muhammad would clearly not let his daughter move to the slums. In addition, Haider had kitted out the apartment with a refrigerator, television and air-conditioning unit – the type of electrical goods that had flooded into Basra since the invasion and were the trappings of new wealth. The rest of the money Haider had spent on gold jewellery for Nora; in a land plagued by wars that gold was Nora's insurance policy.

But then, having prepared his offer, he did nothing. William Kearney, his boss at ArmorGroup, who had been following the story of his relationship, uncovered the reason for Haider's reticence. In Iraqi society it was customary for the suitor's father and other male relatives to accompany him to the woman's house. This allowed the bride's father to eye up the suitor's lineage. Haider had no one but a younger brother to take, and being shy and self-deprecating, he had not asked for anyone's help. Kearney decided to take matters into his own hands and asked another Iraqi translator, Hassan, a former Iraqi army general with an impressive paunch, to speak for Haider at the meeting.

One mid-morning in early April 2005 the two men sat down on the couch opposite Muhammad in his reception room. Nora's family lived near the power station in northern Basra, in a relatively affluent tree-lined suburb. Haider wore a paisley shirt that hung off his gaunt frame. His trousers were neatly pressed, and his square-toed shoes, the latest fashion in Basra, were black and shiny. Hassan was more conventionally dressed in a suit. Muhammad wore, slightly self-consciously, a long tribal dress that showed his chest hair.

The centrepiece of the reception room was a small chandelier, beneath which was a coffee table conspicuously bereft of tea or Arab sweetmeats. At a wedding proposal the offer of drinks depended on whether the offer was accepted. The tone was clinical and officious. Haider sat quietly beside Hassan as the general responded to Muhammad's questions. The grilling was a clear sign that Nora's father was interested. Hassan pretended to be Haider's employer, a wealthy builder who had won some contracts with the British.

'Don't mention the British in this house,' Muhammad had declared early on. 'What have they done for Basra? It's become a den of thieves.' A few months before, militiamen had assassinated Muhammad's cousin, a lecturer at the technical college, as he lay in bed, alleging, falsely, that he had belonged to the Ba'ath Party. Hassan wisely changed the subject to Haider's extensive wedding preparations.

Muhammad eyed Haider up and took a deep breath. 'Would you like some Sharbat?' he asked.

Haider's heart leapt. The appearance of Sharbat, a sweet cordial made from black grapes, was usually a sign that the proposal would proceed. Nora appeared in the doorway with a large pitcher. She was dressed in a black *abaya* that exposed her face. She was clearly trying to suppress a smile.

'Is it true, my daughter, you want to marry this man?' Muhammad asked.

Eyes downcast, Nora replied that she did.

'Then you shall,' her father pronounced and waved her away.[1]

The meeting adjourned after a few minutes more small talk. Haider left feeling elated, but that evening, sitting on his rooftop in the slums, his customary nervousness returned. He had slowly and patiently built a life for himself after all the years of isolation, but now he was on the verge of achieving his dreams, he felt worried by the fragility of his creation. His fate was pinned to that of the British occupation, and yet everything he had seen of the past two years suggested their grip over the city was waning. He could see the difference in the police officers he was helping Kearney mentor. Conscientious officers like Major Shaqir had been sidelined, and the newer recruits were insolent and uninterested. Some even asked prying questions about who he was and why he worked for the occupation. He had tried using a Kuwaiti accent to hide his origins, but he didn't know if it was effective.

Why didn't the British do more to enforce security in the city, restart the ministries and factories, and return life to normal? he wondered. Like most Iraqis he thought Britain and America wielded almost limitless power, and there must be a reason for their failure to rebuild Basra or take on the militias. Some believed the British wanted to keep the city permanently downtrodden, but Haider blamed his fellow residents for failing to take advantage of the reconstruction projects the British had offered.

Either way, as Haider sat on his rooftop, he realised his feelings of insecurity were so much worse because, for the first time, he had a lot to lose.

Chapter 18

DON'T MENTION IRAQ

REG KEYS' FIRST SUSPICIONS ABOUT THE British army's investigation into his son's death were raised after a meeting at the Royal Military Police headquarters in October 2003. Keys and the families of the five other Red Caps murdered by a mob in Majar al-Kabir police station in June 2003 had travelled to the base in Bulford, Wiltshire to hear details from Lieutenant Colonel Jeremy Green from the Special Investigations Branch of the Royal Military Police.

Through the media, disturbing details about the Red Caps' lack of ammunition and a satellite telephone had emerged. The families had also learned from military contacts further details that pointed to an army culture of carelessness: the riot two days beforehand that should have warned senior officers of the dangers posed in Majar al-Kabir; the subsequent deal between 3 Para and the locals to stop house searches, which the military then jeopardised by aggressive patrols on the day of the attack. The most shocking discovery was that a British army vehicle had been within sight of the police station shortly before the attack. The families wanted to know why the vehicle had not gone to the Red Caps' rescue.

Keys' slim fox-like face, with its greying hair and thick moustache, had an earnest, focused demeanour as he listened to Lieutenant Colonel Green describe the events that had led to his son's death. Keys had long resigned himself to the reality that soldiers are killed in battle, and he had approached the meeting in a spirit of openness, expecting the military to try to make sense of their loss, and make sure that every lesson was learnt. His mood quickly soured as it became clear that Green would not address the major mistakes that had resulted in their sons being stranded in the police station. At one point Green told them they had not been able to identify the British vehicle that had passed near the station just before the mob descended.

'That's a load of bollocks,' said John Miller, a former non-commissioned officer and father of Corporal Simon Miller, the Red Cap whose 600cc Suzuki motorbike still sat at home in the garage.[1] He suspected the military knew exactly whose vehicle it had been and were deliberately covering up. Keys suspected the army would rather not question the aggression of one of its premier fighting units or the quality of leadership in the military police.

Nine months later the army board of inquiry published its findings, and the Red Caps' plight once again dominated the news. The families had not been asked to either attend or contribute. However, on the day of publication they were invited to meet Minister of Defence Geoff Hoon. Keys had asked to see the report the night before but was refused. Instead, the families had half an hour to read the ninety-page document. Most names were blacked out and all the crucial questions ducked. The report concluded that the deaths were 'not preventable'.

In the short meeting with Hoon, Keys struggled to get his points across. 'Why didn't our boys have proper equipment?' he asked. The minister fobbed him off, saying there had been enough equipment to go around. Calm and deliberate from his years as a paramedic, Keys was not quick to anger, but he was fuming when he left the meeting with Hoon, as he told the reporters waiting outside. A few weeks later Keys got in touch with John MacKenzie, a lawyer who had successfully brought a number of cases against the military. Mackenzie offered his legal opinion that the military had broken the law by failing to provide the Red Caps with adequate communications equipment. He also suggested that the British government had a larger case to answer over the legality of the war and the use of intelligence. Mackenzie's views struck a chord with Keys, who had started to ask why his son had been sent to Iraq in the first place. Tony Blair, he believed, had manipulated the case for war and, despite two inquiries, had failed to provide an adequate explanation. Talking to Mackenzie, Keys saw how his personal grief was wrapped up in the bigger issue of bringing the government to justice. There needed to be a new public inquiry into the war, one that did not dodge the difficult questions.In the short term Keys asked Mackenzie to represent him and the other families at the coroner's inquest into their sons' deaths. The inquest would be the next opportunity to press the military. Unfortunately, the Oxfordshire coroner dealing with inquests into soldiers' deaths as their bodies arrived at Brize Norton airbase was hopelessly overworked, with a backlog of eighty cases, and was unable to set a date.[2] Keys was dismayed to hear that the earliest that the Red Caps' inquest could start was the spring of 2006. He feared the strains that leaving the case open would place on his wife, and that the momentum of recent weeks would be lost.

After hearing his thoughts, Mackenzie made a startling suggestion. The only way this government would respond would be under pressure so why didn't Keys gain the maximum publicity for his cause and run as an independent candidate against Geoff Hoon in his Derbyshire constituency in the next general election, scheduled for May 2005? The election was shaping up to be a referendum on the Iraq war, and Keys could become a powerful critical voice.

Keys said he would think about it, unsure how to begin and concerned about further wounding his family. He had recently founded an organisation called Military Families Against War with another bereaved parent, Rose Gentle, whose son Gordon had been killed by a roadside bomb in June 2004. The organisation had quickly evolved into a mixture of support network and grass-roots advocacy for ending the war. Gentle was considering running against Adam Ingrams, the armed forces minister in her home constituency. She encouraged Keys to take the leap.[3]

In early 2005 Keys spoke to Felicity Arbuthnot, a freelance journalist who had covered the impact of UN sanctions before the invasion and worked with Margaret Hassan, the murdered aid worker. She was passionately against the war. When Keys told her he was thinking of running against Hoon, she told him she had a better idea. 'Why not run against Blair himself?'

Arbuthnot was in touch with a group of MPs, writers and artists pushing to impeach Blair for 'high crimes and misdemeanours'. It was a diverse bunch which included Harold Pinter, music producer Brian Eno, author Frederick Forsyth, Alex Salmond, leader of the Scottish Nationalist Party, and Adam Price, the Plaid Cymru MP who had tabled the first questions in Parliament about the scale of the military's abuse of Iraqi detainees over the summer of 2003. The idea of impeaching Blair had first been raised by the academic Dan Plesch, a prominent anti-war campaigner who felt the movement had stalled since the invasion. The last occasion an impeachment motion had been tabled in Parliament was in 1848, when Lord Palmerston was charged with concluding a secret treaty with Russia. Plesch, along with Cambridge academic Glen Rangwala, prepared a report laying out the charges against Blair, which included failing to resign from Parliament after misleading the country, making secret pacts with a foreign power, undermining the constitution and general negligence and incompetence.

In August 2004 Adam Price presented this report to Parliament. The first step was to table a motion on the subject for debate, but with none of the main political parties prepared to agree to the vote, Price and the motion's supporters faced an uphill battle. They had already begun looking for a suitable candidate to stand against Blair – someone in the mould of independent MP Martin Bell, who had defeated the discredited Tory MP Neil Hamilton in 1997. Greg Dyke, the BBC director general who had resigned over the 'sexed-up' dossier complaints was one potential candidate. When Arbuthnot told them that Keys was considering standing, they knew they had found their man. The chances of dislodging a serving prime minister were slim, but the campaign would ensure that Iraq remained a key election issue.

In February 2005 Keys travelled up to London to meet Forsyth, Eno and Price. Don't go for the organ grinder's monkey; go for Blair, they counselled. The group had no funds available, but they would tap their networks for support and advice. Martin Bell would prep Keys on how to run a campaign and deal with the media, and Bob Clay, the former MP for Sunderland South, would be his agent. Emboldened, Keys rose to the challenge.

Six weeks before the election he travelled with Arbuthnot to Sedgefield to test the water, staying at the local Travelodge. Arbuthnot had arranged for *Guardian* journalist Stuart Jeffries to meet them, and Keys fired the opening shots of his campaign. He recalled the words of the widow of a man who had died on the *Kursk*, the Russian nuclear submarine whose crew perished in 2000 after Putin's government failed to rescue them: 'If you betray your country you are a traitor and you will go to prison. But if your country betrays you, what can you do?' Keys told the journalist. 'I think I have an answer to that: we can use our vote to get rid of those people who betrayed my son and other men like him. That's what I want the people of Sedgefield to do.'

A few days later he returned to begin the campaign in earnest. A local pub, the Hardwick Arms, had been volunteered as campaign headquarters, and a member of the Labour Party's executive committee in Sedgefield lent him his garden annex for accommodation. At first Keys provoked a mixed reaction on the streets. Sedgefield was staunchly Labour. 'Sorry, mate, always voted Labour here' was a typical reaction on the doorstep. But Keys plugged away: 'Think how much money has been wasted in Iraq, and what just a bit of that could have built for you in Sedgefield.' Keys was more successful in attracting the media, who found the story of an impassioned Everyman taking on a morally bankrupt ruler irresistible. He soon had a small posse of photographers following him around, including at one stage a BBC film crew who were making a documentary called *Reg vs Blair*.

The press attention transformed the campaign. Keys had begun with only a few team members, but as word spread, the Hardwick Arms was inundated with dozens of anti-war volunteers. With his growing campaign fund Keys rented some offices nearby where volunteers were given a crash course on how to canvass. His posters began appearing in windows, and small crowds attended his soapbox speeches.

But the campaign was taking its toll. Every night was spent in teary-eyed conversation with his wife, who was struggling to cope with constant reminders of her son. Also challenging was the reception from some military families who had lost loved ones. They objected to the way Keys had introduced politics into the sacrifice of their sons and daughters, husbands and wives. Those who had died had answered their country's call. Questioning the war effort with British troops still in Iraq and fighting for their lives seemed to some like a form of disloyalty.

At times Keys felt like his campaign was further dividing the nation, not bringing it together, and he came close to quitting. Martin Bell travelled up and spent a week with him, offering to carry on campaigning if he needed to return home. A call from his surviving son bolstered his resolve. 'You've got to see this through for Tom,' he said.

After a few weeks Keys spotted the first sign that he was making political headway when the Conservative Party candidate, Anthony Lockwood, called him in to discuss folding Keys' campaign into his. Keys chuckled and told Lockwood that perhaps *he* should stand down and join him. Meanwhile the Labour team was doing its best to keep Iraq out of the campaign. As one Blair adviser recalled, the strategy was 'Economy first and second, and whatever you do don't mention Iraq.'[4] Blair was bolstered by the return of Gordon Brown to the fold. After sulking for several months over the prime minister's refusal to step aside, the chancellor's instincts for survival had kicked in. On 2 May the two men ate ice cream together in Blackpool before the cameras in a symbolic patching-up of their rift.

But with over 8,000 troops still on the ground, Blair couldn't make Iraq go away. On 27 April the debate about the legality of the war was reopened when *Channel 4 News* leaked a March 2003 memo from attorney General Peter Goldsmith in which he expressed doubts over the advice he had given to Blair about the legality of the invasion. 'While a "reasonable" case could be made for going to war, I was not confident that a court of law would necessarily agree,' he conceded. Anti-war protesters believed Goldsmith's comments demonstrated that the government knew that the war was illegal from the start.

Then on 1 May 2005 disaster struck in Iraq. Guardsman Anthony Wakefield was riding top cover during a night patrol through the outskirts of Amarah when a roadside bomb detonated next to his Snatch Land Rover. Such attacks were a daily occurrence in American-controlled areas of the country but still relatively rare in the south. Roadside bombs consisted of a mine or artillery round encased in scrap metal to create a deadly blast of shrapnel. The devices were detonated manually through wires or, as was becoming more common, a signal from a mobile phone.

British patrol vehicles carried a transmitter capable of blocking such signals, but the one on Wakefield's Land Rover had failed. Shrapnel struck Wakefield in the neck, severing one of the main arteries to his brain, and in the side of his chest, damaging his heart and lungs. He slumped into the Land Rover, which rumbled on for fifty yards before breaking down. Lance Sergeant Craig Newton took Wakefield in his lap. He had a pulse but was not breathing.[5]

At Camp Abu Naji, Captain Steve Vickers, on duty in the operations room, called up a helicopter to rescue Wakefield. The other Land Rovers in the patrol formed a defensive perimeter around the stricken vehicle, but when the helicopter arrived, the pilots refused to land without a safe landing site and flew off. The patrol ended up having to drag Wakefield's Land Rover back to Camp Abu Naji, and the young soldier from Newcastle upon Tyne was declared dead shortly after arrival. He left behind three children.

The death, a few days before the general election, caused panic in Labour Party headquarters. He was the eighty-seventh soldier to die in Iraq. Wakefield's estranged wife Ann Toward confronted reporters outside her home in Newcastle. 'It's Tony Blair's fault . . . If he hadn't sent them out [Anthony] would still be here.' She had just had to explain to her distraught children that their father had 'gone to heaven'.

Blair now had to defend his rationale for going to war once again. He was also forced to respond to one of Keys' main campaign demands: a new Iraq inquiry. Keys and nine other families had recently taken a letter to Downing Street saying they would apply for a judicial review if an inquiry was not announced within fourteen days. Blair tried to dismiss the idea, telling *Channel 4 News*, 'We have had inquiry after inquiry. We do not need to go back over this again and again.' But the issue wasn't going away, and neither was Keys.

On the night of the election Blair nervously awaited the results in his Sedgefield home with his inner circle. It would take a 22 per cent swing for Keys to defeat Blair, who in 2001 was elected with a majority of 17,713. He asked his agent, John Burton, how their prospects looked.

'I think we'll do all right.'

'Do all right, John? Is that all?'[6]

The night got off to a bad start when Labour lost Putney to the Conservatives. Blair could not stand the tension and went out into the garden, muttering, 'It's all my fault,' and, 'Iraq.'[7] In the early hours of the morning he set off for Sedgefield's town hall to hear the result. Before a packed auditorium Blair took the stage with the other candidates. He ignored Keys.

The Solihull man had worked hard for this night. Defeat was likely, but that wasn't the point. After the failure of the Hutton and Butler Inquiries, he was there to call the prime minister to account. At 2.17 a.m. the results were declared. Blair's majority had decreased from 65 to 59 per cent. Keys had won 10 per cent of the vote, just behind the Conservative Party candidate. Blair shook hands with everyone but Keys who would not have accepted his hand anyway. It was a tense moment. With Blair and Cherie Booth standing behind him, Keys told the cameras, 'I hope in my heart that one day the prime minister will be able to say sorry.'

Overall, Labour's parliamentary majority was cut from 200 to sixty-six. Philip Gould, Blair's pollster, later estimated that the Iraq war only cost Labour around 2 per cent of the vote in the last ten days of the campaign with the furore over Goldsmith's comments and Wakefield's death.[8] Despite the headlines, the war's unpopularity had not translated into a major election issue. Some on the left who had opposed the invasion had not been prepared to vote for the Conservative Party. Others had tuned out the conflict after the constant drip of bad news. Then there were those who didn't care; Iraq was a distant war fought by a volunteer army. The economy was booming, fuelled by easy credit, a housing bubble and a buccaneering culture in the City of London.

But the victory still felt like a defeat to many in the Labour Party, who worried about the direction their party and the country was taking. Blair, however, had lived to fight another day and was determined to inject some direction into his premiership. On the domestic front that meant more privatisation of public-sector services; overseas that meant getting out of Iraq and beginning a new offensive in Afghanistan.

Chapter 19

IRAN

LIEUTENANT COLONEL ANDREW WILLIAMS had been in the army long enough not to be surprised by the air of neglect and isolation at the British base outside Amarah, Camp Abu Naji. General Riley had sold him the Overwatch concept on his way through Basra and given the impression that great forces were at work in the province training the Iraqi army and police force to take over from the British. The reality, as his predecessor Lieutenant Colonel Ben Bathurst confessed, was that since the heavy fighting of the previous summer there had been no additional resources or manpower. Instead Bathurst had been left to manage as best he could the political tensions between the followers of Moqtada al Sadr and the Badr Brigade, a militia loyal to one of the main Shia political parties in Baghdad. Since the evacuation of the headquarters in Cimic House, there had been few attacks on British forces. It was a testimony to Bathurst's diplomacy as well as reduced British ambitions in the region. Bathurst had wryly greeted his replacement at the runway near Camp Abu Naji: 'Welcome to the loneliest command in the British army.'[1]

Williams was an altogether different type of character to the genteel Bathurst. His rise through the army had been steady. He matched unforgiving drive with a meticulous eye for detail, one of the qualities that had attracted the attention of Peter Wall, the former commander in Basra who had recently been appointed deputy chief of joint operations in Northwood. Before deploying, Williams made sure his Staffordshire Regiment had carefully practised the counter-ambush techniques that Matt Maer had used the previous summer, such as constantly varying patrol routes. He expected a seasonal increase in violence brought on by the summer heat but thought he would mostly be patrolling in soft-skinned vehicles and working on Riley's security sector reform.

On this first day in command the Marsh Arab leader Abu Hatem paid a visit. Williams knew his treachery had previously endangered the lives of British soldiers. The tribal leader had continued jockeying for power, even setting up a political party to run in the elections, but he had fared badly against the Sadrists. He was losing local support but not prepared to admit it. Gone was any pretence of cooperation. 'You are on my property,' he informed the lieutenant colonel and demanded the British leave. Williams was not going to be intimidated and was equally terse. That night a salvo of rockets slammed into Camp Abu Naji.

The next morning Williams set off for a meeting in town with the Provincial Security Committee to inform them about Overwatch. The Committee was intended to provide a forum for the province's various militia groups and tribal leaders and had had moderate success at keeping the factions at bay – at least as long as the British provided a common source of resentment. Despite successive commanders insisting that British forces wished to hand over responsibility for security to the committee, the Iraqis on the panel remained sceptical and showed little willingness to work with their UK counterparts.

Williams hoped to invigorate the relationship. His column of Land Rovers had just reached the outskirts of Amarah when the lead vehicle containing the UK police adviser and two ArmorGroup contractors was engulfed in a fireball. The convoy screeched to a halt in the shocked silence after the explosion, but the smoke cleared, and the targeted vehicle drove on unscathed. The force of the blast had travelled away from the Land Rover, leaving it without a scratch. Williams got the message clear enough, however. We're the new boys in town, so we're being tested, he thought. He insisted on carrying on to the meeting to let the Iraqis know that he was not about to be cowed. Back at the base later that evening, Williams mulled over his options. There had been a number of attacks on British troops, including the deadly bomb attack that killed Guardsman Anthony Wakefield on 3 May. Any British response risked provoking further attacks, but sitting back had consequences as well. He recalled Matt Maer's advice that the Iraqis didn't see the British army as an impartial peacekeeping force; they were a tribe, and to command the Iraqis' respect they needed to show they were the strongest one around.

The next day Williams pointedly went to see the governor in a Warrior, a larger, more threatening vehicle. It was the first time the twenty-eight tonne armoured vehicles had been seen on the streets since the fighting the previous summer. From its turret Williams could see the reaction of the people. They sense what's coming, he thought.

The Iraqi elections had placed Adel Muhoder al-Maliki in the governor's office. An educated Sadrist who had trained as an engineer, Maliki wore his hair brushed back and thick gold rings on his fingers. Only the twitch under his left eye betrayed the considerable dangers of his position, caught in the uneasy truce between the Jaish al-Mahdi and the British army. However, Williams had little sympathy for Maliki and any pressure he might be under from the local militia leader Ahmed al-Gharawi. The meeting was tense.

'We don't like outsiders,' Maliki said.[2]

'Well, I don't like people killing my soldiers,' Williams returned. He left the meeting angrily.

Williams decided a show of force was needed. He wasn't the first to think that way. During the Shia uprising in 1920 Iraq had come to embody the 'irrational other' that lurked on the edges of the colonial imagination. Optimists like Harry St John Philby were convinced there were rational explanations for the outbreaks of violence in the country, which ranged from local tax disputes to Britain's failure to grant elections. But for the senior British officer in Iraq, General Aylmer Haldane, there was a wearying familiarity to the bloodshed. He noted the comments Ibn Saud, the future king of Saudi Arabia, had made to him:

> The tribal leaders and notables of Iraq from whom you want the improvement of the country . . . do not wish that the people of Iraq should be quiet, and that there should be law and order in the land. It is impossible to change their nature, as this has been their policy of old . . . Their whole point in life is to stir up the people in order to gain profit from the government . . . It will be impossible to manage the people of that country except by strong measures and military force.[3]

THAT SAME EARLY-SUMMER NIGHT IN 2005 WILLIAMS launched Operation Titanium, in which 500 British troops cordoned off and searched four Iraqi compounds looking for militia members. Most of the soldiers didn't know what to look for other than young men with surly attitudes. The intelligence inherited from Bathurst's soldiers was patchy. There was still no electronic database for exchanging information between battle groups, and other than a few names and a family's reputation there was little else to go on. But to Williams's mind the mission was as much about signalling the British presence as taking out known troublemakers.

The soldiers lined families up against walls and rifled, room-by-room, through their homes, stepping in their boots over children asleep on thin mattresses. Young men were hauled off in vehicles, with no information given to their families. In all, thirty-three Iraqis were detained, a 'textbook' operation, division headquarters noted. However, all but two of the thirty-three were soon released for lack of evidence.

The show of force improved morale on the base, even if it left fury and resentment in its wake among the Iraqis. Earlier in the day at the desert airstrip near Camp Abu Naji Wakefeld's coffin had been wheeled out for the long flight back to Brize Norton. Something had been done for a comrade, and the unwritten code of soldiering upheld. At Williams's next meeting with Maliki, the governor was initially cordial, although he clearly had received complaints about the raids. Williams did not have to wait long before he suddenly asked in perfect English, 'Why are you punishing us?'

'We're not punishing you,' Williams responded.

The governor went on talking, his eye twitching compulsively, laying out conditions for further contact between the council and the British.

'Number one, no more punitive raids against Amarah. Number two, advance knowledge of all British operations,' he demanded.

Williams told him the conditions were unacceptable and promised Maliki that if there were any more attacks he would personally lead a raid to arrest him. It felt good to take the gloves off. Maliki seemed to think so too. The meeting broke up.

Five days later Lieutenant Ben Bishop of the King's Royal Hussars led a small column of Land Rovers out of Camp Abu Naji. The camp had taken steady mortaring every night since Wakefield's death, and, like any young troop leader recently graduated from Sandhurst, Bishop was eager to get out of the camp and test his mettle. His column was providing close protection support to Major Nick Hunter, who was mentoring a small

Iraqi border force attempting to staunch the illegal flow of weapons and money into the country from Iran. The week before, Bishop had accompanied Hunter to the border, which in parts was little more than a gully and a line of barbed wire. Hunter and Bishop had climbed out of their vehicles to look around when they were spotted by two Iranian border guards and shot at. They returned fire but beat a hasty retreat, not sure how quickly Iranian backup might arrive.

They were running late for their next trip to the border. Their Iraqi interpreters had refused to go out that day, the type of flakiness that Bishop and the others had grown used to. Then, just outside the gate, Lance Corporal Alan Brackenbury, in charge of communications, announced he had forgotten the satellite phone, and they had had to turn back to get it. Brackenbury was a serious, quiet soldier, but he took the ribbing over his forgetfulness with good grace. Bishop sat beside his driver Trooper Darren Smiles in the lead Land Rover, which was stripped down to the frame. Brackenbury was in the back with Lance Corporal Liz Rawlinson, the medic. The men and Lance Corporal David Simcock were standing up, guns in hand, their upper bodies exposed through the rear hatch. The desert was empty, and the wind buffeting them was hot and dry. Bishop watched absently as they overtook a beat-up camper van.

Suddenly a terrifying rush of metal shook the vehicle, and Bishop's Land Rover careered off the road. As they pitched over the rough ground, Bishop yanked on the handbrake to stop them tipping, and the vehicle skidded before coming to a stop fifty yards further on. Smiles slumped forward against the steering wheel. Looking over his shoulder, Bishop saw a huge dust cloud rising into the air. He scrambled clear of the vehicle with his rifle. Ambushes often followed IED attacks. He scanned the horizon with his rifle raised. The other two Land Rovers had driven past the blast site and stopped 400 yards up the road. He waited thirty seconds before rushing back to the vehicle. Smiles was stirring. Shrapnel had gouged his back and neck, but he seemed calm in contrast to the shrieking from the rear of the Land Rover. Bishop ran around to find Rawlinson hysterical, splattered in blood. Simcock was sitting opposite her nursing a shattered arm. Beside Rawlinson was a bloody mess. All that was recognisable was a leg. 'It's Bracks, boss,' said Simcock.[4] The force of the blast had travelled straight through Brackenbury's body, killing him instantly.

Bishop grabbed the satellite phone. With their radio malfunctioning it was the only communication back to camp. He was patched through to the Abu Naji operations room via London. He sent their grid reference and details of the blast as Hunter and a couple of others came over to administer first aid. The sergeant treating Simcock was so shaken by the scene he injected himself accidentally with morphine. Bishop only realised he was injured himself when he handed the satellite phone over to Hunter and found it was covered in blood. Both his arms felt like they were broken, and the blast had scored his left knee. After ten minutes a Sea King helicopter showed up, but the RAF medics on board initially refused to get out, saying it was too dangerous. When they finally emerged almost ten minutes later, Bishop gave them an earful.[5]

As the injured were flown to safety, Captain Simon Bratcher, the battle group's bomb disposal expert, arrived. His job was to check for further bombs and gather evidence from the scene. A quiet-mannered twenty-nine-year-old from Dorset, Bratcher had joined the Royal Logistics Corps and gone on to complete the army's sixteen-month bomb disposal course in Kineton, Warwickshire. He had only qualified to work in high-risk environments overseas a few months before. He met Hunter beside the parked Land Rovers. Bratcher knew there was a risk of a second bomb further up the road. The first step was to deploy the 'wheelbarrow', a small robot with a camera and arms, to remotely assess the site. In Northern Ireland it had been a favourite trick of the IRA to use secondary and tertiary devices, but the Iraqi insurgents seemed to favour a more direct approach. After checking the site as best he could through the grainy monitor in the back of his vehicle, Bratcher suited up for a manual inspection. Bomb disposal experts wear twenty-eight pounds of Kevlar with a heavy armour plate over the chest. Movement is restricted, but Bratcher had been trained to compensate for this.

Near the crater he immediately spotted something: a white plastic box about the size of a fire alarm with wires coming out of it. Bratcher suspected it was the trigger device, but it was not like the simple wire- or phone-operated mechanisms he had been trained to look for. Walking over to the bombed Land Rover, Bratcher went straight to the rear. The silence was deadening, and he could hear himself breathing heavily. Brackenbury's body had been covered with a poncho, which Bratcher slowly raised. In forensic mode, he scarcely registered the gore. Instead, he traced the entry and exit points of three holes in the vehicle less than a few inches in diameter, one near the driver's seat, a second down the middle, and the third beneath Brackenbury. The force of the blast appeared to have been extremely focused, and the holes were lined with what looked like copper.

Returning to his own vehicle, he was joined by Captain Dave Goddard, a bomb disposal expert from Brigade headquarters, who had flown up from Basra. As he described his findings, both men realised what had killed Brackenbury. Explosively formed projectiles had only been used once or twice before in Iraq. The technology was simple but deadly: a disc of copper sat above the explosive charge in a metal tube that ensured the force of the blast was concentrated, turning the copper into a molten fist that could punch through several inches of steel. Not even the heavily armoured British tanks could offer full protection. The trigger also represented a departure from the usual devices. It looked like a passive infrared system triggered automatically by the heat of a vehicle's engine, making it immune to the army's jamming devices.

The suspicions of Bratcher and Goddard that militia groups in southern Iraq had acquired a powerful new weapon were confirmed a week later. Hunter was meeting village elders close to where the armour-piercing roadside bomb had detonated to find out who had planted it. Second Lieutenant Richard Shearer, twenty-six, was leading a patrol in support. In his lead Warrior, Shearer spotted two Iraqi police cars parked along the side of the road. He drove past and then his hackles began to rise. He ordered his patrol to stop. Only then did he spot, fifty metres down the road, some wires sticking out of the dirt alongside the road.

Bratcher arrived half an hour later with his wheelbarrow and disarmed the device as best he could. He then suited up to retrieve it. If this was one of the more sophisticated roadside bombs then it was vital he disarm it intact in order to examine it. From the camera pictures coming from the wheelbarrow, he suspected there were three more attached to the trigger, but once he started probing the earth he realised there were even more. Covered with sweat, his heart pounding, he quickly stripped off the suit. The armour could not protect him from the number of bombs surrounding him. In all, he uncovered ten devices. Had Shearer not got a sinking feeling and listened to it, the entire convoy could have lost their lives.

Bratcher's findings caused a stir in Baghdad, where the devices were sent along with the fragments from the bomb that had killed Brackenbury. American intelligence analysts discovered that the casings from Amarah were identical to devices used by Hezbollah against the Israeli army. The likely origin was Iran. Intelligence also suggested that training camps in Iran belonging to the Al-Quds Force, a branch of the Iranian Revolutionary Guard, had been revamped in order to train Iraqi militia groups. The name Abu Mustafa al-Sheibani featured prominently in the US reports. Captured Iraqi intelligence documents from before the invasion referring to 'conically shaped bombs' suggested that Sheibani's network in Baghdad may have had access to armour-piercing roadside bombs as early as 2001.[6] The question for the analysts was why these bombs were appearing now.

At a video conference with Blair a few weeks later Bush talked aggressively about 'dealing' with the Iranian threat, one staffer remembered.[7] Yet the Iraqi elections, which had brought Iran-friendly Shia parties into power, seemed only to be cementing Tehran's influence. On 1 March 2005 Ibrahim Jaafari, the leader of Dawa, one of the two main Shia parties, was finally announced as the next prime minister. The rise of this undemonstrative former doctor with strong ties to Iran was a low moment for American and British officials. Jaafari was opaque and distant, and given to flowery speeches about Arab history that could last for forty-five minutes. Anti-American Sadrist politicians had made a strong showing, and Jaafari was dependent on their support. In return, Jaafari gave them control over the health and electricity ministries, which soon became known centres of patronage and corruption.[8]

Equally troubling was the transformation of the Interior Ministry under Bayan Jabr, a former commander of the Badr Brigade, an Iraqi militia originally set up by Iran to fight Saddam. The ministry, which was the equivalent of the British Home Office and oversaw the police and justice system, was rumoured to have set up its own paramilitary units, which were kidnapping and torturing former Ba'ath Party members and Sunni tribal leaders. 'The new government felt that it needed protection from Sunni insurgents, but in unleashing the death squads [the paramilitary units] we fuelled this country's descent into sectarianism and civil war,' recalled Ali Allawi, Jaafari's finance minister.[9] The sectarian violence strengthened Tehran, noted Allawi, by tying down American forces and pushing Iraq's Shia deeper into Iran's embrace.

There had also been a radical shift in Iranian politics. Over the summer of 2005 hardliner Mahmoud Ahmedinejad had replaced the reformer Muhammad Khatami as president of the Islamic Republic, marking the end of Western attempts to negotiate with Tehran over its nuclear programme. The news was a particular blow to the British Foreign Secretary, Jack Straw. Having been a detached figure in much of the Iraq debate since the invasion, he had passionately advocated negotiations with Iran over its nuclear ambitions. In late 2003 he joined the French and German foreign ministers in signing an agreement with Tehran to suspend enrichment activities. The Iranians largely ignored the terms of the agreement, but Straw persisted. The threat of American or Israeli military action was the terrifying alternative. 'After his failure to stop the war in Iraq or influence the aftermath, he [Straw] viewed diplomacy with Iran as a means of salvation,' noted one official on condition of anonymity.[10]

At one stage, in May 2004, Straw learned of a CIA initiative to capture or kill Iranian agents working in Iraq. Targets included Mahmoud Farhadi, a Quds commander in the most northerly of the Iranian training camps, who the US military had identified as a trade envoy with the Iranian embassy in Baghdad. The argument at CIA headquarters in Baghdad ran: if we don't arrest Iranian agents now, we can hardly do so when their Iraqi proxies are in power. Straw quickly intervened, fearing action against the Iranians would escalate the conflict and scupper the nuclear talks. Straw threatened to use Britain's veto to stop the CIA initiative, the first time the UK had threatened to withdraw from Iraq since Greenstock confronted Bremer over his privatisation plans. It worked, and the US operation was called off.[11]

In November 2004 Iran announced the suspension of its uranium enrichment programme ahead of preliminary talks with Straw and the other foreign ministers, but it was soon clear the gap between Iran and the West was too great. Iran wanted to end suspension as soon as possible and allow the International Atomic Energy Agency to supervise. Britain, France and Germany were prepared to give Iran ready-prepared nuclear fuel rods but wanted no more uranium enrichment. At the end of May the Iranians rejected the EU offer, the same month as the first bombs with a likely Iranian origin began appearing in Maysan.

Was it a coincidence? Richard Dalton, British ambassador to Iraq at the time, did not make the connection. As he tried to explain to London,

Iran was not a monolithic entity. The supreme leader, Ayatollah Ali Khamanei, remained in overall charge, but the different power structures in the country – the Council of Guardians, the Revolutionary Guards and the intelligence agencies – often appeared to work independently of one another.[12] The Iranian approach to Iraq often baffled American diplomats, who believed Tehran was acting duplicitously by backing mainstream Shia parties in Baghdad, at the same supporting violence against the government they served. Seen in a different light, however, the Iranian stance was clearer: Tehran was simply backing multiple options in Iraq while waiting to see who really held the cards.

David Satterfield, US deputy head of mission in Baghdad, was convinced that Iran posed the most serious threat to the country's security.[13] The collapse of the nuclear talks along with the appearance of more sophisticated roadside bombs led to senior American officials talking openly about military action against Iran. But for now Jack Straw's cautious approach prevailed. 'We've got to reduce tensions,' Straw told Condoleezza Rice during a phone call. 'There is no other option.'[14]

Nonetheless, the British government had to confront the fact that an Iranian-made bomb had killed Lance Corporal Brackenbury. Ambassador Dalton delivered a strongly worded statement to Tehran. The Americans sent a similar note of protest via the Swiss embassy – there had been no US diplomatic presence in Iran since the 1979 Islamic revolution. When Dalton next saw Hassan Rowhani, Iran's chief negotiator at the nuclear talks, he confronted the Iranian. 'You can't keep saying that Iran wants a peaceful and stable Iraq while supporting violence in the country,' said Dalton. 'This two-track diplomacy has to stop.' Rowhani, a former Iranian ping-pong champion, gave little away.

In Baghdad pressure was also brought to bear on Jaafari to distance himself from Iran, which he duly consented to. Jaafari sent Mowaffak al-Rubaie, his national security adviser, and Barhim Salih, the deputy prime minister, to Tehran to demand an end to the violence. 'The trip was a complete sham,' recalled Rubaie, the scion of a prominent Shia family who had spent his exile working as a doctor in Glasgow.[15] In Tehran they met Qasem al-Soleimani, the commander of the Al-Quds Force, and asked for Iran to end all harmful activities in Iraq. A week later Rubaie saw Soleimani in Jaafari's office. 'Jaafari knew exactly what was going on,' said Rubaie, 'and was happy for the Iranians to take on the Sunni extremists.'

In Amarah the fighting was escalating. Responding to Lance Corporal Brackenbury's death, Williams launched a second raid, this time into what he called 'the heart of darkness', the town of Majar al-Kabir. Since the Danny Boy incident of the previous year the only British troops to have gone near the town were a Welsh Guards patrol, which had gone in on foot with an Iraqi police escort to hand out money in a vain attempt to win over the residents. The town was festering with anti-British sentiment and a likely base for the Jaish al-Mahdi's smuggling operations. Williams's target was militia leader Ahmed al-Gharawi himself.

The plan was to helicopter-drop a small party to arrest Gharawi outside the house where the militia leader was believed to be hiding while the rest of the battle group surrounded the town. But just as the mission was about to start, one of the two Sea Kings flipped over and crashed in the unpredictable air currents of the desert. Incredibly there were no casualties, although Williams was left missing half his arrest party. He decided to push ahead with the operation and send ground forces to raid Gharawi's compound instead. Gharawi was in Majar al-Kabir that night, but he managed to escape into the desert after being tipped off that British troops were on their way.[16]

His revenge was swift and brutal. Second Lieutenant Richard Shearer, the officer who had narrowly escaped a roadside bomb earlier in his tour, was leading a late-night patrol of Snatch Land Rovers through one of Amarah's slums. To counter the new threat of the Iranian-made armour-piercing roadside bombs, Shearer's vehicles had been escorted to the edge of town by Warriors. So far, there had been no bomb attacks in residential neighbourhoods. Just after 1 a.m. the sound of an explosion drew Shearer's patrol back to the edge of town, where bystanders directed them towards the city's football stadium. The patrol had travelled several hundred metres down the darkly lit street when Shearer's vehicle was struck by three armour-piercing bombs. Shearer, his driver Private Philip Hewett and top cover Private Leon Spicer were killed immediately.[17] From then on, no one left the base except in a Warrior.

Chapter 20

JAMIAT

MAJOR RUPERT JONES HAD BARELY STEPPED off the helicopter when bad news hit. Basra's chief of police, General Ahmed Hassan, had told the *Guardian* that the Shia militia had infiltrated half the city's police force. Given the favourable reports coming from General Jonathon Riley and the UK's senior police adviser Colin Smith – and Whitehall's own predisposition towards good news – Hassan's comments shocked the Cabinet Office.[1] Smith was ordered to draw up a list of dodgy Iraqi officers that the British government would present to Iraq's Ministry of the Interior for dimissal – as if a typed sheet of names would solve the deeper problem.

Jones quickly reached the conclusion that the Overwatch scheme to train the Iraqi security forces to take over – in place for only a few months – was going to fail unless the endemic corruption and militia infiltration was addressed. As chief of staff to the newly arrived 12 Mechanised Brigade, Jones was in a position to do something about it. His father was Lieutenant Colonel Herbert 'H' Jones, the most senior officer to die during the Falklands War. Few officers had been able to match H's drive, and his son was equally single-minded.

Jones suspected that much of the violence gripping Basra could be traced to the Jamiat police station. The serious crimes unit headquartered at the station, originally trained and equipped by the British, was linked to scores of Iraqi deaths across Basra. The role of the militia in all this was not clear, but most of the murdered had opposed Moqtada al-Sadr's followers. Jones pushed to create a brigade surveillance team to assist special forces in monitoring all movement in and out of the station. The danger posed by the Jamiat was believed to be so grave that regular British army patrols and police advisers were now barred from the station.

Once again a debate raged between the British on the ground as to whether to adopt an aggressive strategy against suspected insurgents, or a diplomatic approach. William Kearney, who had first uncovered abuse at the Jamiat, felt he and his staff were developing a relationship with the Jamiat's officers, and to arrest or fire them now would only fuel further confrontation. But for officers like Jones that was precisely the point. 'When you've got a cancer like the Jamiat corrupting the heart of your enterprise, your whole mission is corrupted,' said Jones.

Unfortunately, Smith's list of not-to-be-trusted Iraqi officers grew quickly to over a hundred officers, way too many to sack or arrest. Top of the list was Captain Jaffar at the Jamiat. The military had also come up with a broader directory of names, dubbed the Forces of Darkness list, which served as an uncomfortable reminder of just how few allies the British had on the ground. Prominent names on the list included head of the British-trained tactical support unit Major Shaqir, police chief Hassan Sawadi, Governor Waeli and of course the militia leader Ahmed al-Fartosi.

Information was scarce, as few journalists dared to enter Basra for fear of kidnapping or death. Steven Vincent, a thirty-nine-year-old freelance American, was one of the few Western reporters to take the risk, and he wrote damningly about Basra's nexus of police and militia, fuelled by oil smuggling and Iranian support. He interviewed an Iraqi police lieutenant who described how one unit – presumably Captain Jaffar's – was responsible for hundreds of assassinations. 'The British know what's happening but they are asleep, pretending they can simply establish security and leave behind democracy,' the policeman said.[2] Two days later men dressed in police uniforms kidnapped and killed Vincent. His female Iraqi interpreter Nouriya Itais was shot several times in the stomach but survived. When the *New York Times* sent in its local correspondent Fakher Haider, he too was murdered.

At the end of July the UK's senior police adviser Colin Smith delivered the list of policemen to Interior Minister Jabr, along with a letter from the military asking for the officers' immediate dismissal. When nothing happened, brigade headquarters pushed to arrest Fartosi, only to be informed that the Iraqi prime minister had put the Jaish al-Mahdi leader on a 'no lift' list, not wishing to antagonise Moqtada al-Sadr, whose followers were vital allies in the new parliament.

By then Jones was convinced that insurgents would soon be using the Iranian-made armour-piercing roadside bombs in Basra. On 30 July one of the new devices killed two private security guards working for the FCO. Jones rushed through new safety procedures, which included erecting concrete barricades around British bases in Basra, furthering the sense of isolation from the local population. He also required armoured vehicles to lead every patrol. Then, in early September, a blast killed Fusilier Stephen Manning and Fusilier Donal Meade in Az-Zubayr, a few miles outside Basra. A few days later Major Matthew Bacon, an enterprising intelligence officer, was killed by another bomb as he left the palace. Both devices appeared to have been machine-made and mass-produced, suggesting an origin outside Iraq.

The question of what to do about the deteriorating situation was now in the hands of Brigadier John Lorimer, who had been called in at short notice after his predecessor, Chris Hughes, was transferred to a planning job at the MOD in London after just three months. Lorimer was the eighth brigadier to command Basra in just over two years. Powerfully built, with heavy features and a balding pate that glistened with sweat in the heat, Lorimer came across as a man of uncomplicated morals for whom the army's blunt style was well suited. But others recognised a deeper, subtler thinker who had done brilliantly as a planner at the British military headquarters in balancing aggressive instincts with what was politically expedient.[3] His great-grandfather had been a former political officer on the North Western Frontier of the Indian Empire before becoming Political Resident in the Persian Gulf during the First World War. Like his predecessor, Lorimer had studied Arabic, and brought to his headquarters an unusual grasp of tribal behaviour, albeit one flavoured by colourful stories from his family's imperial past.

With the death of three British soldiers in a week, Lorimer decided it was time to arrest Ahmed al-Fartosi. He might be on a Baghdad no-lift list, but this was no time for politics. The Jaish al-Mahdi leader was easy to find. He spent most evenings at his house, apparently convinced he was untouchable.

On 17 September an SAS detachment, backed up by C Company Coldstream Guards, moved into position around Fartosi's house, which was surrounded by a number of water-filled ditches and open sewers. The militia guards were quickly overpowered and Fartosi dragged out into the evening gloom a few moments later, along with his right-hand man, Seyed Sajjad. Dressed in a tracksuit, unshaven, Fartosi screamed abuse as the soldiers held him. His wrists were secured behind his back with plastic cuffs. A search of the house revealed several photographs, one of which showed Fartosi with known members of Hezbollah. Fartosi later tried to sue the British government, claiming he was beaten with rifle butts during the arrest and on his way to the detention facility at the Shaibah logistics base.

The next day Fartosi's followers demonstrated outside the palace, but otherwise Basra was largely peaceful. There was a sense of elation at 12 Mechanised Brigade's headquarters, recalled Major Jones. They had arrested Fartosi easily, and attention could now turn to the Jamiat police station and Captain Jaffar's serious crimes unit.

Two days after the arrest of Fartosi two SAS soldiers called Paul Jenkins and Lee Harris set off on a mission. Like most British special forces, they viewed Basra as a backwater, where commanders were cut off from the flow of resources and intelligence in Baghdad and had little desire to pick a fight with the militia. Since the start of the war, the SAS commander Richard Williams had insisted that British special forces should be strategically deployed alongside the Americans in Baghdad. So far, they had spent most of their time 'fannying around' looking for weapons of mass destruction or ageing Ba'athists.[4]

Either way, Jenkins and Harris were about to have a bigger role in shaping the fortunes of the British occupation than anyone could imagine. Dressed as civilians with their faces disguised behind Arab scarves, they drove out of the palace gates on an surveillance operation. For the past few weeks Captain Jaffar's house had been regularly watched in preparation for a possible arrest. The SAS men were to observe the building and track Jaffar's movements. They left the palace without informing their support team – a routine procedure – a sign they viewed the mission as soft.

The previous year ArmorGroup had set up a facility for members of the Iraqi public to call in if they spotted suspicious activity. The 115 call centre, attached to the Iraqi police headquarters, had proved sporadically effective. That morning the centre received three separate calls from Iraqis in the Hayaniyah district, reporting the suspicious activity of the white

Toyota being driven by Jenkins and Harris. The men were wearing *shemaghs* and taking photographs. The Iraqi police responded quickly, setting up a roadblock to intercept the men.

When Jenkins and Harris spotted the checkpoint they panicked and opened fire, seriously injuring a police officer, before turing round and fleeing.Their old banger was no match for the Iraqi police's brand new pickup trucks. The two soldiers were rapidly boxed in. Jenkins was able to raise the alarm over his radio before they stopped, dropped their weapons and nervously raised their hands in the air as an angry mob of police descended. They were clubbed to the ground with rifle butts, handcuffed and shoved into the back of a police car.

Lieutenant Colonel Nick Henderson, commander of the Coldstream Guards battle group in Basra, was the first to respond. Darkly handsome, with olive skin and heavy eyebrows that gave him a brooding look, Henderson had grown increasingly unhappy with the army's failure to recognise the deteriorating security situation in Basra over the past few months. On learning from brigade of the SAS men's detention, Henderson's first thought was that they had been kidnapped by militia elements within the police. He assembled the whole battle group. Brigadier Lorimer was in transit to the palace for a scheduled meeting with the governor, so effective control of the response had passed on to his hard-driving chief of staff Major Jones. The SAS squadron based in Baghdad was preparing to fly down to lead the rescue, but the flight would take over an hour. Initial reports claimed one of the Brits might be injured. Henderson ordered his force to set up vehicle checkpoints at major junctions to close down the city. 'The first moments after a kidnap are vital.'[5]

The British officer's worst fears were quickly realised. Captain Jaffar had also been informed about the incident, and had ordered the police cars with the SAS men inside to the serious crimes unit at the Jamiat. If he didn't realise it at once, when Jaffar saw their British military ID cards he knew he had achieved a major coup. Jenkins and Harris could be powerful bargaining chips for securing the release of Fartosi.

Shortly after the two men were thrust into a cell at the Jamiat, Major Jones at brigade headquarters located them. He hastily arranged for Major James Woodham, head of the brigade's surveillance unit and a frequent visitor to the Jamiat, to chopper over to the station. He took with him an interpreter and six soldiers from his unit. Henderson also diverted a company attached to the Coldstream Guards commanded by Major Andy Hadfield to form a cordon around the station. If the SAS men were in the Jamiat, then it was vital they stay there during negotiations.

A small Iraqi crowd had already gathered outside the station when Woodham arrived, and the police inside were on edge. The thirty-seven-year-old Woodham felt his breath quicken as he over marched to the office of the station head, Colonel Ali al-Sewan. Gazing down his long thin nose at the Iraqi, Woodham knew he would get nowhere demanding the men be released. After two years of British efforts to create a functioning police service, Woodham had to pay lip service to the local authorities, underlining the delicate balance of power between the British forces and their Iraqi counterparts. Furthermore, Sewan did not have the power to go up against Jaffar, and to push the issue might force him into taking a harder line. Woodham waded through the elaborate introductions that Arabic courtesy demanded before quietly enquiring about the two soldiers, trying to hide his frustration. Sewan confirmed they were holding two Westerners who worked for Israeli intelligence.

'Well, I think actually they might be British soldiers. Can I see them?' Woodham asked.[6]

Sewan prevaricated before finally granting permission. They were being held in the Jamiat's other compound, the one controlled by Jaffar's men. Surrounded by shouting police officers, Woodham was taken to the soldiers' cell on the second floor. Both Jenkins and Harris were handcuffed to chairs. They had clearly been beaten, their bruises clumsily dressed. Woodham ordered the blindfolds to be removed. The soldiers were obviously relieved to see Woodham before he was ushered out of the room.

At least Woodham was able to confirm to Jones at brigade headquarters, via mobile phone, that the men were there. With Lorimer arriving at the palace to confront the governor, and the ambassador in Baghdad, William Patey, approaching the Ministry of the Interior, the full weight of British diplomacy could now be brought to bear. By then Hadfield's men had also set up a security cordon outside the station. Shortly after midday the Iraqi police's city reaction force arrived, lights flashing, outside the Jamiat. The British had created the paramilitary unit for just such occasions, but instead of coming to Hadfield's rescue, they took up positions around the buildings opposite, their guns trained on the British soldiers. The police on the station walls now cocked their guns too.

A desperate Hadfield called Henderson for permission to withdraw.

'Stay put,' ordered Henderson. 'If you move, the men in the station will be gone.'

The ensuing stand-off lasted for twenty minutes before Hadfield noticed some of the Iraqi policemen starting to smoke and look bored. Henderson instructed him to walk across to the Iraqi commander and issue an ultimatum: if they did not leave there would be a bloodbath, and it would not be British blood in the dirt. Hadfield bravely walked across no-man's-land with his sergeant major in support and cowed the police commander into pulling back. Efforts to dislodge Hadfield's men were only just beginning, however.

Inside the station James Woodham's negotiations had reached an impasse, but he did not give up. Colonel Sewan had introduced him to Raghib al-Mudhaffar, chief justice of the Basra anti-terrorism court. Woodham tried to explain to the judge the legal justification for releasing the men – the Iraqi government had agreed that Coalition forces were not subject to Iraqi law. Judge Raghib appeared unable or unwilling to accept his explanation. Over the phone Woodham requested headquarters to send the brigade's lawyer, Major Rabia Siddique, who had worked closely with Raghib establishing the city's courts. Siddique was an oddity in the British Army. Raised in Australia by her Indian Muslim father and Australian mother, she had worked as a federal criminal prosecutor before emigrating to the UK to join the army's legal team. Smart and attractive, she had thrived in the male-dominated world of the military.

Major Jones spoke to her at brigade headquarters shortly after 11.30 a.m. Siddique had already been doing what she could to release the men, phoning William Kearney, the ArmorGroup manager, to pool contacts. His police mentors had already been in touch with the Iraqi police inside the Jamiat about handing over the SAS men, but in the confusion this line of communication was ignored.

'You're going to the Jamiat. There's a chopper waiting,' Jones instructed her.

Jones was one of the few officers Siddique did not get along with. She considered him overbearing and she was not easily intimidated. Nor was the female Muslim lawyer thrilled at the prospect of being choppered into the centre of such a potentially violent situation. To Jones's irritation, Siddique asked to call Brigadier Lorimer, who told her to stay put.

Jones erupted, overriding his commanding officer: 'Listen to me. The brigadier doesn't know what's going on. I'm telling you to go to the Jamiat. Stop asking questions and get on with your fucking job.'[7]

Siddique asked what protection she would have, although she knew that if there was any chance of helping the hostages then she had to go. She called her colleague Lieutenant Colonel Alex Taylor in the divisional legal department and left a message. Taylor was interrogating Fartosi with SIS officers at the Shaibah logistics base. But as she sat in the chopper in her body armour, SA90 rifle between her knees, she began to doubt her decision. She had not trained for close combat, let alone fired a shot in anger. As the helicopter plunged down on a combat descent she glimpsed the Hayaniyah district wreathed in black smoke. There looked to be hundreds of Iraqis protesting before a tiny cordon of British troops at the entrance to the Jamiat. Her heart sank.

The helicopter landed with a thud 200 metres from the police station, and Hadfield's men pushed forward to clear a passage through the crowds. She dashed towards the compound gates, where Woodham was waiting for her with his British Army interpreter. The interpreter, perhaps hoping for British reinforcements, took one look at the solitary woman approaching and announced, 'I can't do this.' He dashed through the open gates to the cordon, where he took shelter in an armoured vehicle. Woodham grabbed Hadfield's interpreter.

'I hope you're up for this,' he said to Siddique as they entered the compound.[8] Siddique looked up and saw the walls lined with Iraqi policemen, their guns pointing down at them. They moved cautiously on to Sewan's office. There was the usual crowd of subordinates behind him, but this time Captain Jaffar was standing casually to his right. The Iraqi judge, Raghib, stepped forward to greet Siddique.

'You must release these men at once,' Siddique told Sewan. The police chief repeated that they were Israeli spies. Siddique insisted they were British soldiers who would be tried if they had done anything wrong. The conversation appeared to be going nowhere fast when Siddique hit upon a formula. If Sewan received a letter from the Iraqi government ordering their release, would he do so? The police chief glanced nervously at Jaffar before nodding. Sewan also agreed they could see the men again, provided they went without their security detail.

'Well done,' muttered Woodham under his breath.[9] He relayed the terms by phone back to Jones before they were led back up to the cells with a crowd of Iraqis, including the judge. Jenkins and Harris again had their blindfolds removed as the judge declared he thought their release was imminent. Woodham was speaking to Jones when the tremor of an explosion and the roaring of the crowd cut him off. It sounded like the station was being stormed. The policemen in the cell quickly bundled the two SAS men to the floor, and Siddique and Woodham were forced outside.

'I'm sorry, there's nothing I can do,' said the judge, who promptly disappeared. All pretence of British control now disappeared as they were hustled downstairs in the middle of a shouting gaggle of policemen. The sound of fighting was more intense in the open-air compound. Woodham managed to call Jones on his mobile phone, and tell him that the cordon outside was aggravating the situation. He and Siddique were hurried towards one of the portacabins. Inside was a group of agitated clerics and militiamen with AK-47s. One of them shoved his gun in Woodham's face, screaming obscenities. They were now prisoners of the Jamiat as well.

The explosions had come from gunmen within the crowd firing RPGs at the cordon of British soldiers. Shortly after Siddique entered the station, the crowd outside had swelled to almost 500 and turned violent. Rocks and bricks were lobbed at Hadfield's men as the protesters surged forward. The company was soon locked in hand-to-hand combat as the crowd attempted to isolate individual soldiers in the cordon and drag them away. Hadfield's sergeant major had to leap into a Warrior at one stage and drive forward to rescue one soldier who had been grappled to the ground out of reach of his colleagues.

'We need to pull back; we're driving the crowd wild,' shouted Hadfield down the radio to his operations room. Another RPG slammed into the wall behind Hadfield, and the crowd cheered. From his headquarters, Henderson again insisted they stand firm, although he dispatched 3 Company, led by Captain James Bradford, to the rescue, with another company on standby. Henderson was also coming under pressure from Jones at brigade to pull back to ease the tension within the compound.

But Henderson was sure if he did, the hostages would be spirited away. In the heat of the day it was impossible to launch a Sea King surveillance helicopter. In service since the Falklands, the Sea King was the brigade's only way of tracking where they might be taken. Compared with the American unmanned Predator drone, which could hover over battlefields for up to forty hours in most conditions and carried two Hellfire missiles, its capabilities were limited.

When Jones ordered Henderson to pull back, he refused, hoping to avoid censure as the order had technically come from Jones and not from his superior Lorimer. The rescue company led by Captain James Bradford was just arriving at the Jamiat, and Henderson wanted to use them to draw the crowd away from Hadfield's men. That didn't prove difficult. As soon as they got out of their vehicles 300 metres from the station, the crowd surged towards them. Bradford and his men fought with rifle butts and fists to force back the mob. After a Warrior charged forward to scatter the crowd, the British remounted and reversed up the street under a steady barrage of petrol bombs. The strategy of pulling the crowd away from the cordon was working.

Out of the narrow streets of Hayaniyah, Bradford's company reached the raised highway that ran along the western edge of the city and formed a defensive line. Henderson was advancing to reinforce the position, but before his Warriors arrived Bradford was hit in the face by a brick and knocked senseless. A flurry of petrol bombs rained down as Henderson's force pulled up. He had already made the decision that a static location might be overwhelmed; now he dashed between the burning Warriors to find Bradford, who was groggily coming round.

As the two men spoke, one of the iconic moments of the war took place. Henderson's second Warrior, commanded by Sergeant George Long, was struck by a petrol bomb, and he and Private Karl Hinnett in the front gunner's seat were engulfed in flames. They leapt desperately from the vehicle, a moment caught on camera by a local Reuters cameraman. Colleagues doused the flames, and both soldiers survived. For the next two hours, as both companies withdrew, there was barely a moment when one or other of the armoured vehicles was not in flames, put out as best they could by Henderson's men as they slowly withdrew north.[10]

By midday the Cabinet Office in London had realised the scale of the calamity. James Tansley, in for a final briefing before heading out to take over as consul general, was collared in the newsroom.

'What the fuck is going on?' one alarmed staffer asked. 'Basra's burning.'

'It looks like it,' Tansley said uncomfortably.[11]

In the drab rooms beneath Downing Street a Cobra meeting was hastily assembled. Cobra – referring to Cabinet Office Briefing Room A – was only convened in times of national crisis, chaired by the prime minister or a senior cabinet member. The former defence minister Geoff Hoon was in that morning. He focused on the political dimensions of the Jamiat crisis, much to the exasperation of some of the officers present.

'If we take action won't we be undermining the police and everything we've tried to build in Basra?' one FCO member asked.[12] Another senior civil servant was horrified to discover that a Muslim female British officer now appeared to be one of the hostages.

Hoon was encouraged by news from Baghdad that Ambassador Patey had strong-armed the Interior Ministry into ordering the two SAS men's release. In Basra Brigadier Lorimer had sought to get a similar order from the governor, but the official refused to cross the likes of Jaffar. So instead Lorimer wrote a letter to staff confirming that the detained soldiers were members of the British army. The Iraqi judge inside the Jamiat had suggested that this might be enough. Perhaps a negotiated settlement was still possible.

By then, the SAS squadron from Baghdad had reached Basra and were anxious to save their colleagues. Standard operating procedure was to launch an immediate rescue, without consulting their chain of command. This time, they were instructed by Brigadier Lorimer to hold off while negotiations were conducted. A furious squadron commander contacted Richard Williams, the senior SAS officer in Hereford. The reason why they were being stopped was clear to both men. 'A rescue operation would expose the flaws in the UK's withdrawal plan, that the police they were mentoring so carefully were working for the "other side"', said Williams. He ordered the SAS squadron commander to prepare for a rescue, irrespective of Whitehall's concerns and what his bosses might say.

However, before the SAS rescue could begin, Major Jones had arranged for a chopper to deliver the Lorimer's letter to the Jamiat. The letter was meant for Hadfield at the Jamiat, but instead the pilots flew over the station and followed the line of the highway to Lieutenant Colonel Henderson's besieged position. By then Henderson had retreated a mile up the road, with a large storm ditch filled with water on the eastern side of the highway channelling the rioters onto the open ground to the west. The riot had taken on an oddly festive air. 'They realised that we weren't shooting at them, so this was just an opportunity to come and give us a good kicking,' said Henderson.

Henderson ordered his men to shoot only if their lives were threatened. Half an hour before the helicopter's arrival, a soldier had fallen from the back of his Warrior as it charged forward to push the crowd back. A young Iraqi had rushed at the prone man with a large stone raised above his head before being shot and killed, prompting a surge of aggression from the crowd.

When the helicopter landed, Henderson assumed it was to pick up the injured soldier, lying on a stretcher beside the road. Instead the pilot rushed out and handed him a letter, dashed back and took off without picking the man up. By the time Jones realised the pilot's mistake, the helicopter had been tasked to another mission. With no more helicopters available, and realising the letter had gone to the wrong place, Jones told Henderson that he would have to deliver the letter to the Jamiat in person. You've got to be kidding, thought Henderson. He could not drive back through the crowd, as this risked drawing them back to the Jamiat.

There was only one option. During a lull in the fighting he stepped beyond the defensive line of Warriors with his sergeant major, interpreter and signaller. As they approached the line of protesters, a large and volatile crowd engulfed them.

The British officer quickly spotted an older Iraqi man and tried to explain that he had a letter from Lorimer which he needed to pass on to the police officers at the Jamiat. The jostling intensified. Henderson withdrew hastily as the crowd surged forward again. Unsure of whether the old man would take the message, Henderson tried twice more in subsequent breaks to find an intermediary. Finally, in the late afternoon, his efforts were rewarded when Captain Jaffar himself appeared at the edge of the crowd. Henderson stepped forward awkwardly to speak to him.

'If I take the letter, will the British forces withdraw?' Jaffar asked through an interpreter.[13]

Henderson looked up at the waning sun. By now his men had been fighting for hours, but more importantly the air temperature had fallen sufficiently to allow the Sea King surveillance helicopter to get airborne. It was time to remove the soldiers from outside the Jamiat.

Ten minutes later Jaffar appeared in the portacabin where Siddique and Woodham were being held. A rescue attempt had been attempted by Woodham's security detail, who had finally tracked them down and barged inside, but that was as far as they got. With so many armed militiamen around, they couldn't shoot their way out. Instead they had joined the captives, now totalling eight. The atmosphere remained tense, but Jaffar ushered them back to the offices of Colonel al-Sewan. Siddique immediately asked about the SAS men. Sewan was at first evasive, but finally threw up his hands theatrically and said the men were no longer under his control, although Siddique and Woodham were free to go.

Siddique flashed Woodham a worried look. 'I think they've moved the men,' she said.

Woodham got on the phone to Jones, but he was one step ahead of them thanks to a live video feed coming from the hovering Sea King helicopter. As soon as Henderson's men had withdrawn from the Jamiat, Jenkins and Harris had been smuggled out of the rear of the building. The men, suspecting they were under British video surveillance and realising this could be their only chance, staged a fight, running at each other and falling to the ground. The scuffle drew the attention of the surveillance team watching the Sea King feed. The British were able to follow the car with the men on board as it left the station and drove through a maze of streets before pulling up outside a house.

The pace quickened at brigade headquarters. The lifting of the cordon had also put pressure on London to take action. John Lorimer was finally given permission to raid the Jamiat using the recently arrived SAS squadron from Baghdad. The Special Forces team had been on the verge of launching their own unilateral operation, and set off at once with the 2nd Battalion The Royal Welsh in support.

At 8.30 p.m. Siddique heard the grating sound of approaching Warriors. Since the SAS men's departure the station had rapidly emptied. She and the other soldiers were able to stand near the exit unmolested as the first armoured vehicles appeared, and they clambered inside safely. Major Steve Manning, commanding the rescue effort, was not taking any chances. The Warriors ploughed straight through the wall of the station into the main courtyard, crushing an aluminium building and a dozen British-supplied police cars. The destruction of the very supplies that the British had paid for was an irony lost on the major in the heat of the rescue. An SAS search team quickly confirmed the building was empty, and the armoured vehicles reversed over the rubble and back onto the street.

The rescue party's next stop was the house under Sea King surveillance. The SAS burst into the house to find it deserted except for Jenkins and Harris, the two men dressed in just their underwear. Jenkins got into the same warrior as Woodham and Siddique. Having seen the bodies of assassinated Iraqis at Basra's morgue, also stripped of their clothes, she suspected that they had been rescued just in time.[14]

The Jamiat affair was over, but the fallout was just beginning. Arriving back at the Shatt al-Arab Hotel, Henderson wanted to give a press conference early the following morning to make sure they got their side of the story out first. The MOD press team rebuffed the idea, preferring to release a written statement through Lorimer. The instinct in London was to hunker down and try to work out what had happened. It was clear some sort of disaster had occurred, but the question was how big.

They found out the next morning. The British press uniformly led with the startling Reuters photo of a British squaddie on fire leaping from a tank. SOFTLY SOFTLY ARMY TACTICS SHATTERED BY DAY OF CHAOS, declared the *Guardian*; ARMY'S HOLD ON BASRA LOOKS TENUOUS AS TENSIONS WITH MILITIA BOIL OVER, announced the *Daily Telegraph*. A slew of editorials described how Britain was losing its grip on the south. With so little media coverage from Basra other than saccharine MOD press releases, the Jamiat affair had finally revealed the city's dark undercurrents.

Worse was to come when Governor Muhammad al-Waeli gave a press conference in which he accused the British of behaving like terrorists. The Iraqis were well within their rights to detain two suspicious Westerners who had openly shot at Iraqi officers, he insisted. Cameras were invited into the Jamiat to see the gaping hole in the wall and crushed police cars. He would have no more to do with the British, said Waeli, unless they apologised, offered compensation and promised not to enter the city again. He also wanted Brigadier Lorimer sacked. Britain's political partner for any future handover had just turned his back.

No one, it appeared, was more shocked by events than Tony Blair. At a Cabinet meeting the next day he laid into Jack Straw and the new defence minister John Reid, accusing their ministries of misleading him into believing that Basra was under control. Blair's anger did not stop there. He penned a stinging letter distributed to every relevant government department, singling out the FCO for its failures to manage the Iraqi police. Dominic Asquith, head of the FCO Iraq team and shortly to become British ambassador to Baghdad, was humiliated.[15]

At the prime minister's request, Ronnie Flanagan, the recently retired RUC chief constable, was asked to conduct an inquiry into Iraqi police training. The MOD promised its own inquiry. The army head Mike Jackson visited Basra in October. He had been irate when he learned that political considerations had initially taken precedence over the lives of the SAS hostages, fuming, 'I cannot believe that they were concerned about the political consequences of taking on a corrupt, penetrated and fucked-up police force.'[16] For the first time Jackson got the unvarnished truth from his commanders in the field. In his report on the trip he wrote dryly, 'Though there was no sense of defeatism in theatre, the possibility of strategic failure was mentioned in earnest on this visit more than on any before.'[17]

The first reference to 'strategic failure' by a senior British general should have set alarms ringing at the MOD given the intimate connection between withdrawing from Iraq and the approaching deployment to Afghanistan. The British plan to train Iraqi security forces was clearly failing, but the MOD inquiry into the Jamiat incident was limited to decision-making during the affair, which it duly found to have been exemplary. The military was to return to its role training Iraqi security forces and preparing to withdraw as if nothing had happened.

Everyone was relieved except Nick Henderson. The Coldstream Guards commander resigned his commission at the end of his tour following the death of another of his soldiers. On 18 October, near midnight, thirty-year-old Sergeant Christian Hickey was leading a convoy of lightly armoured Snatch Land Rovers through Basra. The spotlight on his vehicle was not working, making it difficult to see the roadside. Halting the convoy, he dismounted to check the route. As he stooped to inspect a suspicious lump in the road he triggered a passive infrared device. He bore the full brunt of the explosion, which ripped off his limbs and dumped his body on the road. His bravery certainly saved the lives of the other men in his vehicle.[18]

Henderson believed Hickey's death was a result of his men not having sufficiently armoured vehicles, and represented the military's woeful effort to come to terms with the violence in southern Iraq. Before departing, he warned his superiors that if the MOD did not do something soon they would have a lot more blood on their hands.

Chapter 21

HELMAND

On 7 July 2005 the trains were packed as usual for the morning rush hour on the London Underground, making the explosions all the more deadly. Three British Muslims carrying backpacks each containing ten pounds of high explosives boarded separate trains. At 8.50 a.m., within minutes of each other, three blasts ripped through commuters, tearing open carriages and knocking out power. Grainy mobile phone footage shows injured survivors streaming out of station exits from tunnels filled with acrid smoke. As rescuers rushed to the scene and administered first aid, and the government struggled to piece together what had happened, a fourth bomb exploded on a Number 30 double-decker at Tavistock Square, taking off the roof and rear of the bus. The first suicide bomb attacks on British soil had killed fifty-two and injured over 700.

Blair, attending a G8 summit at Gleneagles Hotel in Scotland, rushed back to London to deliver a defiant statement outside Downing Street. He immediately connected the attacks with the 'war on terror': 'When they try to intimidate us, we will not be intimidated. When they seek to change our country or our way of life by these methods, we'll not be changed. When they try to divide our people or weaken our resolve, we will not be divided and our resolve will hold firm.' The possibility that British foreign policy had actually provoked the attacks was not raised by the prime minister, although the bombers themselves made the connection. In a prerecorded video, one of the bombers described his motivation as vengeance for the British role in Iraq and Afghanistan.[1]

However, the attacks underscored for Blair the importance of sending

British troops to southern Afghanistan.[2] As details of the attackers emerged, the role of terrorist training camps in Pakistan close to Afghanistan, where two of the bombers had trained, suggested that British high streets would not be safe until tensions in the region had been quelled.[3] Plans were now well under way for Afghanistan. Since Blair's May 2005 election victory the task of preparing the British mission to Helmand had fallen to the new minister of defence, John Reid, a belligerent Scot with an ear for emotional sound bites. There were two main issues that needed to be resolved: should British forces deploy to Helmand in the first place and, if so, what size force should be sent?[4] Reid immediately recognised the crucial factor in any mission to Helmand: the pace of the withdrawal from Iraq. The last thing he wanted to do was commit to fighting on two fronts.

Chief of the General Staff Mike Jackson reassured him that while it was risky, the military could handle a protracted withdrawal from Iraq. 'At a pinch, at a considerable pinch,' Jackson added.[5] There was no reason to suggest that plans drawn up for an autumn 2005 handover of the Iraqi provinces of Muthanna and Maysan could not be met, Jackson insisted, demonstrating the gulf in understanding between the military's leadership and its commanders in the field.[6]

Still sceptical, Reid asked for a more detailed plan for a force of 3,000 troops to be sent to Afghanistan. Both he and the defence chiefs knew that the further down the planning route they went, the harder it would be to pull back.[7] Over the next few months the MOD set about winning Treasury approval for a three-year mission costing £1 billion. Several military officials knew the mission's costs were likely to exceed this figure but that Whitehall would not approve a higher budget. 'We would have asked for double the money if we thought we'd have got it,' said one officer.[8]

Lieutenant General Rob Fry, the man who had conceived switching the British effort from Iraq to Afghanistan, was instrumental in pushing the deployment through Whitehall. The growing violence in Iraq over the past year had highlighted the need to end an unpopular campaign before the military was drawn into a protracted counter-insurgency campaign. Afghanistan, in contrast, was a mission the British public supported, where UK forces would be tackling vital strategic issues for the UK such as terrorist training camps and the opium trade.

In September 2005 Rob Fry presented Reid with the MOD's plan for an Afghan deployment of 3,150 troops, mostly drawn from the Parachute Regiment. John McColl, the former deputy commander in Baghdad, was present along with members of the intelligence community. Fry reassured Reid that the Helmand mission would primarily be a peacekeeping operation, with a limit on time and resources. That appeared to mollify the still-doubtful Reid. The subject turned to details of the deployment. British forces would take over a base under construction by the Americans called Camp Bastion, located in the desert to the north-west of Helmand's capital, Lashkar Gar.

'Won't British troops be isolated and exposed?' Reid asked.[9]

Fry assured Reid that the Taliban were in their base in Quetta, Pakistan, and if they ventured into the deserts of southern Afghanistan, they would be easy to pick off. The senior SIS man in the room rolled his eyes. His operatives in the region and an initial SAS reconnaissance mission to Helmand had reported that an increase in Coalition forces in the region might have the effect of provoking a fight, especially if the British troops interfered with the lucrative opium trade. Fry dismissed their concerns.

One British officer present at the meeting was appalled by Fry's characterisation of the situation in Helmand. It was true Camp Bastion was unlikely to be attacked – it really was in the middle of nowhere – but what was the point of being in Helmand unless British forces were deployed throughout the province? The officer confronted Brigadier Chris Hughes, who had left Basra to become Fry's deputy, and asked him to provide Reid with the truth. 'We have no idea what we will find on the ground,' he warned. Hughes disagreed. A formal letter of concern was also delivered to Reid by the SIS operative, but by then Reid had made up his mind.

Brigadier Ed Butler was chosen to lead the force in Helmand. The 16 Air Assault Brigade commander had found out about the plan in March when the Canadians were placed in overall charge of the mission, with a headquarters in Kandahar. A wiry intense officer with a career in special forces and the scars to prove it, Butler was immediately suspicious of the arrangement with the Canadians. Having served in Bosnia, he had seen the lengthy and bureaucratic decision-making processes which plagued multinational operations.

When Butler was briefed in more detail in September, the command structure in Helmand was more tangled than he had feared. As the Canadian headquarters in Kandahar would be commanded by a brigadier, the same rank as Butler, Coalition etiquette demanded that he step aside and allow a more junior officer to command his men. That meant Butler would have to oversee operations from Kabul, while a colonel from another British regiment would be drafted in to take his place. Butler was further dismayed to learn that his headquarters would not be doing the operational planning for the British deployment; instead a staff officer from the British military headquarters in Northwood would be making the initial reconnaissance of Helmand province and drawing up the crucial operational plan. Rather than have the freedom and authority to run the Helmand mission as he saw fit, Butler was going to be constrained by the dictates of Northwood.

But the mission's gravest failure was its lack of resources. Butler would have just four Chinook helicopters at his disposal, barely enough to support one offensive mission a month.[10] Following a fiery briefing at Northwood, Butler put his objections down in writing to the deputy chief of joint operations, Peter Wall, although he knew he had little choice but to 'crack on'.

A few weeks later the hostage crisis at the Jamiat police station rocked Whitehall. The hand-wringing that followed offered the opportunity for a fundamental reappraisal of policy on Iraq and Afghanistan. The commander-in-chief at Land Command, Richard Dannatt, responsible for generating the Helmand force, considered questioning whether going to war in Afghanistan was sensible. The mission in Iraq was far from over, and if the situation worsened there and troop levels remained high, the military would be desperately overstretched. Dannatt later told the Iraq Inquiry he saw a 'perfect storm' heading the military's way as early as the autumn of 2006. But although he briefed visitors to his headquarters in Wiltshire on the dangers, he never pushed the defence chiefs to take action, even after the Jamiat incident clearly raised the unlikelihood of a smooth transition of power back to the Iraqis.[11]

Meanwhile, in Helmand the British mission was starting to take shape with the arrival of Brigadier Gordon Messenger and a small planning team. Messenger had accompanied General White-Spunner to Kandahar the previous year as his chief of staff. Since then he had made a number of trips on behalf of Northwood to assess British plans for Helmand. In his early forties, Messenger was heavyset with the granite jaw of a bruiser, which contrasted sharply with his soft and intellectual manner.

Messenger knew the military needed to take a different approach in

Afghanistan than they had in Iraq, one that emphasised building up local leaders. Military defeat of the Taliban would not be enough to secure stability; Helmand needed a provincial leadership capable of sustaining itself. The British ambassador in Kabul, Rosalind Marsden, had already made an important intervention on that front by insisting that the Afghan president, Hamid Karzai, remove the current governor of the province, Sher Muhammad, the leader of the region's dominant Akhundzada tribe. Sher Muhammad was notorious for rape, murder and his links to the drugs trade. In June government forces raided his compound and discovered nine tonnes of opium, the largest find since 2002. This was not a man who could champion Tony Blair's counter-narcotics policy, and Marsden, backed by Whitehall, insisted that Karzai replace him before British troops arrived.[12]

'It was a little naïve to dismiss Sher Muhammad because he had links to the drugs trade. Anyone with any power did. Karzai's brother was the biggest drug dealer in town,' noted one official.[13]

Nonetheless Karzai named Muhammad Daoud, a well-meaning if bland technocrat whose absence of ties to Helmand was viewed as a strength, although his lack of a power base was to have crushing implications later on. He was due to take up his position in November 2005, a few months before the British deployment. Messenger was delighted to have a governor whose views mirrored his own concerning the future of the province and the British mission. Here was an historic opportunity, Messenger realised, to combine military and civilian objectives into a single campaign plan.

Messenger was not alone in deriving a more holistic approach from the lessons of post-conflict disaster management. In the summer of 2004 Brigadier Nick Clissitt had served on a US Department of Defense investigation into what had gone wrong in Iraq in the first year led by the American general Karl Eikenberry, who was later to become US ambassador in Kabul. The report criticised the slowness of the American response to events and its failure to coordinate civilian and military planning. Clissitt developed his own ideas for what the UK needed in the immediate aftermath of an invasion: a dedicated pool of reconstruction experts ready to deploy to the war zone. He also wanted a cross-departmental team in place in Whitehall that would end the bureaucratic politicking and turf battles between ministries that had led to so many mixed signals being received on the ground.[14]

In Whitehall Clissitt won agreement from the MOD, the DFID and the FCO, with the DFID providing £20 million of funding and offices for the nascent team, which would be called the Post-Conflict Reconstruction Unit, or PCRU for short. The MOD's Paul Schulte would head the unit, and day-to-day operations would be run by his deputy Colonel Gil Baldwin, one of the officers who had raised concerns about the military's abuse of Iraqi prisoners eighteen months earlier.

However, the mere idea of the PCRU put some government bureaucrats on edge. The fact that Schulte and Baldwin were mid-ranking officers was a reflection of the caution many in Whitehall felt at a cross-departmental organisation with no clear lines of responsibility. Baldwin further stoked bureaucratic anxieties by insisting on seeing every slip of official reporting, from the intelligence reports at SIS headquarters in Vauxhall Cross to the MOD's daily briefings. At the same time the PCRU's own reports would bypass the usual ministerial committees and go straight into the Cabinet Office morning briefing papers. The idea that reports would be circulated that had not been carefully vetted by the departments was received in Whitehall 'like a red rag to a bull', Baldwin later ruefully admitted.[15]

Despite its admirable ambitions, steps were instantly taken to rein in the PCRU. Jim Drummond, in charge of Iraq for the DFID, was one of three overseers who made sure the unit did not throw up any surprises, much to Baldwin's annoyance. The departments themselves refused to recognise PCRU-headed reports; in an extraordinary display of pettiness, Baldwin's team was instructed to print its reports on paper individualised with the heading of each respective department or no one in that department would read it.

However, he was having more success in establishing a ground team for the next disaster zone. He had 200 contractors with combat zone training, equipped and ready to deploy, a secure war room in the DFID with access to classified reporting, ten armoured Toyota Land Cruisers, satellite phones, electrical generators and battle rations. Baldwin also established a permanent staff with experts on post-conflict reconstruction including Minna Jarvenpaa, a smart former special adviser to Finnish President Ahtissari, who had worked with him on weapons decommissioning in Northern Ireland and with the UN in the Balkans; Hugh Walker, a diplomat who understood the workings of Whitehall; Matt

Baugh, one of the DFID's rising stars, who did not mind breaking with development orthodoxy; and Babu Rahman, a free-thinking FCO employee.

The question was what to do with the unit's brainpower. Baldwin, through a contact at the British military headquarters in Northwood, had succeeded in getting his team deployed within days of the Indian Ocean tsunami on Boxing Day 2004. His three-man team arrived in Banda Aceh, the worst-hit region, with high-speed internet access and phone lines, making them a fulcrum for other EU countries reporting home. The result was a remarkably co-coordinated relief effort, which also negotiated a truce between the Indonesian government and a long-standing rebel group in Banda Aceh.

The PCRU won plaudits in Whitehall for its performance but did not assuage the lingering concerns of government departments. 'The unit was held in very high regard – by itself,' noted one senior officer.[16] The tensions came to a head as the Helmand force took shape over the summer of 2005. The mission was an ideal opportunity for the PCRU to deploy, but to Baldwin's intense frustration it was not being considered. Hostility towards the PCRU was epitomised one evening by Margaret Aldred, deputy director of the Security and Defence Committee, who unknowingly remarked to a PCRU employee, 'The Cabinet Office has been managing disaster for a hundred years. What do we need the PCRU for?'[17]

Aldred marshalled consensus over Afghanistan in the Cabinet Office, a role similar to the one Rob Fry had adopted at the MOD – only with none of the operational experience of the general. She seemed to view officers like Baldwin as naïve and unable to see the bigger picture. Baldwin had sent her a number of messages in which he said that the proposed number of troops was a recipe for disaster. Three thousand was neither small enough to avoid antagonising the Taliban nor big enough to fight back.[18] She ignored him.

Baldwin was not surprised, and was sick to death of the petty infighting in Whitehall. By then he had sat in on a number of planning meetings of the new Afghan steering group chaired by Nigel Sheinwald. Like its counterpart for Iraq, the group was often mired in turf warfare. The FCO needed to send a special representative to southern Afghanistan when the mission began, Sheinwald said. Edward Chaplin, who had returned from his stint as ambassador in Baghdad to become Middle East director, announced he did not have the budget to do so, unless DFID returned an official they had recently lent them in Nepal. The Maoist insurgency in the Himalayan kingdom had sparked the latest round of inter-departmental scrapping. The DFID thought the Maoist rebels should be supported. The FCO sided with the monarchy, which it had supported for years.[19]

'This is Tony Blair's next foreign policy priority, and you're quarrelling over £60,000,' said an exasperated Sheinwald. 'What planet do you people come from?'[20]

Baldwin knew that time was running out. The next day he stormed over to Edward Chaplin's office with Paul Shulte. What were the FCO's plans for Afghanistan he wanted to know, and who did they intend to send? Chaplin admitted they had none.

'Well, that's why we're here!' declared Baldwin.

Chaplin reluctantly acquiesced. It still took a further three months of sitting around, but by early November the half-dozen permanent staff, with their supplies and communication equipment, were finally ready to deploy. To lead the mission, Baldwin turned to Mark Etherington, the former governor of Kut. The two men were old friends from their time together in Kosovo. Etherington's insistence on standing his ground in the British outpost in Kut against an Iraqi mob and not evacuating even after the Ukrainians had scurried back to their bases had won him heroic status in the corridors of Whitehall. Haunted by the failures of the CPA, Etherington had taken a few months off before managing an election-monitoring programme in Afghanistan. He was in the country working on a new project for an NGO headed by the former Finnish president Martti Ahtissari when Baldwin contacted him.

As winter approached, Etherington and Jarvenpaa flew to Kabul, where they learned that enmity towards the PCRU extended to the field. Their first call was on the British ambassador in Kabul, Rosalind Marsden, who promptly said she was too busy to see them. What could be more important than this? wondered Etherington. Instead he met the DFID representative, who ended the meeting by warning him, 'You will be reported upon.'[21]

While Etherington flew on to Kandahar, Jarvenpaa remained in Kabul to gather information on Helmand and get a sense of how the UK mission would be perceived. Jarvenpaa was shocked by how little the British government knew about Helmand, which she attempted to remedy. In London she had managed to track down an elderly engineer who had worked in Helmand in the 1960s, when the province had been showered with US aid money as part of the Cold War battle against the Soviet Union.

The US had spent millions damming the Helmand river, with the dual aim of irrigating the desert and promoting capitalism by turning the local tribesmen into farmers with their own parcels of private property. The locals had proved reluctant small-holders, preferring their centuries-old tradition of herding the region's distinctive fat-tailed sheep. Further plans were abandoned following the Soviet invasion in 1980. The legacy of the US-built irrigation system, however, was poppy-growing, helping to turn the region into one of the world's top opium-producing regions.[22]

Most of the British reports Jarvenpaa could find on Helmand referred to the 1880 Battle of Maiwand. Few on the British side were interested in re-examining the battle, but that was not true among the Afghans she met in Kabul, for whom Britain's colonial role was neither forgotten nor forgiven. Poems about the victory at Maiwand had even passed into Afghan folklore. In one legend a woman called Malalai, seeing the Afghan forces falter, had used her veil as a banner and stirred the courage of her compatriots by shouting, 'Young love, if you do not fall in the battle of Maiwind/ By God someone is saving you as a token of shame.'[23] Ashraf Ghani, the chancellor of Kabul University, warned Jarvenpaa that if the British arrived like occupiers, there 'would be a bloodbath'.[24]

Apprehensively, Jarvenpaa flew down to Kandahar airbase a few days later to link up with Etherington. She found that he had struck up a rapport with Brigadier Messenger. The two had known each other as junior officers in Northern Ireland, and both advocated empowering the locals to fend for themselves. By early November Messenger had drawn up a basic plan for the region. It was textbook counter-insurgency stuff: establish security in one area first, in this case the triangle demarcated by the towns of Lashkar Gar, Nad-e Ali and Gereshk, before expanding. Messenger had developed a number of so-called military effects, ranging from Taliban interdiction to developing the Afghan security forces.

Messenger instinctively liked the PCRU's approach, and agreed to introduce a broader set of goals for areas like the rule of law, counter-narcotics and alternative livelihoods. Northwood was expecting the Helmand plan the following week. After an awkward conversation, Messenger managed to persuade his headquarters it was worth spending more time incorporating the reconstruction unit's ideas and presenting a single plan.

Before beginning the joint planning, Etherington and Jarvenpaa paid a quick visit to Helmand's provincial capital, Lashkar Gar. The town had been the centre of the US aid effort in the 1960s, even earning the nickname Little America. Its tree-lined streets were laid out in a neat grid on the eastern side of the Helmand river, the houses made of brick and lacking the traditional compound walls. Driving around the leafy town, the two experts struggled to get a sense of the challenges. The outgoing head of the American Provincial Reconstruction Team was enthusiastic about British plans to transform the province, although he had rarely ventured beyond the town. More illuminating was meeting the province's education director, who turned out to be illiterate. Reversing that sort of backwardness was going to take a generation, thought Etherington, as they returned to Kandahar.

Over the next month military planners and the PCRU – which had been joined by Hugh Walker, Babu Rahman and Roy Fleming, a police expert – met each morning in the main tent of the British enclosure at Kandahar airbase. Jarvenpaa led the first session, and listed the various objectives set by the Afghan Steering Group on a whiteboard. On governance they were to provide an 'effective and transparent provincial government . . . free from illicit influence and accountable to the Government of Afghanistan and the people of Helmand'.[25]

It was clear to everyone present that in the three-year timeline they had been given this would be a challenging aim, such was the corruption and poverty they faced. Part of their task would be to lower expectations in London. Etherington hoped that Whitehall would take a pragmatic view of the timeline – once the operation started they were not going to pull funding or withdraw suddenly – but he worried the aims of the mission betrayed a deeper naivety. He had heard several officials in Whitehall worrying that things might go 'too well' in Helmand, thereby exposing the faltering reconstruction efforts in other parts of the country.

On 28 November Etherington and Messenger issued an interim report warning about the lack of information on Helmand and asking for time to do deeper field research before settling on a long-term strategy.[26] The following day they held a video conference with George Fergusson, the Head of the Cabinet Office Foreign Policy Team, who was not impressed by the interim report. With the deployment date looming, he was under pressure to present a plan to the Cabinet early in the New Year. Instead the PCRU was telling him to hold off. 'Just get on with it,' he insisted. Whitehall would need their report before Christmas.

Jarvenpaa was able to make one more trip to Lashkar Gar with Hugh Walker, meeting a broader range of locals. She carried out straw polls with them on Britain's likely reception, but the main sounding board for UK plans was the newly installed governor, Daoud. He offered vague assurances that he would be able to deliver the Afghan side of the deal, improved governance and a competent police force. It struck Jarvenpaa later that Daoud, a technocrat who had spent most of his life in Kabul or in exile in Pakistan, knew little more about the region's tribal dynamics than she did. After a month in country, the PCRU team flew back to find the Christmas decorations up in Oxford Street, and Whitehall preoccupied by office parties. The prospect of deploying troops to Helmand had failed to rouse much public debate. Unlike Iraq, the rationale for the mission in Afghanistan was still widely supported; in fact, the unpopularity of Iraq seemed to unite support for a new effort in Helmand.

Jarvenpaa had taken responsibility for pulling together the report, and the last few days in Afghanistan were spent hunched over her laptop, desperately compressing six weeks of discussions into a twelve-page document. The paper, entitled 'UK Joint Plan for Helmand: Final Report', offered some stark advice. First it said that the British effort needed to assume a ten-year goal rather than three, before advocating much more detailed study of Helmand and how the plans for policing, job creation and governance would be received and implemented.[27]

Etherington and Messenger presented the plan to the Cabinet Office just before Christmas. The PCRU man already knew words of caution would prove unpopular. Even so, he was surprised to encounter a bad-tempered atmosphere when he got up to speak. The room was packed with representatives of the various departments and chaired by Margaret Aldred, wearing her habitual look of arch disapproval. The presentation did not last long. After running through the findings, Etherington concluded by saying they simply did not know enough about the area and needed to conduct further reconnaissance. 'What do you mean we don't know anything?' snapped Aldred, 'We've been there since 2001.'[28] She then hurriedly thanked Etherington and ended the meeting.

'The need to deploy into Helmand quickly drove everything,' reflected Etherington. 'It was clear that this was going to be very difficult, but we got no sense this had been understood in London or accepted. We'd had the temerity to tell Whitehall what they didn't want to hear.'

Baldwin, who had run the reconstruction unit since its inception the year before, announced his resignation from both the PCRU and the army the following day. He was taking up a well-paid job with Norwich Union. He had, by his own admission, 'spilt too much Whitehall blood' setting up the PCRU, and viewed the dismissal of the unit's suggestions as a personal rejection of all he had been trying to achieve. 'It beggars belief that we would go into Afghanistan even worse prepared than we were for Iraq,' noted Baldwin.

Jarvenpaa struggled to keep PCRU involved in Afghanistan, gamely writing a paper explaining how the reconstruction unit would set about further planning, but the idea did not go far. Whitehall was happy for Jarvenpaa and the other members of the unit to return when the deployment began, but only under the auspices of separate government departments. The unit's work, at least as far as Afghanistan was concerned, was over. The PCRU was later stripped of its mandate to bring together government departments for post-conflict planning, having succeeded only in entrenching bureaucrats in their respective ministries. Rebranded the Stabilisation Unit, the group effectively became a government think tank with a limited ability to prepare contractors for rapid deployment to the front line. It had cost £20 million to set up.

Etherington's contract also expired over Christmas, leaving the former paratrooper deflated and questioning whether Britain and America would ever stage a successful nation-building exercise. He knew from history that British colonialists had been no experts, frequently blundering into countries and upsetting the delicate balance of centuries of indigenous life. But at least they had staying power and a long-term vision – however myopic and colonial – for a country's development. Current nation-building efforts were governed by the twenty-four-hour news cycle, which reduced planning to politically expedient short-term 'projects'. The PR-driven optimism that fuelled ministers all too quickly turned to defeatism.

Clear-headed thinking was unlikely to endear Etherington to possible future employers, but he was determined to get to the bottom of the mess. Taking a break in Croatia, he began writing a chronicle of his experiences and wondered whether he would find a job again.

With Etherington gone, Messenger was left to return to Afghanistan alone to complete the planning. He did not have much time to dwell on the disappointment he felt. In early January the defence minister John Reid chaired a Cabinet meeting at which a non-binding vote was held on whether to continue with the deployment. Several senior ministers attended, including Hilary Benn, Jack Straw and Des Browne, the chief secretary at the Treasury, a former lawyer with no military background. Reid ran the meeting imperiously, recalled one member who was present.

During the debate Des Browne asked Reid if enough was known about Helmand to send troops in. Reid waved away the thought, and the matter was dropped. 'Are you voting against the deployment?' Reid asked him.

'I'm abstaining,' said Browne.[29] A few months later Browne himself was promoted to defence minister, a case, noted one insider, of removing a possible opponent to the war by placing him in charge of the strategy.

Having won the Cabinet's approval, Reid announced the deployment in January 2006, just in time for an Afghan donors' conference in London, which Tony Blair was chairing. Reid declared, 'We are in the south to help and protect the Afghan people construct their own democracy. We would be perfectly happy to leave in three years and without firing one shot because our job is to protect the reconstruction.'[30] The statement summarised the prevalent wishful political thinking – for a simple deployment, occupation and withdrawal. It perfectly reflected the type of war the British military wanted to fight, but not the one they were going to get.

Chapter 22

PIKE FORCE

THE BRITISH BASE IN HELMAND, AFGHANISTAN sat in the middle of a dreary desert ringed by a distant line of snow-capped mountains. Will Pike's troops were the first to arrive at Camp Bastion, which would be their home for the next six months. Pike was the commander of 3 Para's A Company and had only just finished a gruelling tour in Iraq. With thinning blond hair and a ruggedly handsome face, he was considered a fierce officer who never shirked the hardships of the field. He came from a military family: his father, Lieutenant General Hew Pike, had commanded 3 Para during the first, desperate battle to establish a beachhead on the Falklands. Rather than overawed by his father's illustrious career, Pike sought to emulate it.

There was the usual grumbling among Pike's 130 men about deploying so soon to another godforsaken desert. In Iraq there had at least been a veneer of civilisation: Saddam's palace with its marble dining rooms and riverside views; Shaibah logistics base with its fast-food outlets and military-run supermarket, where soldiers coming or going from the front lines could stock up on cigarettes and DVDs. In Afghanistan there was nothing more than a few tents flapping in the harsh desert wind. Food consisted of flavourless battle rations, and the toilet was a pit dug in the ground.

Pike was also struck by the challenge his men faced. Under the scheme drawn up the previous year, the British mission was broad-ranging. Pike's men were to conduct patrols and provide basic security to enable reconstruction work and the creation of accountable local government in a province that had only ever known tribal rule. At the same time they were to reform local security forces infamous for their corruption, interdict the barons who controlled the lucrative opium trade and push back a resurgent Taliban, which, over the previous six months, had killed four of Helmand's twelve district police chiefs and seized control of the northern district of Bagram.

In Iraq his unit had been part of a 9,000-strong deployment, with a dozen outposts scattered across the south of the country. The plan in Afghanistan was to slowly trickle in British forces, a few units at a time, to give the locals time to adjust to their presence. But the result was that the first troops of the 3,500 scheduled to arrive in Helmand would have to shoulder a disproportionate share of the burden.

To make the task more manageable, the British plan stipulated that 3 Para was to focus at first on just a small area of Helmand – the triangle of land between Camp Bastion, the provincial capital Lashkar Gar and the commercial hub Gereshk. Both towns and the villages between them were relatively safe and well developed. If the British could demonstrate progress there, they would have a sound base of operations for expanding the mission north up the Helmand river to the mountains, where the Taliban were strong.

The problem was that there was no reconstruction team ready to start work. The Post-Conflict Reconstruction Unit was the most obvious candidate for the job, with its cadre of experts, war-zone training and armoured vehicles. Unfortunately, it had fallen victim to entrenched bureaucracy, and the organisation had been barred any further role in Afghanistan. Whitehall had kept on a few experts like Minna Jarvenpaa, but otherwise insisted individual departments send their own representatives. Furthermore, the DFID considered it too dangerous for its representative to travel beyond Lashkar Gar.

Pike's first mission, a fortnight after arriving, was to establish a British presence in Gereshk, some thirty miles away to the east on the Helmand river. With the highway connecting Kabul and Kandahar running past, Gereshk was a bustling trading town of 50,000 inhabitants, whose civic life was focused around the market place. His company's home for the next ten days until more troops arrived to relieve him was a base on the outskirts of Gereshk that had been established earlier by the Americans. Pike had a rough idea what to expect from an earlier tour of Afghanistan. He knew that the country was poorer than Iraq, with 90 per cent illiteracy and a powerful tribal system that had never accepted central rule from Kabul.

He noted the mud-walled houses, unpaved roads and the piles of uncollected refuse on the streets of Gereshk upon which herds of goats grazed.

Despite the thousands of bureaucrats on the payroll at the DFID and the MOD, Pike realised he was going to have to invent Gereshk's reconstruction plan as he went along. 3 Para's medical officer, Harvey Pynn, had already taken it upon himself to survey health provision in Gereshk. During a visit to the town's hospital, he found the conditions filthy, with bloodstained sheets littering the floor. USAID had donated a washing machine the previous year, but it was still wrapped in plastic because the hospital did not have a regular water supply. Pynn thought it would be a simple job for military engineers to sink a well and get the thing running – an exemplary quick-impact project to win over the locals.[1] When Pike sent the suggestion back to headquarters, the DFID scuppered the idea. An Italian NGO was meant to be in charge of the hospital, he was informed. Neither the DFID nor the Italian NGO wanted the British military involved for fear of tainting their work. Pike pointed out that there had been no sign of the Italians since their arrival, but the DFID representative held firm.

The story reached Ed Butler, commander of the British effort in Kabul, who was incensed that the DFID would put such concerns over practical solutions. 'DFID isn't some fucking NGO; it needs to be working in the British national interest,' he declared.[2] But although the issue of the washing machine in Gereshk made it all the way to the Cabinet Office, where Nigel Sheinwald was apoplectic, nothing happened. Pike left Gereshk for Camp Bastion after ten days, having achieved little.

This underwhelming start to the British deployment had failed to dampen expectations among the Afghan leadership. The newly installed Helmand governor, Muhammad Daoud, implored Butler to deploy British troops outside the safe cities and into Taliban strongholds in the north. 'Anywhere the government of Afghanistan's flag was not flying was an assault on its authority,' Daoud urged.

Butler did his best to temper Daoud's expectations. Early on he warned the governor that if British forces spread themselves too thinly over the province they would be able to do little other than protect themselves. Three thousand British troops in Helmand did not translate into the same number of front-line soldiers. In fact all but 600 or so were support staff – engineers, drivers and cooks. Surely, Butler suggested to the governor, he would prefer them to build security around Lashkar Gar and Gereshk? Daoud would agree but the next day would return to his default setting – pushing the British to do more. The governor frequently used President Karzai to apply pressure through the British ambassador and the Americans.[3] A compromise was finally reached: the Afghan army would occupy a bleak desert camp deep in Taliban-infiltrated territory near the town of Sangin. There they would be routinely monitored by British troops, who would patrol the surrounding area.

British officials like Minna Jarvenpaa greeted this first deviation from the plan to first develop areas around Lashkar Gar and Gereshk with alarm. The tribal dynamics in and around Sangin were complicated even by Afghan standards. From what she could make out, the town was divided between two main tribes, the Noorzai and the Ishakzai, both heavily involved in the opium trade. The Noorzai, however, were supported by the Afghan government; Western diplomats believed Karzai's brother Ahmed Wali to be an infamous drug lord.[4] For protection the Ishakzai tribe had turned to the Taliban. Jarvenpaa feared the Brits might be about to take sides in a vicious drugs war that they didn't understand.

Jarvenpaa raised the issue with Butler, but her concerns were soon overtaken by events. In late May Colonel Charlie Knaggs, the British commander in Helmand, received a panicky message from Daoud that Naw Zad, the next town along the river valley, was in danger of falling to the Taliban. Gunmen had shot at the police station in the town centre, and the town's leader, appointed by Karzai, had disappeared. Daoud prevailed on Knaggs to commit a small force of Paras to guard the local garrison. The Paras were subsequently relieved by Royal Gurkhas, whose arrival conicided with an American request to arrest or kill a Taliban commander at his mud-walled compound on Naw Zad's outskirts.

Will Pike's company was sent on the mission. Four Chinook helicopters dropped Pike's men on either side of the compound. They immediately drew fire from Afghan gunmen in the surrounding buildings. Breaking through the compound gates, Pike found some weapons and boxes of ammunition. His next order was to look for the Taliban commander, whom surveillance now reported to be in a house some three hundred metres away.

With little information and no picture to identify the man, Pike knew it was a fool's errand. Worrying news was emerging from Naw Zad, where a squad of Gurkhas, sent to meet Pike from the district centre, had been ambushed and pinned down. Nonetheless, Pike gamely sent 2 Platoon on the mission. The terrain on the outskirts of the city was a maze of walls, ditches and dense pomegranate and apricot orchards. Pike's men were engaged with machine-gun fire as they crossed a stretch of open ground. The soldiers quickly thinned out to flank the position, in the process encountering more gunfire. The result was a long shifting line of gun battles with neither side gaining the upper hand.

As the shadows began to lengthen, Pike ordered his men to disengage and fall back. The 3 Para commander, Lieutenant Colonel Stuart Tootal, overseeing the operation from a temporary command post nearby, put in a request for Chinooks to come and pick them up, only to be informed through the Helmand commander, Knaggs, that Governor Daoud did not want to lose control of the 'gains' made that afternoon. They were to stay in Naw Zad overnight. After what they had just been through, Tootal feared that if they stayed they would be drawn into a longer battle for the town, overstretching his already limited resources.[5]

Knaggs insisted they stay. He had only been drafted into command of Helmand after the Canadians had objected to having a more senior officer in the province. As one witness noticed, the pressure of having Butler breathing down his neck was not helping the Irish Guards officer adjust to a new regiment in the heat of battle.[6] Butler's chief of staff agreed with Tootal's assessment but was unable to reach his boss. He ended up countermanding Knaggs's order, a decision that exposed the tensions within the command structure. Knaggs was soon removed from battlefield command and confined to dealing with Governor Daoud, placing Butler in charge of operations.[7]

Worse was to come a few weeks later in Sangin. On 18 June the Taliban ambushed fighters belonging to the district chief, Dos Muhammad Khan, killing forty of them. Khan's nephew was shot through the stomach, but managed to escape with other family members. Whereas the Brits had expected to first meet the police chief to discuss training, joint missions and offering ancillary peacekeeping support, Butler was instead being prevailed upon to lead a dangerous rescue of the district chief's family and the local chief of police, who the town's residents accused of raping a little girl. The charge was entirely in keeping with the notorious behaviour of the local police, and suggested why many locals from the Ishakzai tribe had sided with the Taliban.

Realising how contentious the issue was, and the real possibility of another lengthy firefight with the Taliban, Butler pushed the decision on to Peter Wall, the deputy commander in Northwood, who was visiting him. The increasing violence had alarmed the defence chiefs. Since the fighting at Naw Zad there had been three further serious engagements with the Taliban, and the first British soldier had been killed. The Americans were also requesting they man Kajaki Dam, the hydroelectric project that the US had funded in the 1960s, making this the fourth base outside Camp Bastion that the British were committed to defending. Far from being a routine peacekeeping mission, it suddenly seemed as though British forces were in danger of being sucked into a full-scale war with the Taliban as they were spread ever more thinly across the province.

The problems faced by the mission had yet to be reported in the British media, and the MOD was happy to keep it that way while Wall tried to rein in the fighting. But Wall now found himself in a vortex of political pressure over whether to rescue the Sangin chief's family. On 20 June he met Butler at Camp Bastion, where the two men agreed to hold off the deployment. It helped that the Americans were strongly against the idea of manning isolated outposts because of the strain it placed on air assets.[8]

A few hours later, after an intervention from President Hamid Karzai, the two men changed their minds and asked Stuart Tootal, 3 Para commander, to draft a last-minute risk assessment of the operation.[9] Butler told him, 'Stuart, we have got reports coming in that [Sangin] district centre is about to fall. If we are going to reduce the risks to the helicopters we need to use the cover of darkness and go before first light. Given that dawn is less than three hours away, I need to know whether we can launch the mission in the next ninety minutes.'[10]

After days of discussions and months of preparation, it was extraordinary that an important strategic decision to expand the mission should be turned into a tactical query as to whether troops could be deployed in time. Tootal, unflappable, said he needed twenty minutes to think about it. He was against the deployment but understood the political pressure Butler and Wall were under.

An hour later he was airborne with Will Pike's A Company. Dubbed Pike Force by the other company commanders, this time Pike's men arrived on the outskirts of Sangin ready for a fight. They landed near the district centre without incident, where they hastily deployed. The nephew was brought to them in a stable condition, and Pynn, the medic, administered a shot of antibiotics. Tootal left shortly after daybreak, but Pike and his men were ordered to remain after Butler came under more pressure from Afghan leaders.[11]

Fearing a Taliban attack, Pike began beefing up the defences of the district centre, which consisted of two small buildings surrounded by a low mud-brick wall that was unlikely to survive much of a pounding. The centre's rear was guarded by a bend in the Helmand river; the rest of the building faced the town's cramped bazaar, with its narrow streets and stalls with brightly coloured awnings. He put a mortar team on the roof to watch for Taliban fighters, while the rest of his men filled sandbags or stacked breeze blocks purchased from a nearby brick factory around the compound wall and placed a pickup truck in front of the main gate to guard against suicide bombers. Governor Daoud had promised to send Afghan forces to guard the centre, but they hadn't shown up. The locals were not enthusiastic about their introduction to British troops. Upon learning that Pike had no reconstruction funds, a delegation of town elders asked him to leave, fearing the devastation that a firefight would cause local homes and businesses. They warned that a Taliban attack was inevitable. For a group of soldiers risking their lives to supposedly maintain the peace, the request was disheartening and ironic.

A few days later the mortaring and harrying by small-arms fire began. The following day the sniping continued. Unlike the firefight in Naw Zad, when the adrenalin was followed by the chance to wind down back at camp, Pike's men found this non-stop attritional warfare draining. The base had no amenities. They lived on battle rations and used an open ditch as a toilet. With the shops and houses near the district centre disintegrating under the bullets from both sides, the elders came daily to plead with Pike and his officers to depart.

'It was a mockery of the idea that we were there on a peacekeeping mission,' said Pike.

Then on 1 July, a week into their deployment in Sangin, disaster struck. The day got off to a bad start when Pike went to visit a British outpost several miles outside town. There was only a quad bike in the district centre, so he had to wait for a helicopter to transport him. On his return, however, the Chinook pilot mistook another compound for the Sangin district centre, dropping Pike off at the wrong end of town. He realised the mistake only after the helicopter had flown off. The landing had alerted the locals, who swarmed out of nearby buildings. Pike did not wait to find out if they were armed, but began jogging towards the river as he desperately tried to call the operations room on his radio. He reached the riverbank coated with sweat and feeling desperate, and ducked down behind the embankment for cover, certain he was about to be surrounded, but the Chinook returned with its Apache helicopter escort, and he was plucked from danger.[12]

That night he was relaying the incident to his company sergeant Zak Leong when an explosion rocked the operations room.

'What the fuck was that?' he shouted. Leong dashed outside. The roof of the main building was glowing like a volcano. Pike had just shifted five members of the Signals Corps onto the roof, which offered better protection than the exposed top of the other building in the compound. Dashing up the two flights of stairs, Leong was confronted by a scene of horror. A 105mm rocket had struck a concrete pillbox at the top of the stairs, blasting a massive hole in its floor and coating the remaining sandbags along the perimeter with molten lead, still burning from the heat of the explosion. One of the signallers, Lance Corporal Jabron Hashmi, was on his side with a gaping hole in his head. Corporal Peter Thorpe was motionless on his back beside an Afghan interpreter, Daoud Amiery. Both were clearly dead. Other soldiers on the rooftop had suffered shrapnel wounds.

Leong rushed over to to check for signs of life, and was soon joined by other soldiers, who carried the dead and dying men downstairs and over to the small medical facility Pynn had set up a few days before. On one of the makeshift beds Pynn, covered in blood and cursing, was trying without success to pump some life into Hashmi.

Pike's deputy, Captain Martin Taylor, came over to find out what they needed.

'Fuck off. Just get me some helicopters,' Pynn shouted.[13]

Headquarters was reluctant to launch a night-time rescue with helicopters to take the men to a proper medical facility, given the danger of Taliban attack, and once it became clear that the surviving men were not critically injured, the decision was taken to wait for dawn. It was a grim night for Pike as he sat up in the operations room, within earshot of the men's groans. Even though he had sent them to the rooftop precisely because it offered superior hard cover to the roof of the other building, he couldn't help but feel guilty. He had been excited by the deployment to Helmand; it was just the sort of seat-of-your-pants mission he had stayed in the army for. Now he wondered how anyone back in London could have thought such a cavalier approach would work.

The next morning Stuart Tootal flew up from Camp Bastion in the helicopters to pick up the injured and the three bodies of those who had died. He asked Pike whether the district centre at Sangin was worth holding.

'If we stay here any longer we will have more casualties,' said Pike. 'It will only be worthwhile if I have enough resources to defend the base and support from the local security forces.'

In Kabul Butler took a different view. Urged on by Governor Daoud, Butler decided that a withdrawal from Sangin would cause a catastrophic loss of face. The British media had now picked up on the unfolding disaster after *Sunday Times* reporter Christina Lamb was caught in a harrowing ambush during an embed with troops. Once again it appeared that the army's belief that a small detachment of soldiers could have a far greater strategic impact than their size suggested had come undone. However, Butler was adamant the British were to remain in Sangin.

Tootal spent the following night at Pike's headquarters. The A Company commander and his men clearly needed a break after almost two weeks of bombardment, so he made arrangements for another company to take over. That night the British retaliated with artillery barrage, and an American Spectre gunship, with a 105mm cannon slung beneath its fuselage, flew in slow circles firing at suspected Taliban gathering spots in Sangin.

But the fighting was only just beginning. At British military headquarters in Northwood planners were starting to fear what many had forewarned: that UK forces were now dangerously overcommitted. No one was ready yet to contemplate what impact a war on two fronts over 1,500 miles apart might have on the British military, but the pressure was mounting for – yet another – change in approach.

Chapter 23

NEW COIN, OLD ROPE

If the British military was searching for a model to transform its fortunes, it needed to look no further than the Americans, who were coming to terms with the prospect of defeat against a deadly Sunni insurgency in Iraq. By the autumn of 2005 the US military was suffering their worst year yet, with an average of sixty dead and 400 wounded each month. The British, in comparison, would lose twenty-three soldiers that entire year. Despite their mistakes in Basra, the British prided themselves on their knowledge of counter-insurgency, honed in Northern Ireland, and blamed the Americans for being too aggressive, insensitive to Iraqi culture and overly reliant on technology.

But that was about to change. US General David Petraeus contacted his British colleague Brigadier Nigel Aylwin-Foster with an unusual request. The two men had worked together for the past year training Iraq's security forces and had developed a rapport. Petraeus had left Iraq in September to take charge of the US Army's training and doctrine at the Combined Arms Centre in Fort Leavenworth, Kansas. He was all too aware that the US military risked defeat in Iraq without a comprehensive rethink, but that change in such a large institution would be slow without provocation. Petraeus asked Aylwin-Foster to submit an article to Fort Leavenworth's magazine, the *Military Review*, giving vent to British frustrations with the American style.

Aylwin-Foster's views proved controversial, as Petraeus had hoped. The article, published in January 2006 said American soldiers were 'incapable of effective counter-insurgency' and 'at times their cultural insensitivity, almost certainly inadvertent, arguably amounted to institutional racism'[1] The piece was quickly picked up by Tom Ricks, Pentagon correspondent at the *Washington Post*, and became front-page news on both sides of the Atlantic. 'We were more successful than we had imagined,' noted Petraeus.[2]

Some American officers greeted this latest claim to British superiority based on their past history with annoyance – Basra was hardly turning out well – but most viewed it as a challenge. The following month Petraeus held a 'big tent' gathering of counter-insurgency experts to herald a revolution in thinking. It was time to learn from the British, said Petraeus, and for that matter the French, and from America's own dismal record in Vietnam and Iraq. The American army had failed so far, but the real test was whether they could learn from their mistakes.

At the heart of Petraeus's effort was the creation of a new counter-insurgency doctrine. Lieutenant Colonel John Nagl led the small team of drafters. Holding a doctorate from Oxford, Nagl had written an influential book called *Learning to Eat Soup with a Knife* – a phrase coined by T. E. Lawrence to refer to the difficulties of counter-insurgency – about Britain's defeat of ethnic Chinese insurgents in Malaysia in the 1950s. British success in Malaysia had rested on a light and responsive counter-insurgency force capable of adapting to the needs of the local population. The book had already proved instrumental in General George Casey's decision in 2005 to set up a counter-insurgency warfare centre in Iraq modelled on the British army's jungle warfare school established in Malaya in 1949 at the beginning of the fighting there. But Casey's training centre had failed to broaden the US military's approach, which remained preoccupied with targeting insurgents rather than protecting the community.

The new doctrine was encapsulated by the mantra 'Clear, hold, build,' which bridged the gap between clearing areas of insurgents and maintaining security, and providing development assistance. A first draft was published in March 2006. The doctrine was greeted with a degree of self-satisfaction by some British commanders, who believed US thinking was only now catching up with their own. But the American transformation also raised awkward questions. With security deteriorating in Basra, and the fighting escalating in Helmand, some were beginning to wonder whether Britain needed its own doctrine rewrite.

Responsibility for answering that question fell to Major General Graeme Lamb, a former commander in Basra. On leaving Iraq in 2004 Lamb had become director general of training and support at the Land Warfare Centre at Warminster, a post roughly analogous to Petraeus's at Fort Leavenworth. In the hands of some generals such a job would have been an excuse for time serving, but Lamb shared the American general's drive to reform the military after his experience in Iraq. There were other similarities between the two men: a protean quality and elusiveness that allowed them both to shift easily between positions to the consternation of their more conventionally minded colleagues. But where Petraeus was intellectual to the point of scholarly, Lamb relied on a more esoteric reading of moods and motivations. A severe dyslexic, he spoke in a low mumble, in a rough jumble of swear words, military jargon and Estuary English. 'Lambo rarely said a good word without the word fuck preceding it,' recalled one associate.[3] His inner poise took the form of slow and measured movements. Even in his late fifties, with a slight build and mousey looks, Lamb's steely brown eyes exuded a quiet menace. His brutal experiences in the special forces seemed to feed the dark, Calvinistic side of his nature, as if he were both driven and repelled by his deeds. He had told several of his subordinates earlier in his career to stay away from him. 'I'm toxic,' he told one officer, who was not sure if Lamb was referring to his conflicted conscience or his ability to stir up the top brass – usually with a string of expletives.

His latest effort to provoke his bosses included a push to properly teach the current British doctrine on counter-insurgency. Written and updated in 2000 by Brigadier Gavin Bullock, this drew heavily on the Malaya experience. Bullock had focused on a few clear principles: the need to articulate political aims in London and coordinate across government; and an effective information campaign in the field to win over locals and generate intelligence. The police were also to have primacy over the military to ensure that the justice system was fair and transparent. The doctrine had its shortcomings, as Bullock readily admitted. There was little on how information technology and the twenty-four-hour news cycle had transformed the way the military sold its operations to the public, or the particular ideological challenges presented by Islamic fundamentalist insurgencies. But the core principles of counter-insurgency had not changed.[4]

The problem for Lamb was that the doctrine had been sidelined in the British army's main training academies. At the Joint Services Command

and Staff College at Shrivenham, counter-insurgency had been restricted to one day in the forty-two week course for mid-ranking officers since the late 1990s. At Sandhurst the third term was devoted to doctrine, but the examples given to cadets were drawn from an air-brushed history of Malaya that discussed Templar's hearts and minds campaign without analysing the brutal methods – like forced relocation of villages to internment camps and summary executions – that the army had employed but were considered no longer acceptable.

'Everyone thought that peacekeeping operations like the one in Bosnia had defined a new paradigm in conflict,' said Colonel Alex Alderson, a member of Lamb's team at the Land Warfare Centre. Alderson had led the first 'lessons learned' exercise for the first and second rotation of troops in Iraq and had been struck by the military's arrogant and complacent attitude. When Casey made his counter-insurgency school in Iraq compulsory for all new US staff, British officers were not allowed to attend because it was deemed 'unnecessary' for the situation in southern Iraq. 'We just seemed to think we could show up in Iraq with soft hats and smile,' said Alderson. 'Politically, we weren't able to recognise we had an insurgency on our hands, as that would imply the Iraqis did not want us to invade their country, and that this wasn't a peacekeeping operation but a bloody war.'

As Graeme Lamb was pushing for better application of the existing doctrine, the organisation overseeing all British military philosophy, the Directorate General Doctrine and Development, had come to the opposite conclusion: British thinking was out of date and more suited to the withdrawal from empire than the type of kinetic, urban battles the Americans were fighting in Iraq. This body pushed successfully for a full rewrite of the doctrine, a time-consuming exercise that would set back efforts to realistically implement it.

The arguments over doctrine did little to address how counter-insurgency was being taught. The task of actually applying existing counter-insurgency thinking to the British military effort had fallen to a lone instructor at Sandhurst. Daniel Marston was a rarity at the academy in that he was American. His PhD, focusing on how the Indian Army had rebuilt itself after initial reverses against the Japanese during the Second World War, was later published as *Phoenix from the Ashes: the Indian Army and the Burma Campaign*.[5] As he began teaching in 2004 Marston was convinced an insurgency had begun in Iraq – although few in the military would openly admit it at that time – and the current curriculum at Sandhurst was leaving cadets unprepared for the challenges there. During his first term he threw out the old fluff and made sure his students grasped the grim realities of waging a counter-insurgency campaign. Marston brought in recently returned Iraq veterans to brief his classes. One of the first was Major Justin Featherstone, one of the defenders of Cimic House in Amarah. Marston's students listened enthralled to Featherstone's stirring account of the Jaish al-Mahdi assault.

Marston's direct and earnest approach provoked some snootiness. A few British officers did not take kindly to being told they were falling behind intellectually by a Yank. But Sandhurst's head of war studies, Duncan Anderson, was delighted. 'We were in our comfort zone talking about the successes of Malaya and Northern Ireland,' said Anderson. 'Dan did his best to shake us up.'[6]

By the autumn of 2005, Marston had completely rewritten the counter-insurgency course for lieutenants and begun work on one for captains, but he was growing increasingly concerned that his lessons were not getting to where they were needed most: the front lines in Iraq. The following year Marston persuaded Anderson to let him fly to Basra to teach his course to officers in the field. Marston was also keen to see the battleground for himself. He got in touch with James Everard, the newly arrived commander of 20 Mechanised Brigade, to ask if he could speak to his officers about counter-insurgency and spend time with units in the field. A charismatic and thoughtful leader, Everard welcomed suggestions for combating the deteriorating situation in southern Iraq.

British commanders had recently been ordered to 'lean into the risk' – that is to start pulling out troops even if Iraqi forces were less than prepared. It was the first sign that the army leadership was preparing to dispense with its commitment to create a competent Iraqi security force in the name of political expediency. 'It was naïve to think we could wait for a benign environment to hand over. We needed to start driving the process,' recalled Andrew Kennett, the assistant chief of staff for planning at Northwood.[7] In provinces like Muthanna that meant manipulating the military's 'traffic light' system for assessing the ability of local security forces to take over without any improvement in quality.[8]

Unfortunately, the rush to hand over to Iraqi security forces had coincided with a spate of fresh attacks. In Maysan, scene of the 2003 murder of the Red Caps, a period of calm was about to end. On 12 February 2006, the *News of the World* ignored MOD warnings and released a video showing British soldiers beating protestors during riots in 2004.[9] The day of publication twenty-two rockets and mortars slammed into Camp Abu Naji, the British base outside Amarah. Later that week a further sixteen fell. Despite the worsening security, BBC reporter Jane Corbin was invited to assess the military's progress training local security forces, including the work of Captain Richard Holmes, a charismatic twenty-eight-year-old who had taught himself Arabic and insisted on sleeping at the Iraqi police station where he worked during local elections. Returning from the station with Corbin, Holmes and Private Lee Ellis were driving a Snatch Land Rover that was struck by a massive roadside bomb. Both men were killed instantly. The Iraqi police cars escorting them had conveniently disappeared moments before the explosion, suggesting they knew of the impending attack. The British rescue team was set upon by a mob of Iraqis.[10] Lieutenant Colonel Ben Edwards, the British commander in Maysan, blamed the Sunday tabloid for their deaths. In addition to attacks on British forces, Basra's small Sunni community was being subjected to brutal reprisal killings following the destruction in February 2006 of the Al-Askari shrine in Samara, an important pilgrimage site for Shia Muslims. As sectarian tensions escalated, a prominent cleric was gunned down in April and on average a fresh corpse was showing up at the city's morgue every hour.[11] Hundreds of Sunnis were fleeing the city. The US commander George Casey flew down to urge Everard to take action but, as Everard pointed out, he lacked the intelligence network to combat the death squads, and even then could not arrest militia leaders without permission from the newly sworn-in government in Baghdad.

After weeks of politicking Nouri al-Maliki had been appointed prime minister of Iraq following the latest set of elections in January 2006. A former intelligence chief of the Dawa, he owed his position to the Sadrist bloc in the new parliament. He had an intense dislike of the British. His grandfather, a prominent cleric, had been arrested during the 1920 Shia uprising, and he saw the UK's current operations in Basra as a continuation of its colonial ambitions for the country. During his first trip as prime minister Maliki had visited Basra to declare a state of emergency in the city and a new security plan. He had told the British he did not want to meet them, and he was not forthcoming on what the plan entailed.[12]

Brigade headquarters was in a sombre mood when Marston addressed them in July. The Sandhurst instructor gave a harsh assessment of what he had seen so far. The army was failing to apply some of the basic working practices developed in Northern Ireland, such as keeping a database of knowledge to pass from one brigade to the next. The problems were exacerbated by the use of six-month tours of duty, which were too short for commanders to gain a proper grasp of the situation. In Belfast tours could last up to two years, and the US currently insisted on year-long deployments to Iraq. The army's rationale was that the strains of Iraq were too great for longer tours, but the result was a lack of coherence in the overall campaign.[13]

As Marston warned his audience, each new commander arrived in Iraq seeking to burnish his career and transform the situation. The reality was that he often repeated the mistakes of his predecessor, and by the time he had built up relations with the Iraqis and worked out the best course of action, it was time for him to depart. 'We are caught in a cycle of repeating the same mistakes,' he concluded.

Marston's courses in Iraq were mostly well received, although he recalled how one young officer sat sullenly through an early discussion. He asked the captain what was wrong. 'Why is this the first time I've heard about counter-insurgency?' asked the officer resentfully. While Marston's efforts were helping the military to come to terms with the intellectual challenge posed by the insurgency in southern Iraq, the MOD was still struggling to meet the basic challenge of supplying soldiers with the equipment they needed. One issue had particularly captured the UK public's anger over Iraq: the use of Snatch Land Rovers. Earlier that year Pauline Hickey, mother of one roadside bomb victim, delivered a letter to Number 10 demanding to know why British troops were driving to their deaths in 'fibreglass Jeeps'. Hickey had learned from her son's commanding officer Nick Henderson about a request to use the antiquated but armoured Glover Webb patrol vehicle. The request had been turned down, although seventy-nine of the vehicles were given to the Iraqi police instead. Why wasn't her son given better protection, she wanted to know.

Since learning of her son's death, Hickey had been in contact with Military Families Against the War, the organisation Reg Keys and Rose Gentle had set up. After Key's campaign against Blair, MFAW had become a focal point for the British anti-war movement, which was setting up a political party called Spectre to field candidates in every constituency at the next general election. Keys made sure Hickey's visit was well publicised, and the issue of poor equipment, which had lain dormant since the invasion, was picked up by the blogger and researcher Richard North. North thought the idea of replacing the Snatch Land Rover with the Glover Webb patrol vehicle a poor one. Its steel armour offered no better protection against roadside bombs and made the vehicle top-heavy and prone to tipping in off-road conditions. The Snatch was still the best option for a light patrol vehicle. What the army lacked was a medium-sized armoured vehicle capable of withstanding most blasts but mobile enough to use on narrow streets or irregular terrain. The failure to supply such a vehicle in Iraq was an altogether sorrier story of bad equipment procurement.

Prior to the Iraq war, the army had already spent a whopping £179 million developing two medium-sized vehicles, neither of which saw the light of day. The 1998 Strategic Defence Review had called for a vehicle with sufficient armour to offer protection from high-impact rounds, but light and small enough to be carried in a Hercules transport plane. The Tracer armoured vehicle had initially been developed with the Americans, only to be discarded in 2000 after incurring £131 million in development costs, once it became clear its chassis and components were not compatible with other European armies – a prerequisite for the European defence force that was being touted at the time. The Boxer, a replacement for the Warrior armoured personnel carrier, was developed with several European countries, only to then be cancelled at a cost of £48 million after it was deemed too heavy at around thirty-one tonnes to be deployed by air.

After scrapping them both, the military opted for a light, readily deployable vehicle that would form part of the so-called Future Rapid Effects System or FRES programme of armoured vehicles. This would not have enough armour to offer protection from many of the roadside bombs currently being encountered in Iraq, but it would compensate with a futuristic array of communication and surveillance equipment, which would allow soldiers to spot and target the enemy with precision weapons before they could pose a threat. In 2002 FRES was first endorsed by the Executive Committee of the Army Board, the panel of senior generals that governs the institution, and was included as part of that year's supplement to the Strategic Defence Review. On Michael Jackson's appointment as chief of the general staff in 2003, a tentative date of 2009 was set for the deployment of the vehicle, a highly optimistic forecast given that the technology needed for FRES did not yet exist, and new and more speculative electronics were constantly being proposed.[14] 'The army was preparing to fight the type of war it wanted to fight in the future, not the one it needed to fight in Iraq,' noted North.[15]

Mine-resistant vehicles were readily available, but the army didn't want to buy them for fear of undermining the rationale for its high-tech FRES programme. In May 2004 Lieutenant General Robert Fulton, deputy chief of the defence staff for equipment capability, told the Commons Defence Committee that buying a stopgap armoured vehicle for British troops was 'not a very attractive option because it would divert much-needed funds from FRES or from some other programme . . . It would be a stopgap but it would be a dead end'.[16]

Spending on the FRES programme was about to go through the roof, and ground troops still did not have the equipment they needed. By 2005 the cost of the programme had increased from £6 billion to £16 billion. Commanders were told that the earliest any vehicles would be ready was 2015. It later transpired that money for FRES that had been saved by cancelling the earlier medium-sized armoured vehicles – the type of vehicles that would be saving lives in Iraq and Afghanistan – had been siphoned off by the navy to pay for aircraft carriers. Not only was FRES delayed but the army had been robbed as well.

Some members of the Army Board like Richard Dannatt blamed bureaucrats at the MOD for overcomplicating the process.[17] Others pointed their fingers at William Bach, the parliamentary under-secretary of state for defence procurement, and his successor since May 2005, Paul Drayson. While the army was responsible for setting the broad parameters for future weapons systems, Bach and then Drayson were responsible for overseeing their delivery. Neither man had much experience of running a multi-billion-pound procurement programme: Bach was a former barrister and government whip in the House of Lords and Drayson the co-founder of a pharmaceutical company that in 2002 won a £32 million government contract shortly after he had donated £50,000 to the Labour Party. At the time, questions were raised over whether other bids had been fairly considered. After being made a life peer in 2004 he gave the party a further £500,000.[18]

For their part, the politicians blamed soldiers like Dannatt for failing to identify their needs and wasting hundreds of millions of pounds of taxpayers' money. There had clearly been an 'almighty cock-up' and by the summer of 2006, as bereaved families began to raise questions about soldiers who had died in soft-skinned vehicles, the military was scrambling to find a quick fix. Richard Dannatt, by then commander in chief at Land Command, still saw FRES as the long-term answer to the army's needs, but for now he rushed to the front lines an antiquated armoured personnel carrier called the SE432, recently taken out of storage and refurbished.[19] In addition, one hundred Mastiffs, a heavy armoured truck with a V-shaped hull that could deflect roadside bombs, were ordered.[20] An earlier version of the Mastiff had been enthusiastically adopted by the US Marine Corps the previous year, and though it wasn't impervious to roadside bombs, it offered vastly improved protection. As Dannatt later reflected, if such pragmatic decisions had been taken earlier by the army, 'We might well have saved ourselves a lot of pain and agony and death.'[21]

The issue of poor equipment wasn't the only area where the MOD was struggling. Troop morale was also undermined by shortcomings in the provision of care for soldiers returning from Iraq. Injured troops returned to the UK were treated on the National Health Service at Selly Oak hospital in Birmingham, where there were just fourteen designated military beds separated from a public ward by curtains. If more soldiers needed treatment they were put in the public wards. Soldiers were frequently left unattended, sometimes in grave pain, with nurses distracted by civilian patients. For those injured fighting for their country, the failure to recognise their sacrifice with adequate medical care was a betrayal.

But a potentially greater problem was the treatment of mental health problems. The Defence Medical Services Directorate, the MOD department responsible for soldiers' welfare, had only a limited understanding of post-traumatic stress disorder. Symptoms include flashbacks, nightmares, insomnia, depression and anger. This was partly a reflection of the army culture of 'cracking on', in which emotional distress was perceived as weakness, but the directorate was also wary of acknowledging a problem that it lacked the resources to address. Between 2003 and September 2005 just under 1.5 per cent of the more than 100,000 military personnel who served in Iraq between 2003 and September 2005 were referred for psychiatric treatment. Such a low percentage should have rung alarm bells. In 1982 the US government conducted a study of Vietnam war veterans and found that approximately 15 per cent of men and 9 per cent of women had PTSD at the time of the study, figures that correspond with similar studies after the First Gulf War.[22]

The Defence Medical Services Directorate insisted, 'We do not have a time bomb waiting to go off,' and, 'PTSD is not a major concern.'[23] Meanwhile, soldiers like Sergeant Christopher Broome were suffering but not encouraged to seek help. Broome had fought with the Princess of Wales's Royal Regiment during its violent tour in Maysan in 2004. For his actions during the Danny Boy incident, which included loading the dead bodies of Iraqis into the back of his Warrior, he had been awarded the Conspicuous Gallantry Cross. But a year after returning to the Army Training Regiment in Winchester he was experiencing the short temper and agitation which are the common early symptoms of PTSD.

'There were loads of things I couldn't tolerate,' said Broome. 'I couldn't stand my wife Lynsey twittering on about everyday rubbish. "Look what I've seen in the Littlewoods catalogue. Aren't those curtains nice?"' He found home life grating. He had just become a father, but when his wife complained about the stresses of looking after the baby, he would snap, 'That's fuck all compared with what I've been through.

'I was never an emotional guy, but I dwelled on having killed people, and I felt bad about it. Really bad. And you feel like you can't talk about it,' reflected Broome.[24] It was no better in the pub with his mates. They would ask him about Iraq, but he knew they didn't really want to hear about it. Iraq meant bad headlines on the television.

Broome had complained about stress to a doctor immediately after returning from Iraq but then, left to his own devices, he started taking it out on the recruits he worked with. If one arrived on the parade ground without a full water bottle, he would bellow, 'What about if your mate's on fire next to you? What if you're on fire?' Then he snapped altogether, forcing recruits to eat dirt or lick candle wax off his marching stick. He hit one recruit around the head with the stick so hard he knocked him to the ground. It was only after he was court-martialled and fined £1,000 that he was forced to seek help for PTSD.

Iraq was also taking its toll on the families of soldiers, especially those who had lost loved ones. Inquests were often slow and cumbersome, and many of the bereaved found the MOD defensive. When the inquest into the death of the Red Caps finally took place in March 2006, Coroner Nicholas Gardiner reached the extraordinary conclusion that the deaths could not have been prevented despite the clear failure of superior officers to follow basic standing orders like ensuring satellite phones were carried by patrols. After listening to Lieutenant Richard Philips, the Red Cap officer in charge that day, repeatedly declare he 'could not recall' the key moments of the incident, Reg Keys confronted Philips's commanding officer, the recently promoted Lieutenant Colonel Bryn Parry-Jones. 'My son! My son!' Reg began to break down, but he steadied himself. 'My son died, Bryn, because you left that man in charge!' he said, pointing at Philips.[25]

There was little understanding in the MOD that PTSD could affect family members as well. In February 2006, after her brother Lee died in Maysan province, Karla Ellis sank into a deep depression. She had shared a particularly close bond with Lee, with whom she had grown up in a council house on the Wythenshawe estate in Manchester. A former professional footballer in the lower leagues forced to retire prematurely by injury, Ellis had not wanted to join the military but felt he had little choice if he did not want to flip burgers all his life.

Lee's ex-partner worked nights, so his daughter Courtney spent most afternoons with Karla, who had just fled from an abusive relationship. Karla quickly became nervous and prone to tears, and had to leave her job. When she asked the MOD for help, she was given a number for a grief line and put on a six-month waiting list to speak to a counsellor.

Taking matters into her own hands, she decided to track down the soldier whose film of Iraqi protesters being beaten in 2004 had stirred up the tensions in Iraq that led to Lee's death. The soldier in question – Corporal Martin Webster – wasn't hard to find. The footage recorded Webster shouting, 'Oh yes! Oh yes! You're gonna get it. Yes, naughty little boys! You little fuckers, you little fuckers. Die! Ha ha!'[26] After the film's release in February 2006 Webster was detained, but ultimately no charges were brought against him or any of the soldiers involved in the beatings.

Ellis found Webster on Facebook. She thought she would be angry with him, but found Webster in a worse state than herself. He had recently left the army and was deeply depressed, living in a caravan in Cornwall. They ended up sharing their grief online.

'How do you get over this?' she asked him during one web chat.

'You can't. You've just got to live through it,' he said.[27]

Chapter 24

THE SHERIFF

When General Richard Shirreff arrived in Basra to take charge of British troops in southern Iraq he was the eighth general to take command of the war in less than three years. Like every general before him, he had his own specific ideas on how to run things and didn't care if they were a complete reversal of the departing general's strategy. Raised in a tough colonial family in Kenya, Shirreff commanded a room like a lion on the veld. His exploits as a young tank commander in the First Gulf War had earned him the nickname Rommel, along with the reputation of being one of the most instinctive, aggressive officers of his generation. By the time of the invasion of Iraq, Shirreff was in his early fifties and still physically imposing, his broad shoulders usually thrown back, and his hawk-like nose lending him a predatory air.

His arrival was greeted by some officers with the sort of anticipation that precedes a fight.[1] Others, including the chief of joint operations in Northwood, Lieutenant General Nick Houghton, were nervous. 'No rushes of testosterone,' Houghton warned him, although he would usually defer to his commander in the field. Shirreff already had plans that would ensure Basra would never be the same again.

Prior to deployment from the UK, Shirreff visited Basra on a scouting mission and unsurprisingly criticised what he saw as the UK's timid stance towards the militia. The outgoing general, John Cooper, was due to hand over Muthanna province to Iraqi control the following month, kick-starting the transfer of power to local security forces, as outlined in the Overwatch scheme, and the eventual withdrawal of Britain's 8,000 remaining troops. Cooper had regularly deployed the division's reserve battlegroup for missions alongside Iraqi units to prepare them for the handover, but questions remained over whether they could handle the steadily deteriorating security. Cooper's own forces were struggling to have an impact in Basra. Despite having 3,000 troops, Cooper could only spare 200 at any given time for patrols or strike operations; the rest were needed for defending and supplying British bases in the city.

As Shirreff saw it, the British army was not withdrawing with honour, but about to cede territory under attack. Unless something was done to take on the Shia militia, the British army's reputation would suffer. 'We had a strategy that involved extraction rather than necessarily achieving mission success. It was, in a sense, an exit strategy rather than a winning strategy,' he later said.[2]

The UK stance was also damaging the relationship with US forces. The British insistence that the south was ready to be handed over was greeted with growing scepticism. Cooper had already had a couple of run-ins with the top American commander, George Casey. Since the destruction of the revered Al-Askari shrine in February 2006, the country had edged closer to civil war between Sunnis and Shias, with 200 bodies a week arriving in Baghdad's central morgue as government-sponsored Shia death squads and Sunni suicide bombs terrorised the capital. Casey could not afford to risk British forces withdrawing and leaving a power vacuum in the south.

Shirreff felt that the British stance of withdrawal come what may was missing the point of the mission in Iraq. It was not about building a prosperous Iraq or edifying politicians back home, but about fostering Britain's key strategic relationship with the US. After Basra, Shirreff flew to Baghdad to reassure the Americans that British forces would not be leaving Iraq with a whimper.[3] He found Casey planning Forward Together, a massive operation designed to smother the sectarian bloodshed in Baghdad. Iraqi troops would be deployed in force, making the operation a test of Casey's transition strategy.

Shirreff liked the aggression and grand scale of the American approach, and back at the 3rd Mechanised Division's headquarters in Bulford, Wiltshire he began conceptualising the game-changing event for Basra. In order to reverse the current strategy, he would need permission from London, but he felt confident he could browbeat them into it, even if they had only just agreed to slipping out quietly. Shirreff envisioned British forces sweeping into districts to take on the Jaish al-Mahdi. 'I will see tanks back on the streets of Basra,' he told his planning team. Although past clashes with the Jaish al-Mahdi had always led to deadly reprisals that set back British goals, Shirreff felt this time would be different both in the size of the operation and the use of local Iraqi army units to hold newly won territory. Shirreff's plan also emphasised development, an area that had often been neglected in military thinking. As each district was secured, mentors would revamp police stations, and contractors would follow with quick-impact projects to provide jobs, installing power lines and sewage systems, ensuring that the militia would struggle to regain a foothold.

Northwood greeted Shirreff's plans with dismay, but he ploughed on regardless, bolstered by an experiment in how aid was delivered in Basra with the creation of the first British Provincial Reconstruction Team. In 2002 Nick Carter, who was to become a brigade commander in Basra, had come up with the PRT idea while serving in Afghanistan as chief planner to the American mission. He saw how a fragile local government and the small pool of local contractors had stymied reconstruction, and wanted teams of soldiers and contractors to work with local governors to deliver reconstruction projects. This meant expanding the role of the military to play a humanitarian role.

The concept of military PRTs was not without its critics. Aid agencies complained that the teams retarded development by doing the tasks for the locals; militarising reconstruction efforts also endangered the work itself by blurring the line between aid and military intervention. But then, as Carter and others pointed out, aid agencies were all too often absent from post-conflict zones when something needed to be done.

By early 2005 Condoleezza Rice believed the moribund reconstruction process in Iraq needed reinvigorating. On her first visit to Baghdad as secretary of state she announced a PRT roll-out to a large gathering of officials, including Tim Foy, the DFID's man in Baghdad. There would be some alterations to the model used in Afghanistan, explained Rice, in recognition of Iraq's more established local governments – State Department rather than Defense would have the lead – but she wanted a PRT in every province by the end of the year. In southern Iraq the British were expected to supply their own teams.

Foy was horrified. They might have worked in Afghanistan, where government institutions were extremely weak, but in Iraq the teams threatened to supplant a key function of the provincial councils. Others in

Whitehall feared the teams would inevitably lead to an expanded military effort in the south instead of a drawdown.[4] But after six months of artful stalling on the UK side, the British ambassador in Baghdad was handed an American ultimatum: either establish a provincial reconstruction team or 'We'll do it for you.' To back up the threat a reconnaissance team for a possible American-led team was sent to Basra. British officials in the south were disturbed to discover an already worked-out blueprint for the province 'drawn up by some two-bit official in Maryland'.

Realising establishing a British team was inevitable, Simon MacDonald, the head of the Iraq desk at the Cabinet Office, negotiated some concessions: there would be a single team in Basra for the whole of the UK-controlled south, which would preserve the schedule for handing over Muthanna and Maysan provinces to the Iraqis; the Americans also agreed to give $40 million to the British team, although they expected the UK to provide the start-up costs. Over Christmas £5 million was scraped up, just enough to fund operational costs for the first few months.

It wasn't exactly a ringing endorsement of the new approach, but the search for the first team leader duly began. The lead candidate for the job was Mark Etherington, who in the last few years had witnessed his share of failed efforts in Iraq. After his forced abandonment of the British outpost in Kut in 2004, reconstruction efforts in the province had essentially disappeared. The following year Etherington had led the Post-Conflict Reconstruction Unit's team in Helmand, but his calls to take more time planning the deployment were ignored by Whitehall. The former paratrooper and Cambridge graduate had also just published a book documenting early failures in Iraq and made public his criticisms of the inadequate planning before the Afghanistan deployment. Still, it was clear that Etherington knew more about the situation than most.

After witnessing so many reconstruction projects flounder, first in Iraq and then Afghanistan, Etherington did not believe a PRT consisting of forty contractors would transform southern Iraq, but he did promise himself one thing before accepting the job: if he saw something wrong he was not going to keep quiet about it. 'Too many times in the past I figured I'd just let it go; someone else would do something about it,' said Etherington. 'And I'd always lived to regret it.'

If that vow betrayed a hint of idealism, then the next few weeks of planning rekindled some of Etherington's hopes of ending the infighting between Whitehall departments. Etherington knew that while the bureaucrats were squabbling, soldiers on the front lines were making it up as they went along, without benefiting from any of the expertise of civil servants trained in reconstruction. His efforts had not worked in the poisonous atmosphere of Whitehall, but perhaps in the field the PRT might finally achieve that elusive cross-departmental focus. He successfully lobbied to create a southern Iraq steering group to co-ordinate the civilian and military reconstruction efforts with his own.

By the time Etherington was ready to deploy in April, disaster had struck the Queen's Birthday Party in Iraq. With Basra palace under partial evacuation orders, Whitehall questioned whether it was foolhardy to send more civilians. The mission was rescued in part by a new set of political calculations at Number 10, where Tony Blair was contemplating what might be his final year in office. In April, he summoned to Chequers his closest advisers, with Iraq was high on the agenda.[5] Blair accepted that a 'relatively stable situation' was the best that could be hoped for, but what would there be to show after four years of British occupation?

The need to starting building legacy projects in Iraq helped allay the lingering concerns about deploying Etherington and his team. In May the Cabinet Office asked the new consul general in Basra, Robin Lamb, to come up with a reconstruction plan for the city that would represent Britain's final bequest to the Iraqi people. Lamb, who had only just arrived, was given three days to deliver. He gathered the remaining consular staff and the newly arrived members of the PRT together at the palace for a brainstorm. 'Basically anything they could come up with got put on the list,' recalled Lamb. 'We didn't have time to vet them or draw up a proper budget.' Some of the ideas, including planting date palms and boosting the tomato export business, had been kicking around since 2003, a reflection of the failures of the reconstruction effort so far and the challenge Lamb and his team would face implementing them in a matter of weeks. These ambitious projects were balanced against quick-impact projects like painting schools and laying football pitches.

The three-day deadline, the rough figures and the whole concept of politically driven legacy projects immediately aroused the suspicions of the DFID employees in Basra. They viewed the projects as undermining the very work that New Labour had set up the department to do: build up local governance capacity. One representative warned Lamb, 'Unless you're building up local capacity you might as well pour money down a drain.' But the consul general had little choice but to follow the whims of his political masters.

At the end of an exhausting weekend he had a twenty-page list of projects with a £100 million budget. The last thing to do was name it, and after a little thought he found just the sort of phrase that Whitehall loved: 'Better Basra'. Lamb sent it back to the Cabinet Office with a few hours to spare. He received a short congratulatory note and was told to start preparing more detailed plans. He was dismayed therefore when an irate Des Browne, the recently appointed defence minister, arrived at the palace on the warpath a few weeks later. 'Why the fuck isn't there any money being spent here? Where are the quick-impact projects?' he yelled. Lamb quietly asked Browne's private secretary if he had read the 'Better Basra' paper. 'Apparently not,' came the response.

Etherington arrived a few weeks after the plan had been submitted with his own ideas on how to approach reconstruction. However, his first task was to manage the relationship between himself, General Shirreff, and Lamb's recent replacement, Rosalind Marsden. The former ambassador to Kabul had assiduously ignored Etherington the year before when he had been working for the Post-Conflict Reconstruction Unit, but this time they met without undue awkwardness. The bigger challenge was the pairing of the retiring Marsden, with her nervous gestures but unshakeable convictions, contrasted starkly with the swashbuckling General Shirreff. But it might just prove an inspired grouping, he thought. Shirreff would not be able to bulldoze Marsden in the way he might have done a male diplomat; Marsden would be compelled to apply her undoubted intellect to the campaign issues at hand; and he would act as a facilitator. Still, it was a big gamble to take.

Shirreff began their first meeting together in typical fashion, setting out his ideas for his big push into Basra, now called Operation Salamanca. He had already created a stir in the city. Since arresting or killing insurgent leaders had failed in the past, the Brits had avoided aggressive and public crackdowns on local militias, which tended to scuttle diplomatic relations with Iraqi officials and prompt mortar and IED attacks that killed soldiers and wounded civilians. However, within hours of arriving Shirreff had given James Everard, 20 Brigade commander, approval to arrest the current Jaish al-Mahdi leader Sajjad Badr al-Sayeed. In a midnight raid British forces stormed his compound. No one was killed, but since then the palace had suffered a hail of rocket and mortar attacks, and relations had deteriorated with the Basra Security Committee, a group set up by Iraqi Prime Minister Maliki several months earlier in his only real effort to establish control in the city.[7]

Shirreff was unapologetic over Sayeed's arrest and the reprisals. He told Marsden and Etherington that the British army was facing 'full conventional war in built-up areas' similar to what the Americans had experienced in Sunni towns to the north. The army would take back control of the streets through a series of 'pulses' through Basra's troubled neighbourhoods. British and Iraqi troops would then seal off one district at a time and create a pocket of security. In order for it to work, Shirreff wanted Marsden and Etherington to provide the economic engine to transform Basra's districts with quick-impact projects and job-creation schemes. The important thing was to offer Basra a powerful alternative to the Shia gangs.

Etherington could feel his hackles rise at the mention of quick-impact projects and the tokenism they often led to, but he appreciated a military leader who pushed for reconstruction efforts instead of wanting to blow things up. The British commander's next step was to find the resources for Operation Salamanca. To secure each district, Shirreff wanted at least three battle groups: two to hold the city and the third to conduct the operations. The outgoing general, John Cooper, had already reassigned the battle group dedicated to training the Iraqi army to Basra. Efforts to forge the Iraqi 10th Division into a coherent fighting force had not been a success. Under orders not to partner with their Iraqi counterparts at junior levels and lacking resources, there was little British troops could do to counter militia influence over the locally based force, and an entire battalion had recently mutinied after being asked to deploy to Baghdad for a routine operation.[8]

With two battle groups in the city, Shirreff still needed a thousand troops to form his strike force for the 'pulses'. He saw no alternative but to call on the theatre reserve in Cyprus. Shirreff knew that British military headquarters would take some persuading of the need for more troops with a withdrawal supposedly imminent, but he had a bold plan for securing Northwood's support. He believed he could succeed where earlier commanders had failed and persuade the Americans to approve a British exit from southern Iraq's most troubled province, Maysan. In the process he would signal the start of the final confrontation with the Jaish al-Mahdi in Basra before a complete withdrawal.

Chapter 25

BREAKING POINT

EVER SINCE THE APPEARANCE OF Iranian-made precision bombs the year before, Maysan had festered. Over the winter of 2005, the province's illusion of calm had been shattered by the deaths of Captain Richard Holmes and Private Lee Ellis. The conflict that emerged was a raw and bloody battle for the control of Maysan's 200-mile long border with Iran, over which equipment and supplies for the Shia militia groups across the country were flooding. British efforts to stop the trade provoked a furious backlash.

When Major General Richard Shirreff arrived in Iraq, the fighting in Maysan had reached a new level of intensity. Every time British soldiers entered Amarah they were ambushed. One night in June the base outside Amarah, Camp Abu Naji, was hit by sixty-seven mortar rounds. The commanding officer Lieutenant Colonel David Labouchere had taken shelter under a table with the other officers in the operations room while the base's distressed pet goat Ben 'pissed and shit everywhere'. Tim Wilson, the police adviser for Maysan, emerged after the all-clear to find portacabins ripped to shreds and the ground pockmarked with craters.[1]

Labouchere, who was now forty, had joined the army as an enlisted soldier at the age of nineteen without a university degree, and had worked his way up through the old Etonians and double-barrelled surnames by proving himself on the battlefield. As a cavalryman, he had been drilled in the virtues of manoeuvre warfare. 'Don't go static' was the watchword, which meant not staying still long enough for the enemy to surround you. But that was precisely what had happened at Camp Abu Naji. He felt that handing over the base to Iraqi security forces was long overdue. His forces could then redeploy to the desert, where they would be free to move and fight at the time and place of his choosing.[2]

Labouchere's idea hadn't been popular and seemed impossible to sell to General Casey, but he saw little alternative. His attempts at political reconciliation had taken on a surreal air. One local tribesman implored him to stop punishing his people by switching the power off. Another explained that it was a well known fact that the British were deliberately poisoning the water by dropping the bodies of their fallen into Amarah's water supply. Even Labouchere's own headquarters had been infiltrated. Their resident spook had informed him that the camp barber, a local, was a member of the Jaish al-Mahdi and helped guide in the mortar and rocket salvoes. That did not stop Labouchere visiting the barber's shop in the recreation area a few days later, albeit with a pistol in his pocket, which he cocked once he was covered by the hairdresser's sheet.

By the time Shirreff came to visit Maysan in July, Labouchere was convinced he would never be able to change Maysan's tribal culture and anti-Western bent. If we weren't here, what would Iraq look like? he wondered. Wouldn't it just settle back into the old tribal jungle it had always been?

Labouchere took Shirreff over to see the base which Iraqi contractors were refurbishing as part of a plan to maintain a small British strike force in the province. Standing in one of the sangars overlooking the empty desert, Shirreff asked him what he thought about stationing troops at the base. 'They're just going to be a tethered goat,' Labouchere told him, before sharing his own plan – complete withdrawal save for a light desert patrol.

The British general understood the dangers of withdrawal. The Muthanna handover, which had finally gone ahead, had descended into farce when a crowd of several hundred Iraqis descended on the base a few days later and overpowered the local security forces. They proceeded to strip the base down to its bones. 'A four-million-dollar base that's just gone up in smoke,' fumed one of Casey's staff, for whom the sacking presented an ominous portent for the transition strategy.[3] Fortunately for the Iraqi government and Coalition, the television cameras had long since departed, and the event scarcely registered.

But Shirreff was not a man to be dissuaded by risk. The more radical plan for Maysan reflected his own desire for action, as well as some savvy political reckoning. By relocating half of Labouchere's battle group to Basra he could make the case to the Americans that the withdrawal from Camp Abu Naji was part of his scale-up for the big fight. If he could deliver Basra, then a reduced presence in Maysan would be worth it and he would be able to deliver an even more comprehensive withdrawal than anyone at the headquarters could have imagined.

Unfortunately, the defence chiefs did not share Shirreff's optimism. When the subject of Shirreff's plans for Basra came up, the newly appointed chief of the general staff Richard Dannatt opposed increasing troops levels in Iraq. Any spare soldiers were needed in Helmand, he argued, where UK forces were increasingly overstretched and under siege. Far from melting away into the Afghan desert when confronted with British fire-power, the Taliban had surrounded the district centres where the UK troops were based, turning towns like Sangin into a warren of staging posts for attacks. With half a dozen outposts needing air support, the shortage of helicopters was taking its toll. Troops were eking out their rations and were forced to send patrols into the town to look for food. Furthermore, helicopters were being hit with withering barrages of fire whenever they tried to land, and it seemed only a matter of time before one was lost.[4] The quiet province that had been observed while planning for the deployment had been transformed into a war zone by the arrival of 3,500 British troops.

A further irony, not lost on Dannatt, was that the military had few additional resources to send until British forces left Iraq. The smart strategic shift from Iraq to Afghanistan that had been envisioned the previous year had turned into a protracted war on two fronts – the 'perfect storm' that Dannatt had warned about.[5]

In Basra, Shirreff was aware of the strains that the failing mission in Afghanistan was placing on the British military and that a direct request for troops was unlikely to succeed. Some 125 additional troops had already arrived in Helmand from the theatre reserve in Cyprus – from the same pool of troops that Shirreff was demanding for Basra.[6] Shirreff, however, had an alternative approach. His predecessors in Basra had discovered that being part of an American-run operation sometimes placed the British commander in Basra in an unenviable position, caught between the demands of London and Washington. But Shirreff chose to take advantage of the fact he had two chains of command, knowing that if he could win American approval for his plan to withdraw British troops from Maysan and divert them to Basra, Northwood would come under pressure to do the same.

Shortly after visiting Lieutenant Colonel Labouchere in Maysan, Shirreff flew to Baghdad to meet the US corps commander, Peter Chiarelli. Despite the escalating violence in Baghdad, the American general was immediately receptive to taking on the militia in Basra, offering Shirreff a battalion of American troops from his own operational reserve, any surveillance assets he needed and £50 million for reconstruction projects. The offer was followed by a personal visit to Basra by Casey. He was concerned that a British withdrawal from Maysan would invite more Iranian weapons and money over the border.[7] Shirreff assured Casey that British troops would continue to patrol the Iranian border in Maysan and any forces released would be used for Operation Salamanca. Britain would not be abandoning its strategic ally.

Casey was mollified, and Northwood hardly needed persuading of the merits of withdrawal. Shirreff was given permission to go ahead with the withdrawal of 1,200 troops from Camp Abu Naji in Maysan, ending over three years of occupation. However, the chief of joint operations in Northwood, Nick Houghton, still stalled when it came to approving the aggressive new approach that Shirreff wanted in Basra. Instead, he asked Shirreff to turn down the American offer of men; the presence of US soldiers in Basra was viewed as a recognition of failure. Shirreff had anticipated this reaction but had hoped the American offer would spur the defence chiefs into approving the release of more British troops for his plan. He therefore ploughed on, hoping that once he had built up momentum, Northwood would fall into line.

As the withdrawal of Camp Abu Naji in Maysan began, and non-essential equipment was shipped out, British forces were engaged in heavy fighting. On 18 August, just after midnight, British special forces backed by C Company of the Queen's Royal Hussars entered the Jamariya neighbourhood to arrest a member of the Gharawi clan. The moment the SAS contingent went in, spotters on neighbouring buildings started firing. The noise alerted the target, who was attempting to escape across the rooftops when an SAS soldier rugby-tackled him. As the Iraqi was loaded into the back of a Warrior, one of the tanks in the cordon was struck by an RPG and its commander, Corporal Stephen Killick, knocked unconscious. The blast also removed one of the tank's caterpillar tracks.

Labouchere was directing operations from his command vehicle in the outer cordon. He quickly ordered in a recovery vehicle only for it too to get bogged down in a sewage ditch. By the time a third vehicle commanded by Staff Sergeant Christopher Lyndhurst arrived some 300 Jaish al-Mahdi fighters had assembled, including a sniper who was picking his shots whenever Lyndhurst tried to move between the stuck vehicles. By mid-morning the sergeant had finally succeeded in attaching a chain to one of the tanks, only for the situation to be further complicated by the appearance of Amarah's residents.

Labouchere was shocked to see markets opening and a woman with a baby in her arms walking past his tank as bullets whizzed by. He was struck once again by how the British presence exposed the locals to danger. Two cars now pulled up, and a posse of tribal elders emerged, demanding that the British withdraw. The parley was cut short by sniper fire and then a 120mm mortar round that landed in the sewage ditch where Staff Sergeant Lyndhurst and another sergeant were still labouring to extract the vehicles. Labouchere watched horrified as the two men flew through the air in a hail of debris and excrement. He was certain they were dead. Instead the soldiers got up, deafened and bruised, and finished hooking up the vehicles so they could be towed from the scene.

After this draining encounter Shirreff gave Labouchere a break, leaving his second-in-command, Major Matthew Cocup, to manage the actual withdrawal from Camp Abu Naji two days later. With the regiment's materiel already moved, Cocup informed the Iraqi commander, Colonel Akil, about the handover. Given the nature of the Iraqi forces who would be taking over – they were the ones who had mutinied just a few weeks before – Colonel Akil was only told a few hours before the British forces were to leave.[8] After a short ceremony Cocup and the rest of the battle group drove out of Camp Abu Naji, leaving Akil's men in position. Knowing all too well the danger of roadside bombs, Cocup led his men on a circuitous route through the desert. The last he saw of the base was a distant line of barricades and an Iraqi flag hanging limply from the flagpole.

Within an hour word of the British departure had spread to Amarah, and a crowd began to gather outside the base. By midday there were hundreds of chanting protesters demanding entry to the base, to which Akil meekly agreed in return for safe passage out. A local Al Jazeera crew recorded startling images of Maysani residents ransacking the base. The rioters didn't just take electronic equipment and tents, they gutted the base in what appeared to be a wholesale rejection of the occupation the camp represented. The images were broadcast a few hours later with devastating impact. After spending over £80 million trying to rebuild the province and at least £2.5 million on the camp itself, the British were leaving the province in the worst possible state.[9]

US Secretary of State Condoleezza Rice watched the images in her office in the State Department. For an administration racked by doubts over the conduct of the war in Iraq, the sacking of Camp Abu Naji provided a shock. As Rice recalled, 'It wasn't clear what victory in Iraq looked like, but everyone could recognise defeat: terrorists swarming over the ramparts like they did at Abu Naji.'

The sacking of Abu Naji threatened to expose the whole premise of Casey's transition plan as false: that Iraqi security forces were ready to take over their own security as Coalition forces withdrew. Casey in his quiet, professional way, accepted that the situation 'stank' and sent down his corps commander, Peter Chiarelli, to decide whether Coalition troops – British or American – needed to be ordered straight back in. Beyond rectifying the horrors of the Camp Abu Naji withdrawal, there was now a real concern that the 'ratlines' from the Iranian border camps would bring in a flood of arms and supplies to the Shia militias, despite Shirreff's promises to continue patrolling the border.

The sacking was a grave embarrassment to Shirreff but it did have one unexpected consequence: the MOD was shamed into approving his Basra plan, and he was given two companies from the theatre reserve, an additional 320 men. It was nowhere near the number he had originally wanted, but he accepted that the military was increasingly stretched, and he was hopeful of making up the numbers with Iraqi soldiers.

Having won London's grudging approval, the next step was to take the plan to the Iraqis. After initial resistance from Basra's security committee, many of whom had close ties to the militia, Shirreff won grudging approval from the Prime Minister Nouri al-Maliki. However, the Iraqi leader objected to taking on the Jaish al-Mahdi, preferring to focus the plan on reconstruction. Shirreff paid lip service to Baghdad, even renaming the operation Sinbad following a request from Maliki. Privately, he vowed to carry on with the strike operations against the Jaish al-Mahdi regardless.[10]

As Shirreff prepared to take on the militia the fragility of the British military's position, stretched over two battle fronts a thousand miles apart, was about to be revealed by events in Afghanistan. In June, British forces had occupied yet another district centre, this time in the town of Musa Qala, northern Helmand, following the withdrawal of a Danish contingent. They were soon under heavy attack. The steep hills surrounding the small town of mud-brick homes meant helicopters were dangerously exposed whenever they tried to bring in supplies or remove casualties from the British outpost in the centre of the town. Soldiers in need of urgent medical attention were forced to wait hours for the cover of darkness, when the choppers were less likely to be hit.[11] Given the danger of British soldiers dying because they could not be airlifted out in time, the British commander in Afghanistan, Ed Butler, was confronted with the stark option of evacuating Musa Qala and conceding defeat to the Taliban.[12]

On 6 September the Helmand mission came the closest yet to breaking point. Afghanistan was one of the most heavily land-mined country in the world, with as many as one million landmines laid in the country during the war with the Soviet Union. Hundreds of thousands had since been cleared, but the dangers remained. That day a small team from 3 Para's sniper platoon, preparing for an operation near the large dam at Kajaki, entered a minefield. Lance Corporal Stu Hale was the first to step on an anti-personnel mine, which ripped off the lower half of his right leg. 3 Para commander Stuart Tootal dispatched a Chinook helicopter to the rescue, but the pilots were unable to land anywhere near, forcing the sniper team to clear a path to a safe landing site. That was when Corporal Stu Pearson stepped on a second mine, shearing off the lower half of his left leg. Worse was to follow. The Chinook kicked up a storm of dust and debris as it landed. Private Mark Wright was caught in the chest by an exploding mine and mortally wounded. A few moments later Private Andy Barlow stepped on a fourth mine.

The multiple injuries at Kajaki would have taxed British resources in themselves, but as the stricken sniper platoon was brought to safety, 3 Para's headquarters at Camp Bastion learned of further injuries following mortar strikes at the district centres in both Musa Qala and Sangin. The task of saving the British mission in Helmand fell to Major Mark Hammond, a Royal Marine Chinook pilot. Flying first to Sangin, he picked up a wounded soldier under heavy machine-gun fire, but despite the work of the medics in the back of the Chinook the soldier died on arriving at

British headquarters. On his next mission – to Musa Qala – the weight of Taliban fire initially forced Hammond back to Camp Bastion. He later learned from the Apache pilots flying escort that two RPGs passed above and below his Chinook, missing either side by ten feet, the closest yet that the British mission in Helmand had come to the loss of a helicopter and potential mission failure.[13]

After a second, successful attempt, to remove the injured soldier from Musa Qala, Stuart Tootal met the flight crew as the Chinook arrived at Camp Bastion. The past few hours had been the most intense of his career. Two men had died with eighteen injured, and but for the heroism of Hammond and his fellow pilots, it might have been far worse.[14] In the base's field hospital Tootal spoke to one of the mine victims, nineteen-year-old Fusilier Andy Barlow, who had had a leg amputated above the knee. He had just come to and managed a joke: 'First time I've been legless on the tour since you banned alcohol!'[15]

Barlow's courage briefly reinvigorated Tootal, although the day was not over yet. Another of the landmine victims, Mark Bright, was receiving his last rites at Camp Bastion's makeshift chapel. His body was laid out in a body bag, open at the top to show the young man's face. One of Bright's best friends, Corporal Lee Parker, stood beside Tootal during the ceremony. As they left the room, Parker stopped to ruffle his dead friend's hair. Tootal would later describe that moment in a book about his time in Afghanistan: 'It was the ultimate act of compassion, love and loss. Witnessing it at the end of that fucking awful day, it very nearly broke me.'[16]

A week later Butler met village elders and Taliban fighters in the desert outside Musa Qala to agree a ceasefire. Butler had come within hours of unilaterally withdrawing from the town before local residents contacted him to broker a meeting. All sides were horrified at the scale of the destruction. Every house within a 200-metre radius of the district centre had been reduced to rubble and craters. The local mosque was on the verge of collapse, and the market was filled with shuttered shops. The meeting took place beneath a military camouflage net to shade the participants, and British soldiers took up positions, out of sight, in case the meeting turned ugly.

A cloud of dust rising into the cloudless sky heralded the approach of the elders and Taliban fighters in a convoy of twelve pickup trucks. The long shirts of the tribal leaders flapped in the wind. The Taliban were clearly recognisable – young fighters dressed in black with stony stares. After climbing out of their trucks, the Taliban stood silently to one side as Butler began negotiations with the elders. Butler suggested that if the ceasefire held for one month, both sides could withdraw and leave the town to its inhabitants.[17] The Taliban, who had only come to Musa Qala to fight the British, agreed.

As an eerie calm descended on the town, the military might have looked forward to a period of respite and introspection, as it came to terms with the painful withdrawals from both Musa Qala and Camp Abu Naji in Iraq. Instead British troops were about to be tested in Basra like never before.

Chapter 26

SINBAD

As Shirreff's planning to salvage Basra gathered pace, the whole basis of the British effort in southern Iraq was called into question by the announcement of Tony Blair's imminent departure from office. Blair had been undone not by the war that had drained the vitality from his premiership, but a fresh miscalculation over foreign policy, this time in Lebanon. On 12 July Hezbollah, the Shia militia in southern Lebanon, kidnapped two Israeli soldiers close to the border. The following day Israel launched raids over the border to recover them and threatened to bomb Lebanon 'for twenty years'. The FCO did not approve of the Israeli approach: they saw its new government under Ehud Olmert as too anxious to prove itself.

Blair, however, was clear that Hezbollah should not be allowed to 'benefit from what they had done'.[1] He was feeling bullish after a recent trip to Baghdad, where the SAS commander Richard Williams had briefed him on his unit's new role fighting alongside American special forces against Sunni insurgents. An early British mission had provided a trove of intelligence that ultimately led to the death of the al-Qa'eda leader in Iraq, Abu Musab al-Zarqawi.[2] Blair was particularly delighted by the campaign badge on Williams's shoulder, depicting a skull and crossbones centred on a Union Jack flag with 'Fuck al-Qa'eda' below. He saw Israel's clash with Hezbollah as an extension of Britain and America's war against terror in the region. As Israel responded to Hezbollah rocket attacks with a massive aerial bombardment, Blair followed the American line, refusing to recognise the response as disproportionate or call on Israel to implement an immediate ceasefire.[3]

But for many in the Labour Party Lebanon was the last straw. 'It felt as bad as Iraq in terms of the passions it was stirring,' recalled one insider.[4] In August one hundred Labour MPs had written to Blair protesting his position. By early September the opposition had developed ominous undertones. Sion Simon, a former Blair loyalist, was circulating a letter among MPs calling on Blair to announce his resignation. Tom Watson, a junior minister of defence, promptly resigned, along with seven parliamentary private secretaries. Later that day, after meeting Gordon Brown at Number 10, Blair announced at a pre-scheduled school visit in north London that the next Labour Party conference would be his last.

The announcement alarmed British defence chiefs. Brown, Blair's likely successor, was believed to be deeply sceptical about the military's handling of the wars in Iraq and Afghanistan. With the fighting still escalating in Helmand, the new chief of the defence staff, Jock Stirrup, felt he had little choice but to plan for a handover in Basra by May 2007, the following year. Stirrup had been a prominent sceptic of the mission to Afghanistan, fearing that British forces might get 'caught in the mangle.'[5] His army colleagues in the defence staff dismissed his concerns as typical of those of an airman, and yet it was Stirrup who was left to bring a new steely-eyed determination to the MOD. Previous exit dates had slipped as a result of Iraq's chronic instability. The Iraqis needed to be pushed into taking control and that meant being prepared to leave Basra no matter what condition the British finally left the city in.

Not everyone was pleased at the prospect of an imminent withdrawal. Officers like Patrick Marriot, a former commander in Basra and currently in charge of operations at Northwood, raised the obvious question when he learned of the scheme: what happens if Basra is in flames when British troops start to pull out? Soldiers had died for the promise to leave behind a secure and prosperous city. The withdrawal plans also cast General Shirreff's current plan for Basra, Operation Sinbad, in a new light. The fate of Basra was no longer dependent on British success in calming the city, but the reputation of the army was.[6]

On 28 September Major Johnny Bowron launched Sinbad in Mufotiyah district near the naval academy in northern Basra. The addition of the division's reserves had freed Shirreff to deploy two companies-worth of soldiers for the operation, or about 360 men. The Iraqi 10th Division had supplied about two dozen men, a paltry number but the best that could be expected given the militia pressure on the Iraqi army leadership. Reflecting the smaller size of both the British and Iraqi contingents, the scale of the so-called pulse had also shrunk from whole districts to neighbourhoods. Each operation would now be focused on a single police station, with a few quick-impact projects on the side.

It was not the overwhelming show of force that Shirreff had once envisioned, but the operation galvanised his headquarters, and the first results from Mufotiyah were promising. Lieutenant Colonel Jonny Bowron set up a stall in a local school, offering short-contract jobs for cleaning streets and painting public buildings. The headmaster was nervous at first at the arrival of a hundred British soldiers outside his school, but it was not long before he was helping police the long queues that snaked around the school's playground.[7] When residents of other neighbourhoods heard that the British were dispensing hundreds of thousands of dollars, they were soon clamouring for pulses of their own. Basra palace was still being mortared regularly, and British patrols ambushed, but Shirreff was confident he was winning the PR battle.

He was unprepared, therefore, for army head Richard Dannatt to publically doubt the British effort in Iraq. In October Dannatt told the *Daily Mail* that British troops needed to leave Iraq 'sometime soon because our presence exacerbates the security problem . . . Let's face it, the military campaign we fought in 2003 effectively kicked the door in. I don't say that the difficulties we are experiencing round the world are caused by our presence in Iraq, but undoubtedly our presence in Iraq exacerbates them.[8]

Dannatt had been forced into granting the *Mail* the interview following the newspaper's exposé of how injured soldiers were being treated on mixed civilian and military wards at Selly Oak hospital.[9] None other than Shirreff had raised this issue back in July, following a pre-deployment tour of Selly Oak. Over the summer Shirreff's wife Sarah-Jane continued to visit the hospital, sending him a series of emails describing soldiers screaming in agony without nurses available to administer painkillers and then served cold sandwiches for supper.[10] Shirreff forwarded his wife's emails to the chief military surgeon at Selly Oak, Rear Admiral Philip Raffaelli, who insisted soldiers had adequate care, a line reiterated by Dannatt until his interview with the *Mail*.[11]

But it was not Dannatt's declaration that mixed wards were unacceptable that attracted the headlines. The *Mail*'s front page announced, WE must quit iraq says new head of army. Shirreff was dismayed when he read the comments, an own goal which many believed could only bolster the morale of the insurgents. Realising the damage he had done, Dannatt appeared on the BBC's *Today* programme the following morning to explain his remarks. Iraq remained his number-one objective, he insisted, a point he reiterated on the phone to Shirreff.

If Shirreff felt undermined by his military bosses, he was growing equally frustrated with the civilian effort in Basra. So far the PRT had contributed little to Operation Sinbad. Mark Etherington, the PRT leader, was spending most of his time simply handling the logistics of his staff of forty, who rotated out of the country every two months for a two-week break. Basra had become so dangerous British civilians leaving the palace feared they would be ambushed, kidnapped and killed, yet they could do little confined behind the palace walls.

The Southern Iraq Steering Group still met to coordinate reconstruction, but it was ineffective. In addition to Etherington's programmes, there was the Better Basra plan, the military's quick-impact projects and half a dozen other reconstruction efforts from other countries, including the Americans. Basra's council often hosted back-to-back meetings of different aid organisations all seeking to improve its governance. The military was so frustrated with the civilian effort that it was soon pushing to incorporate the PRT into divisional headquarters.

Etherington resisted what he saw as a military takeover of the reconstruction effort, but the debate was rendered meaningless when the new foreign secretary, Margaret Beckett, ordered the PRT's evacuation to Kuwait for security reasons. An American contractor had been killed at the palace recently, and a British contractor's vehicle ambushed a few weeks later. Beckett considered the risks too grave. Etherington was not consulted about the decision. How the hell are we meant to operate from a five-star hotel in Kuwait? he wondered.

Shirreff reacted to the decision like it was a personal betrayal. The issue of civilian safety had often grated on the military. Whereas soldiers lived in tents and patrolled the streets daily, diplomats could only be housed in reinforced-concrete buildings, usually air-conditioned. If a single shot was fired they stayed inside. That was no way to run a reconstruction effort in a war zone, and the reason, in Shirreff's opinion, the military should be leading the task. Seeing Etherington's team struggling towards the palace's helicopter landing pad in full body armour, wheelie bags in tow, only reinforced the impression of a civilian desertion. Although no one said it, everyone suspected that the diplomats and civilians weren't coming back, which lent an air of desperation to proceedings as offices were rifled for equipment, papers shredded, and personal items piled in heaps to be trashed.

Shirreff was undeterred by their departure. For Sinbad to succeed he would have to up the stakes further and confront the Jaish al-Mahdi in their stronghold: the Jamiat police station. Since the hostage crisis the previous year, Captain Jaffar had continued to run the Serious Crimes Unit from the station, which remained a powerful symbol of all that had gone wrong in Basra. Shirreff hoped that if Jaffar and his men could be driven out, and the building destroyed, other militia groups would see the writing on the wall and residents rallying behind Operation Sinbad.

Furthermore, by early December, the flaws of the operation were becoming apparent to commanders on the ground. Lieutenant Colonel Justin Maciejewski, commander of the 1st Royal Green Jackets, had taken over from Jonny Bowron in October. He found that residents welcomed the quick-impact projects, but that once his forces moved on there were not enough local security forces with the drive or inclination to keep up the effort.

The Jaish al-Mahdi had also started to get the measure of Sinbad and were attacking with renewed intensity, with mortars and rockets raining down on British bases in the city. The limited signs of progress drew criticism. US Corps commander Peter Chiarelli wrote privately to Shirreff wanting to know why insurgent attacks were up while the British Army 'waltzed' around Basra spending US taxpayer's money on reconstruction projects. He reminded Shirreff that if British forces tried a withdrawal under fire, it was going to look like the Jaish al-Mahdi had 'once again' bombed them out of their bases, a reference to the sacking of Camp Abu Naji. The disastrous withdrawal from Maysan still smarted.

Chapter 27

A MARKED MAN

HAIDER'S PHONE BUZZED WITH A TEXT MESSAGE. The note from an undisclosed sender read, 'We know you are working for the occupation and we are watching you. You will be killed and thrown in the garbage.' Haider should have been celebrating. A day earlier, his first son had been born. He had only just brought Nora back from the hospital with the baby when the text arrived.

His first instinct was to drop the phone in Basra creek; that vile message had sullied his family home. But upon leaving the apartment the crisp winter air restored clarity. This threat was no different to the one he had received outside Nora's uncle Ali's house a few years before when he had been beaten up. It was the same Shia thugs. He should stand firm, and the trouble would pass.

Yet fear gnawed at him. The last year had been a mix of high and low points for Haider. He and Nora had been married twelve months earlier, in December 2005. The wedding went off without a hitch. As was the trend among Iraqi women, Nora had powdered her face a ghostly white, drew dark arches over her eyebrows and painted her lips in a bright ruby red. Haider had also gone to extremes, choosing an oversized suit with shoulder pads that were meant to reinforce his delicate frame. Instead, he looked young and nervous in his wedding photos. Throughout the ceremony, he stood stiffly beside Nora as though anticipating that at any moment his happiness would be snatched away.

The reception was held at Nora's family's home as Haider's apartment was too cramped. The newlyweds greeted the hundred or so guests in the garden beneath a decorated palm tree, receiving presents and congratulations. Nora remained stony-faced, eyes downcast, as any sign of happiness would be considered an insult to the family she was leaving. Haider, however, finally allowed himself a moment of elation. After so many years of waiting, banished from the university and confined under house arrest, he had a well-paid job, a new wife, and a future. He had come remarkably far in just two and a half years, all thanks to the British and American invasion.

Yet notably absent from the ceremony had been any of his Western friends or colleagues. Haider had kept his position with the foreigners a secret from Nora's family for fear they would disapprove; and for all the encouragement he received from the police mentors he worked with, none would have considered attending given the security situation. British civilians were already restrained on where they could travel around Iraq, and needed a full military escort for most jobs.

Haider's anxiety had only intensified since, as more bad news arrived. Several months after the wedding, William Kearney, the ArmorGroup contractor and Haider's closest British friend, took Haider to a shooting range at Basra airport and handed him a hand pistol. Kearney had grown increasingly disillusioned with efforts to train Iraqi security forces and blamed the British and American attitude towards the locals, either treating them like bungling children or as dangerous militia sympathisers, and generally failing to understand local culture. Kearney had come to the conclusion that the British occupation was in terminal decline and he had started to make preparations to leave the country. Already, the palace was partially evacuated and ArmorGroup forced to relocate to the airport.

'I'm going to be returning to England soon and I need to know you can take care of your family,' he told Haider.

The news hit Haider like a missile. Nora was already pregnant, and the costs of maintaining a household were demanding. Security had steadily worsened in Basra, and a number of Iraqi interpreters working for the British had already been killed in brazen attacks intended to drive off the Iraqis who still worked with the British. Kearney's commanding presence in the office had always given Haider a sense of confidence. Over the years, some of the Wirral man's swagger had rubbed off on him. He had recently shaved off his moustache, and taken to wearing luridly patterned shirts from Turkey. Now Haider looked crestfallen.

At the shooting range, Kearney tried to make a show of presenting

Haider with his gun, a Glock hand pistol that was standard issue for Iraqi police officers. He demonstrated the safety catch and how to load and clear the barrel, and they spent half an hour shooting, Kearney yelling out advice like a drillmaster.

'Keep it with you at all times,' said Kearney when they had finished. He did not have to add why. 'Listen Haider, mate,' Kearney told him, giving him a brief hug. 'If you ever need anything just give me a call.'

Shortly thereafter, Kearney left for a job in ArmorGroup's London office. The number of police mentors was also being scaled back from over three hundred to less than fifty following the withdrawal from Muthanna and Maysan provinces. Kearney's successor considered Haider's twenty-minute commute from Basra to ArmorGroup's new offices at the airport too dangerous. Haider was employed instead as a roving fixer, although he wondered how long that job would last, and whether the militia would get to him first.

Chapter 28

LAMBO

BY THE AUTUMN OF 2006 THE PROSPECT OF Coalition defeat in Iraq loomed. The US commander George Casey was nearing exhaustion. Operation Forward Together had failed to arrest Baghdad's slide into sectarian bloodshed. Iraqi security forces had once again proved unable or unwilling to protect the neighbourhoods cleared of insurgents by American troops, who after each operation retired to their big bases and rapidly lost momentum and the confidence of the locals. Casey had launched a second, even bigger attempt at joint US–Iraq operations in September, but the same flaws resurfaced: the Iraqi police mutinied, afraid to side with the Americans against the insurgents.

'There was meant to be a tipping point, when the Iraqis would stand up. Casey knew security had to be led by the Iraqis, but he didn't know how to make them do it,' said Rob Fry, Casey's deputy in Baghdad at the time.[1]

In Washington DC there was a deepening sense of gloom in the Republican-led White House. The Democratic Party had swept all before it in the mid-term Congressional elections, and the anti-war movement had coalesced around the mother of a deceased US soldier, as it had in the UK with Reg Keys. Even President Bush, so steadfast and myopic in his insistence that Iraq was improving, was beginning to waver. He had commissioned a report from the Iraq Study Group, a high-profile non-partisan group of congressmen and elderly statesmen, a move widely seen as reining in the neoconservative agenda in favour of a more pragmatic approach to the war. In its final report to Congress in late 2006, the group boldly acknowledged the failures of the US approach in Iraq, and called for the phased withdrawal of troops and immediate engagement with Iraq's hostile neighbours Iran and Syria.

With Bush and Blair now forced to confront the realities, the moment was ripe for a dramatic change in strategy. Rising US military star David Petraeus had recently published a new counter-insurgency doctrine, which emphasised getting US forces out of their compounds and engaged in protecting the local population. While promoting the need for a change, Petraeus challenged the sense of defeat in the Iraq Study Group's report, and along with retired four-star general Jack Keane began making the rounds of Washington with a new idea. Rather than cut and run, he urged the US to increase its troops, confident that a final surge in numbers could topple the insurgents and win round the majority of Iraqis who wanted peace.

Under pressure from Congress to do the exact opposite, Bush took the last gamble of his presidency. The President promptly relieved Casey of his command and replaced him with Petraeus. At the end of January 2007 Petraeus arrived in Baghdad with the promise of 30,000 additional men and a tight deadline – to deliver results within six months. The troops would be deployed exclusively in Baghdad and the Sunni areas of the country, with the British expected to 'hold the line' in the south until al-Qa'eda and other extremist groups could be dealt with.

The US 'surge' coincided fortuitously with a change in mood among Iraq's Sunni tribes towards al-Qa'eda and its legions of foreign fighters from Jordan, Syria and Saudi Arabia. Exhausted by the brutal and macabre Islamic code imposed by al-Qa'eda, and fearful of the rising power of the country's Shia population, the Sunni tribes were increasingly receptive to the Americans, especially those offering cash and weapons. Following the example of Colonel Herbert McMaster in Talafar, who had set up a local militia to fight the insurgents, US commanders had begun funding tribes to turn against al-Qa'eda with remarkable success. The Americans nick-named these bands of fighters *Sahwah* – Awakening – councils.

As the Awakening gathered pace in western Iraq, Petraeus faced the challenge of translating the concept from the tribal lands, where clan bonds were strong and the military could easily identify leaders, to the sprawl and fragmentation of urban Baghdad. What Petraeus needed was a key to unlock the Sunni insurgency in Baghdad; what he got was the wily old British general Graeme Lamb.

Lamb had returned to Iraq in September 2006 as the Coalition's deputy commander, taking over from Rob Fry. He was only too happy to leave the Land Warfare Centre and the doctrine-rewriting process, which had got mired in departmental infighting. He had immediately grasped Petraeus's emphasis on protecting civilians over fighting insurgents, but his mind then raced ahead. It was all very well protecting the Iraqis, but that would not be enough to stop the insurgency.[2] The only way out of Iraq was to talk to the insurgents' leaders, even those with blood on their hands.

In August, before their deployment, Petraeus visited Lamb at his quarters, a farmhouse on Salisbury Plain, to discuss the dangerous path the British general had conceived. Petraeus liked the idea of reconciliation with insurgent groups but couldn't accept negotiating with those who had killed American soldiers. Lamb disagreed: the more extreme the insurgent the better, because then the impact would be all the greater when they were turned. 'We were always talking about working with and bolstering the moderates,' said Lamb. 'That's bullshit. In an insurgency it's the extremists who control the political agenda.'

Petraeus was cautious about Lamb's approach, and the British general left for Baghdad knowing he faced a struggle to convince the Americans. Upon arrival even Lamb was shocked by how desperate the situation had become in the city. The civil war – although few on the American side would admit to calling it such – was claiming 200 Iraqi lives a week, with Iraqi security forces either absent or actively participating in the genocide as members of Shia death squads. A city that had once housed large mixed neighbourhoods of Sunni and Shia had divided along sectarian lines, with concrete blast walls now separating the warring communities and armed militias patrolling the streets. The Green Zone was rocketed day and night.

Arriving at Maude House, the British military residence named after the general who had conquered Baghdad in 1917, Lamb cautiously introduced his plan to the outgoing American general, George Casey. The British military's reputation had taken a battering after the botched withdrawal from Camp Abu Naji, and Casey was unreceptive. Gregarious by nature, Lamb soon turned Maude House, a former Ba'athist mansion, into an evening salon for American generals. They would drink whisky in easy chairs beneath the room's grand chandelier late into the evening.

'Americans work until midnight and get up at 5 a.m. They eat their meals at their desks. Lambo was brilliant at getting them to sit together and talk about the big picture,' said Emma Sky, the former governorate co-coordinator in Kirkuk in 2003 who was known for having boldly told the US commander in the region, General Ray Odierno, that the Coalition's approach to Iraqis was all wrong.[3] Odierno had since returned to Iraq as corps commander, the man responsible for the day-to-day running of the Coalition forces in Iraq. The hulking general had come to represent the best and worst of conventional military thinking, which called for the use of overwhelming force to defeat the enemy. Tom Ricks's influential account of the US Army's failings in Iraq, *Fiasco*, unfairly singled out Odierno for blame, Sky believed. 'He comes across as a ball-busting general, but he can be unusually perceptive,' she noted. After Iraq, Sky had gone on a camper-van holiday in New Zealand, before working in Afghanistan as a political adviser to the NATO mission. She had just returned home when Odierno called and asked her to be his political military adviser. Sky was another Briton about to play a key role in transforming the Americans' approach to Iraq.

Odierno soon became a regular at Maude House, bringing the latest on the surge strategy ahead of Petraeus's arrival. He was joined by Martin Dempsey, in charge of training Iraqi forces, and Stanley McChrystal, the wiry US special forces commander. Improved intelligence from Sunni communities had enabled daily strikes against al-Qa'eda leaders up the Euphrates and Tigris valleys. UK special forces were fully integrated into the American operation, and had gone some way to assuaging US doubts about the readiness of the British military to fight after the Camp Abu Naji debacle. With the success of the Awakening councils, Lamb felt that the Coalition had developed a sense of momentum that could transform Baghdad if they could reach out to enough top-level insurgents. But American unease with the concept persisted. During one discussion Odierno declared, 'I will not talk to folks with blood on their hands.'[4]

Lamb kept pushing the Americans to stop seeing the enemy in black-and-white terms: 'It's all about the context. In a certain environment a man takes up his gun and kills an American solider. Change the environment and he's an upstanding citizen. We need to change the perception of how we see them, and they see us.' Lamb accepted it was easy for him to advocate talking to insurgents when they weren't killing British soldiers, and the UK effort in the south didn't help make his case. There had been limited efforts to engage with the Jaish al-Mahdi, which weren't helped by the Iraqi government's hostile approach to British meddling in Shia affairs.

It was going to take a powerful set of arguments to change Odierno's mind. Lamb needed a dedicated team and a degree of secrecy to build his case. The same month that Casey was relieved of duty, Lamb called up James Simonds, an able commander who had served under him when he was commander of the 5th Airborne Brigade and director general of special forces. Simonds arrived in Baghdad just after Christmas, and Lamb gave him his instructions over dinner at Maude House. 'I need you to find the humans that matter in this country. Work out who they are, how we talk to them,' said Lamb. 'Oh, and convince the Americans that this is a good idea,' he added.

Simonds' first port of call was the CIA station chief. He was shocked to hear the American's assessment, which lumped together the disparate insurgents under the heading al-Qa'eda rather than recognising the complicated jigsaw of competing ideologies and histories of the many small groups. Simonds began drawing up a list of all known insurgent groups and tribal affiliations, ranging in scale from the Jaish al-Mahdi and Ansar al-Sunni to small militias limited to a single town or local leader. The Americans had the information but rarely viewed the human terrain holistically. Immediately apparent was the position of Abu Ghraib as a fulcrum of insurgent and tribal groups. The most westerly suburb of Baghdad, which housed the notorious prison, had been built in the 1980s for Saddam's Sunni officer class. The district had retained its tribal linkages to Fallujah and Ramadi, unlike other parts of the city. If the Awakening concept was to be used to transform the Sunni communities of Baghdad, Abu Ghraib represented the key to the city.

'So who pulls the strings in Abu Ghraib?' Simonds asked his CIA counterpart. The answer was Abu Azzam, a forty-five-year-old former officer in the Republican Guard and a prominent commander in the Jaish al-Islami, viewed by the Americans as the conduit between the old Ba'athist regime and al-Qa'eda in Iraq. McChrystal's special forces had hit Jaish al-Islami's leadership particularly hard. Abu Azzam had made peace overtures to the US embassy in June, shortly after the first Awakening councils had begun in Ramadi. Simonds asked the CIA agent what had happened about Abu Azzam.

'Well, the military have been trying to kill him, so we haven't been paying too much attention to talking to him,' the agent responded.[5]

'Can you set up a meeting with him?' Simonds asked.

'Maybe,' came the reply. Simonds had found the man to test whether reconciliation with insurgent leaders would work.

A few weeks later Simonds, Lamb and a small special forces detachment left the Green Zone on one of the more daring and controversial missions of the campaign. Lamb insisted on seeing Abu Azzam face to face; he was not about to stake his reputation on a man he hadn't met. Abu Azzam refused to meet Lamb inside the Green Zone, understandably fearing arrest. That meant Lamb would be Abu Azzam's guest in the heart of Abu Ghraib, a meeting no less dangerous for Abu Azzam, given his al-Qa'eda associates would view the assignation as an act of betrayal.

Lamb arrived at Abu Azzam's house in the early hours of the morning to find the insurgent nonchalant, unshaven and dressed in an old tracksuit open at the chest to reveal thick gold chains around his neck. 'He looked like a fading '70s film star,' noted one of the Brits on the mission.[6] Abu Azzam also struck Lamb as a showman, but he had seen enough of Iraqi politics to know how to gauge the steel of a man behind a facade. Lamb was content to go through the ceremony of welcome and tea drinking in a lavish reception room, leaving it to Abu Azzam, the host, to broach the reason for their meeting.

'We've watched you for almost five years now,' Abu Azzam began. 'At first we thought you had come to destroy our culture and people and we fought against you. But now we've seen where the truth lies. The Americans don't threaten our way of life; al-Qa'eda does. We were mistaken.' That's a lot of lives lost through a mistake, Lamb thought. He suspected that what drove Abu Azzam was not the best interests of his people but rather his recognition that the Shia death squads and American strikes against the Jaish al-Islami leadership had backed him into a corner. But Lamb was happy to be magnanimous. 'Turn against al-Qa'eda and we can help you,' he said. There was not much time to say more, and Lamb left the meeting with Abu Azzam's agreement to meet a second time at Maude House. He had seen enough of Abu Azzam to believe that reconciliation could be real and effective – and to take it to the Americans.

By February Simonds had prepared a PowerPoint slide mapping out the 'strategic human landscape'. Lamb invited to Maude House Petraeus, Odierno, McChrystal and the chief of strategic operations for Multi-National Force – Iraq, Major General David Fastabend, along with their aides. Simonds presented before supper. In the centre of the slide were written the words 'Future of Iraq', with friendly and neutral groups from the Awakening councils inside a first circle, and further out irreconcilable groups like Jaish al-Islami and Jaish al-Mahdi. No reference was made to the recent meeting with Abu Azzam.

'This is what we're up against. Now we need to talk to them,' said Simonds. He made the case for reaching out to the extremist groups. Special forces raids had decimated their leaderships, and now was the time to present them with a stark offer: 'Reconcile or die.'

The reaction from Odierno was instant: he pushed away his coffee cup, shaking his head.[7] Just the week before, on 20 January, five US soldiers had been killed in Karbala. Shia insurgents disguised as security guards had entered the local provincial headquarters, killed one soldier and abducted four others. They were later found dead with bullets in the backs of their heads. The incident had evoked the horrors of the Blackwater guards strung up under the bridge in Fallujah.

Simonds looked around the room. Only Petraeus and McChrystal seemed unmoved. Simonds could not decide whether they had bought into the idea or not. It was not exactly an endorsement, but Lamb decided the only way to win round the Americans was to show signs of success.

For his second meeting with Abu Azzam in Maude House, Lamb decided to raise the stakes. The two men had just sat down at a table when Lamb took out his knife and placed it in front of him. 'You and I are both bad men,' he told Abu Azzam, who looked unfazed and rocked back on his chair. Then the power cut out and the room was plunged into darkness. There was the sound of screeching chairlegs, then the lights came back on. A startled Abu Azzam had pushed his chair back against the wall, as far as possible from Lamb's knife. The power cut was a coincidence but had worked well in unnerving Abu Azzam. 'Let's keep talking,' said Lamb with a smile.

The mixture of threat and conviviality worked perfectly. By their third meeting – in March – Abu Azzam had agreed to a deal. If he could maintain a ceasefire for fourteen days, Lamb would arrange for all 1,738 members of Abu Azzam's group to join the Iraqi security forces. The deal was a reminder to Lamb of just how low the threshold could be between implacable insurgent and bulwark of society.

'At the end of the day he just wanted to be an Arab sheikh taking care of his own,' said Lamb.

Over the course of Lamb's meetings with Abu Azzam, Emma Sky had also been chipping away at Odierno. She saw how Petraeus's doctrine was revolutionising military tactics but failing to address American perceptions of Iraqis. Since arriving in the country, the petite and bird-like Sky had waged a small battle to ban the use of the umbrella phrase 'anti-Iraqi forces' adopted during General Ricardo Sanchez's period as US commander in the country immediately after the invasion. 'It's too easy,' she would say to Odierno. 'You've got to ask who they are and what they want.'

Colleagues who observed the relationship between the two sometimes likened them to Laurel and Hardy. Odierno was measured, patient, bemused; Sky was sharp, edgy and badgering. But by the end of February her persistence in reframing the debate had paid off. Simonds gave a second PowerPoint briefing reiterating the need to talk to the insurgents. He used the example of Northern Ireland, where the British army had worn down the Provisional IRA and created a stalemate. This was only broken when the British agreed to begin talking *before* the IRA laid down their arms. This time Odierno said, 'OK, I get it. Now what do we do?'[8]

The next step was to bring the Iraqi government on board, including convincing them to accept Abu Azzam's offer to lay down his arms in return for jobs. If the American brass had proved sceptical at the prospect of dealing with insurgents, Iraqi Prime Minister Maliki was even more so. Although the Shia leader had approved the original Awakening council in Ramadi, he had become alarmed by the spread of the American-backed Sunni groups across Anbar. Maliki considered the Awakening councils a sham – Sunni insurgents who had temporarily removed their al-Qa'eda badges in exchange for money. He was deeply suspicious of American motives in assisting such groups. 'He thought the US Army was trying to topple him, or at least keep Iraq weak by making the insurgents strong,' noted Maliki's national security adviser Mowaffak al-Rubaie.

Differences of opinion were exacerbated by the fact that both the Iraqis and the Americans had their own intelligence networks, and these were reporting contrasting pictures of the impact of the Awakenings, recalled Sky. 'The Americans were hearing unremitting good news about the

Awakenings. Maliki's intelligence folks were telling him they were former Ba'athists looking to exploit the situation.'

The lack of trust on both sides was exposed in early March by an unusually effective British strike operation. UK special forces succeeded in arresting two of the most notorious Jaish al-Mahdi leaders in the country, Laith and Qais al-Khazali. Files captured on their laptops left little doubt that they had been involved in the recent kidnap and murder of the American servicemen in Karbala, along with a spate of other attacks. Also captured with the Khazali brothers was a man claiming to be a deaf mute who under careful interrogation revealed himself to be a Lebanese Hezbollah member called Ali Moussa Daqduq. Daqduq admitted to training Iraqi militants at camps in southern Iran, the most compelling proof yet of Iranian involvement in southern Iraq's troubles.

The Americans were left in little doubt of the Khazali brothers' links to the Iraqi prime minister's office when, in the early hours of the morning, an irate Maliki called Petraeus to insist on the Khazalis' immediate release. Petraeus maintained his cool, remembered Pete Mansoor, his chief of staff. 'Petraeus understood why Maliki felt he needed Jaish al-Mahdi as a bulwark against Sunni extremists, although he wondered whether Maliki knew exactly what his aggressive associates were up to,' said Mansoor.[9]

Petraeus arranged for Maliki to see some of the captured Khazali material, which included detailed lists of money and weapons supplied by the Iranian Al-Quds Force and the military IDs of the soldiers murdered in Karbala. 'I think that was the first time Maliki realised the depth of our feelings about the Khazalis and the scale of the involvement of Iran in fomenting attacks inside Iraq,' said Petraeus. 'We told Maliki now was the time to take action against the Jaish al-Mahdi to demonstrate his authority.'[10]

The American arguments came at a delicate time in Iraqi politics, with the prime minister feeling caught between their demands and those of the Shia groups that elected him. His own party, Dawa, and its allies largely comprised exiles who had lived in Iran. They had worked with the Americans since the start of the occupation but had steadily lost popularity to the violent anti-Western cleric Moqtada al-Sadr, who had led a revolt against the occupation of Iraq in 2004. Sadr had since withdrawn to pursue religious studies in Iran, but his followers formed an important political party in the parliament, where they used the political process the

Americans had put in place to work against them. In early April the five Sadrist members of Maliki's cabinet resigned over the prime minister's failure to set a timetable for US withdrawal.

The government survived, and the Sadrist withdrawal had the unintended consequence of showing Maliki he could live without them. Maliki's new independence from the Sadrists gave him more room to work with the Americans although dealing with Sunni insurgents like Abu Azzam would prove highly unpopular with his power base.

The deadlock was broken by one of the more extraordinary relationships to emerge during the occupation. Shortly after arriving in Iraq, Sky had noticed that the Maliki inner circle included a female adviser, which was unusual for Shia political parties. Basima al-Jadiri, the prime minister's chief military adviser, wore a multicoloured *hijab*, and was outspoken and unfriendly. Intrigued by the striking woman, Sky asked Odierno's staff who she was, and was promptly handed a file which stated that Jadiri was on a CIA watch list of the most dangerous sectarians in the country. Sky was undeterred – 'Just imagine what my file looks like,' she joked – and arranged a meeting with Jadiri in Maliki's outer offices.

Sky was not surprised to discover Jadiri bore no resemblance to her portrayal in the US file. Over the course of several meetings she learned that Jadiri was a former rocket scientist from Sadr City who had been put forward by the Sadrists in the 2005 elections to fulfill the 25 per cent female quota in parliament. She had insisted on joining the parliament's first security committee, where she met Maliki and impressed him with her passion and diligence. 'Basima was one of those striking Arab women who, contrary to all stereotypes, are dynamic, forceful and able to terrify a roomful of men,' said Sky. Jadiri was fervently anti-American and convinced they wanted to destroy Iraq. She could not believe that the Americans had invaded Iraq without a plan, and assumed they were deliberately encouraging sectarian animosities to keep the country weak.

'There was no plan,' Sky told her. 'At least there was one, but it got thrown in the bin.'

'If the Americans can put a man on the moon then they should be able to fix Iraq,' insisted Jadiri.[11]

Sky recognised the belief in American omnipotence; she had shared it before her first tour of Iraq. She drew on her experience as a governorate co-coordinator in Kirkuk to convince Jadiri that the Americans were guilty of incompetence not vindictiveness. 'The Americans really did fuck up,' said Sky, 'but they're trying to make it right now.'

The two women shared a moment of surprise that so much distrust could have revolved around an elementary misconception. From then on they were in regular contact ahead of the weekly security briefings between Maliki and Petraeus. They briefed each side beforehand on what the other was up to, a policy that quickly delivered results. In late May Jadiri persuaded Maliki to incorporate Abu Azzam's fighters into the government forces. The decision came at a crucial moment in the surge, as the increased numbers of US troops on the streets inevitably led to more American deaths. Meanwhile, the ability of the British to deliver in Baghdad mattered as never before because Anglo-American relations were about to be put under even more strain.

Chapter 29

PALERMO

AFTER SIX MONTHS SHIRREFF'S TOUR WAS coming to an end. Blowing up the Jamiat had provided an emotionally satisfying climax to his command, but it could not hide the growing cracks in the British effort which his attempt to take on the militia had exposed. Operation Sinbad had fulfilled the military's desire to do something – anything – in Basra to turn the situation around, but under-resourced and without real support from the Iraqis, it had succeeded only in stirring up the militia.[1] With the violence in Basra mounting, Consul General Rosalind Marsden remained isolated and under fire at the palace, the provincial reconstruction team had relocated to Kuwait, and the local provincial council was once again refusing to talk to the British. Shirreff addressed the security committee one final time. 'The only way you will ever have progress in Basra is if you tackle the militia,' he urged them, but he didn't imagine a turnaround anytime soon.

A few days later, as Shirreff prepared to leave, he penned a valedictory letter to Blair lacerating the efforts of the FCO and DFID and calling for a single military authority to take control of the situation. Both departments soon heard about it. Martin Dinham, head of the DFID's Iraq desk at the time, was hauled before Simon Macdonald, Blair's new foreign policy adviser, to make a point-by-point rebuttal.[2] But the ranks soon closed against Shirreff. 'No one really wanted to tackle Iraq at the end of the day. They just wanted to move on and forget about it,' he reflected. Shirreff was about to take up command of the Allied Rapid Reaction Corps, although conspicuously he was one of only two commanders in Basra not to receive a medal.[3]

The inconsistency of the British campaign in southern Iraq, brought about by each new general seeking to impose his own view of the situation, was never clearer than at the handover between Shirreff and the incoming Major General Jonathan Shaw. Where Shirreff was a bullish cavalry officer who had pushed the military into taking a more aggressive approach to insurgents, Shaw, thin-lipped and balding, had the clipped air of a City lawyer and was convinced that politics rather than force was the answer to Basra's problems.

By the time Shaw took over the city in early January 2007, Basra palace had become the most attacked location in Iraq, with a dozen rockets and mortar rounds falling a day. The rockets rarely killed anyone, but they drained morale and restricted the defenders to narrow corridors of sand-filled barricades linking barracks and offices. The remaining civilians moved out of their trailers to sleep on mattresses in the corridors of the offices, where the concrete roofs offered greater protection. The situation wasn't much better at the other three remaining bases in the city, cut off from each other by roadside bombs and the threat of ambush.

Shaw and his commanders were confronted with a dilemma: they could continue with Sinbad and risk the city erupting further, or they could pull back their forces to the British military headquarters at the airport on Basra's outskirts and risk leaving a power vacuum at the centre. In the end Shaw opted to stick to the high tempo of operations and the narrative that the British military had conceived – that Sinbad was laying the groundwork for a British exit.[4]

Shaw hoped that a staggered withdrawal from the bases, starting with the smallest outpost and culminating with the palace, would steadily erode support for the militia. Unlike in Baghdad, where the Americans and the Iraqi government were locked in a battle for survival with Sunni extremists, there was no threat to Shia hegemony in the south other than Britain's presence. British troops were the recipients of 90 per cent of the attacks in the south. Without a British enemy to unite them, Shaw suspected some of the armed groups would revert to petty gang activity. Others would seize the opportunity to represent Basra's poor and stand in elections. The British needed to step back from the fighting and try to reconcile the militia groups with the provincial and national governments.

Shaw approached his task with an intellectual rigour befitting an officer who had read politics, philosophy and economics at Trinity College, Oxford.

He set his headquarters to work unpicking Basra's patchwork of tribal and criminal connections. Before deploying, Shaw had set his staff an extensive reading list which included the reflections of former SIS agent Mark Allen on Arab culture. Allen drew the conclusion favoured by romantics: that the Arab mindset is antipathetic to ideas of statehood, preferring tribe and religion as units of identity. Hence there is often an official state and the 'unofficial state of clanship and faith'.[5]

Shaw was struck by the idea of the two different forms of power, which seemed to offer a handle for understanding the relationship between Governor Waeli and the militias. He was surprised to discover that the local SIS station chief, James Proctor, a former army officer, had no good interlocutors among the insurgents. 'Find me some leads,' Shaw told him. As with the US Army, the idea of talking to insurgents responsible for killing British soldiers would no doubt prove controversial, but Shaw saw little option if he was to broker a peace deal in Basra.

Shaw needed to have his arguments in place before making the case to the Americans that Basra was ready for handover. In November he had flown out to Fort Hood, Texas to see General Ray Odierno, about to take charge of day-to-day operations in Iraq. At that stage withdrawing Coalition troops to big bases was still part of the transition plan envisioned by the US commander George Casey. The British plan clearly sat within that strategy, although Shaw still felt the need to explain that UK forces were leaving Iraq in order to bolster the effort in Afghanistan.[6]

But by the time of Shaw's deployment in January 2007 Casey's replacement by General David Petraeus had been announced and the transition plan jettisoned in favour of a troop surge. For the first time since the invasion, the UK's desire to withdraw – to 'reposture' as the British military headquarters in Northwood put it – would be fundamentally at odds with US policy. There was now a danger that the Americans would 'think we were abandoning ship', warned Shaw's chief of staff, Colonel Ian Thomas.

The issue came to a head in early January within a few days of Shaw's arrival in Basra. The new US defense secretary Robert Gates was making his first tour of the country with the outgoing General Casey. They arrived in Basra on 15 January looking impatient. Gates had met Tony Blair on his way to Iraq. Blair had been due to brief the defense secretary on the British plans for a pull-out from the city by early May, but much to the chagrin of the Cabinet Office had only skirted around the issue. 'Blair was keenly aware that the decision to withdraw unilaterally would jeopardise Britain's close relationship with the US, which he wanted to be regarded as part of his legacy,' said a Blair associate.[7]

'He bottled it,' one of Shaw's staff noted.[8]

Shaw laid out the UK position to the Americans before a full house of British officials. He told them that Operation Sinbad had been a success in putting the Jaish al-Mahdi on the back foot and securing areas for quick-impact projects. Clearly Shaw, in an awkward position, was trying to sugar-coat the situation, and neither Gates nor Casey bought his claims. Shaw might have insights into the insurgency in Basra, which given time would produce results, but there was no disguising the fact that Britain was trying to get out of Iraq quick. Shaw's grandiloquent manner further riled the Americans. After the meeting Gates told Lou Bono, the top US official in Basra, that he felt the British had been trying to sell him damaged goods dressed up with fine words. 'Those who are in the know understand what's really going on down here,' said Gates.[9]

A few days later Shaw learned what the Americans really thought. The current UK plan was to hand over all bases in the city to the Iraqis in May. However, the US headquarters in Baghdad informed Shaw that they considered the Iraqi forces the British had trained to be so infiltrated by the militia that they might as well 'hoist a Jaish al-Mahdi' flag over the city if they handed security over to them. If the British persisted, the US would have no choice but to send its own troops to Basra, clearly an outcome neither side wanted.[10] The Americans were focused on battling al-Qa'eda, while the British simply wanted to withdraw on their own terms. When Defence Minister Des Browne visited Basra he was irate to discover that the American threat to send troops was putting pressure on the handover timetable he had agreed to before Christmas. Shaw did his best to soothe the agitated minister while privately wondering why his headquarters was caught in the middle of the fraying relationship between London and Washington.

Eventually the Americans confronted the British. In February the incoming Petraeus contacted Blair directly and asked him to delay handing over the palace until after the summer. It was a difficult decision for the prime minister, who had hoped to conclude his premiership in a few months' time with the handover. There were no direct benefits for British forces if the surge succeeded, and every chance that, as with earlier

American efforts, the operation might make matters worse in southern Iraq. But Blair was equally sensitive to American criticism of how Basra was being handled. There was also growing incredulity about the British deal in Musa Qala, Helmand. General Dan McNeill, the newly appointed head of NATO forces in Afghanistan, thought Butler had surrendered the place to the Taliban. He was proved correct when in early February Taliban fighters seized the district centre and took control of the town. After the bloody fighting the previous year, the British military had no intention of battling to take it back.[11]

Blair agreed to delay the palace handover by several months, a promise that would only last while he was in office.[12] The decision was unpopular among some commanders. If Patrick Marriott, chief planner at British military headquarters, had objected to withdrawing too quickly from Basra, then James Bashall, the incoming brigade commander in Basra, considered changing the departure date at the behest of the Americans equally questionable.[13] His soldiers would be risking their lives for a few extra months with the underlying reality of an imminent British withdrawal unchanged.

In mid-February Shaw flew to Baghdad to present the amended plan for the British withdrawal from Basra to Petraeus and other US commanders. Operation Sinbad had officially ended on 18 February 2007, the same day Tony Blair stated in a televised address that in the coming months 3,000 of the more than 7,000 remaining British troops would be leaving Iraq. The contrast between the British withdrawal and US surge could not have been starker.

Anticipating a frosty reception from the Americans, Shaw had prepared a paper setting out the intellectual framework for British thinking, entitled 'Palermo not Beirut'. Shaw argued that there was no insurgency to speak of in the south. The armed groups in Basra were more like the Mafia gangs found in the Sicilian city of Palermo than the sectarian militias of Beirut. The solution in the south, Shaw told the packed hall in Baghdad, was political and not military. It was a bold argument that captured Shaw's strengths as an orator. He knew he might not win over a room of hardened US commanders, but felt that as long as his case was intellectually sound, he would be able to defend the military's approach and ultimately the reputation of the British army.

Petraeus did not agree with Shaw's characterisation of Basra – given the scale of the violence, the city was more like Palermo and Beirut rolled into one – but from his conversation with Blair, he knew that the UK's patience was wearing thin. Britain had proved the most steadfast of America's allies in Iraq, but Petraeus had to accept that political considerations in London were overwhelming the UK's ability to continue its operation in southern Iraq. 'The main effort is in Baghdad, defeating al-Qa'eda,' he told Shaw. 'If you can secure the south, I'm happy for you to do what you need to do.'[14]

Back in Basra, the British commander in the city, Lieutenant Colonel Justin Maciejewski, began preparations for the handover of the first of the four remaining British bases. Maciejewski's priorities were twofold as the handover date of 20 March loomed for the smallest British outpost, a three-storey building in the centre of the city which housed a single company. His men must maintain an aggressive presence on the streets, but at the same time the Iraqi soldiers who'd be taking over needed training and vetting. The image of Camp Abu Naji being sacked by a mob was on everyone's minds. The troubled Iraqi 10th Division was considered too flaky, so Maciejewski was relying on an Iraqi military police company, recently returned from a successful tour in Baghdad, to guard the building after the British departure. Its commanding officer was an energetic captain 'intensely proud to be reclaiming a little sovereignty', recalled Maciejewski.

Maciejewski's men, along with their sister Staffordshire battle group in Basra, kept up a frenetic pace of operations. The same week that British special forces captured the Jaish al-Mahdi leader Qais al-Khazali, the Staffords raided an office of the Iraqi Ministry of the Interior, disturbingly uncovering a massive cache of weapons and freeing a dozen prisoners who had clearly been abused. It was not an operation designed to ease tensions between the UK and Prime Minister Maliki, who continued to hold a dim view of British efforts. Maliki labelled the raid an 'unlawful and irresponsible act', but Shaw felt it set the right tone. British forces were going to withdraw, but not meekly.

Maciejewski insisted on a ceremonial handover of the outpost. On the day itself Governor Wa'eli showed up for a military parade and feast to celebrate the occasion, and Maciejewski left with Iraqi troops manning the sangars. That night the palace was struck by a two-hour barrage of mortars, but at the outpost all was quiet. Preparations began for the handover of the next British base, located at the Shatt al-Arab Hotel.

The sense of relief did not last long. Two days after the handover Iran took its revenge for Khazali's capture when fifteen British service personnel were taken hostage while patrolling the Shatt al-Arab, the river that flows along the eastern edge of Basra and marks the border between Iraq and Iran a few miles to the south of the city. The sailors and marines were in two rigid-hulled inflatable boats several hundred metres inside the Iraqi side of the river border when they were detained by two larger Iranian vessels.

Few incidents brought into focus more clearly the Anglo-Iranian rivalry in southern Iraq. After a series of allegations and counter-allegations, with Tehran dominating the exchanges, the Britons were released unharmed fifteen days later. Further embarrassment followed when Defence Minister Browne allowed two of the sailors to sell their stories to the *Sun* and ITV's *News at Ten* for £100,000. Browne wanted to use their stories and allegations of mistreatment to continue the propaganda war, but by lifting the long-established ban on servicemen selling their stories, the MOD only reopened the issue of how little serving men and women on the front line earned.

The hostage affair was an unwelcome distraction for Shaw, who had finally found a Jaish al-Mahdi interlocutor with whom he hoped to meet and reach a political settlement. Since Shaw's initial instructions in October James Proctor's SIS team had trawled through their network of contacts without success. In the end all it had taken was one SIS case officer to look in the most glaringly obvious place: the high-security detainment facility at Basra airport, where dozens of high-level Jaish al-Mahdi leaders were held. The officer found his man at once: the key to the British withdrawal from the south, none other than the former leader of the Jaish al-Mahdi in Basra and the man responsible for the deaths of many British soldiers, Ahmed al-Fartosi.

Chapter 30

THE DEAL

FOR ALMOST THREE YEARS Ahmed al-Fartosi had been locked away in a tiny cell at a desolate logistics base ringed with concrete blast barriers and barbed wire, kept under constant surveillance by armed guards. The Jaish al-Mahdi leader was considered so dangerous he was not allowed contact with his fellow prisoners and restricted mostly to solitary confinement. Fartosi's commitment to driving out the foreign infidels had only hardened since that summer evening in 2005 when British special forces had dragged him out of his house in a tracksuit and arrested him. The reprisal attacks had led to two British soldiers from the SAS being held hostage at the Jamiat police station and the unravelling of the occupation, much to Fartosi's delight.

Since 2005 he had watched the prison population at Shaibah logistics base swell, and by 2007 the British-run facility held one hundred detainees considered the south's most hardened insurgents and too violent to be turned over to the Iraqi judicial system. On the rare occasions Fartosi was allowed to mingle with fellow inmates, he was heartened to hear of the continuing loss of British lives and the steady mortaring of their headquarters under the current Jaish al-Mahdi leadership of Wissam Abu Qader.[1] Intermittently the Red Cross visited, but Fartosi had not been tried or sentenced, and had every reason to believe the British would hold him indefinitely. He was both surprised and relieved when British SIS station chief James Proctor arrived and made him an offer. Negotiate a ceasefire with the Jaish al-Mahdi in Basra, Proctor promised, and the British would stay off the streets of the city and steadily release detainees, culminating with Fartosi.

British intelligence had interrogated Fartosi back in 2005, and knew him to be a wily customer, self-possessed and convinced of his cause. But Fartosi looked ill, and had lost almost two stone in prison. Proctor suspected his hatred for the British had been worn down by imprisonment.[2] The thought of dealing with a killer like Fartosi left a bitter taste in Proctor's mouth, but in the cold realities of war it was deemed a win-win situation. Under the newly signed status of forces agreement between the Iraqi government and the Coalition, the British would hand over their prisoners to the local courts anyway following their departure. Given the Iraqi government's close ties to the militias, that was tantamount to letting the prisoners walk free. The British might as well get some advantage before their inevitable release.

Nonetheless, the offer came on the heels of weeks of intense emotional debate between Proctor, British commander Jonathan Shaw and the handful of other officials in the know. With Shirreff's efforts to take on the militias only provoking them, Shaw needed a strategy to calm the region in advance of the British withdrawal. Unpleasant as it was to negotiate with Fartosi, there seemed to be little choice but to open talks with the militia, even if they had British blood on their hands. In Northern Ireland peace had only come when the likes of Jerry Adams and Martin MacGuiness were co-opted into power, albeit over ten years, not a few months. It was a high-risk game. If Shaw couldn't find a route to a peaceful withdrawal, the British army would forever be described as having been driven out of Iraq in defeat.

The pressure to withdraw with honour was intensifying with each month. In late spring Tony Blair had visited Iraq and been openly depressed at the course of the war. Shaw, who did not mention the possible deal with Fartosi, tried to reassure the prime minister, but his briefing was interrupted by two rockets slamming into the airport. 'Blair looked shit-scared,' said one senior officer.[3] The visit was hastily wrapped up.

A few weeks later Blair held his final dinner at Chequers, the prime minister's retreat in Buckinghamshire. He invited a number of celebrities, as well as Richard Williams, the combative SAS commander who had briefed him about his exploits killing Sunni insurgents the previous May in Baghdad. The prime minister was torn between a desire for reassurance about the US surge and his desire to forget the horrors of war. Blair accepted that the aggressive stance favoured by the US and British special forces had been right. The two men ended up drinking a toast to the SAS with a glass of brandy, while Williams brandished one of Oliver Cromwell's swords.[4]

In Basra efforts to crack down on the current Jaish al-Mahdi leader had worsened the situation. The brigade surveillance team had been watching Wissam Abu Qader for weeks, and had finally deduced a pattern to his movements to and from the Sadr's political headquarters in Basra. On 25 May, as the militia leader attended a meeting, the brigade's strike force led by Lieutenant Colonel Mark Kenyon moved into position along Qader's expected route out. Suspecting an ambush, Qader changed his route, but not enough to stop Kenyon intercepting him. A high-speed car chase ended with Qader's vehicle hitting a wall, the car in flames, and the militia leader dead.[5]

The Jaish al-Mahdi responded a few days later with the most audacious assaults since it tried to overrun Cimic House in Maysan three years before. Lieutenant Colonel Patrick Sanders, commander of 4 Rifles, had taken over from Justin Maciejewski the week before. Sanders' father had been defence attaché in Baghdad in the 1970s, and he could remember playing as a child in the embassy's palm-fringed garden. The memory of the heat and the alluvial green of the Tigris became an exotic counterpoint to the rest of his schooldays in sleepy Sussex and made his return to war-torn Iraq all the more jarring.[6]

Sanders' men were immediately in the firing line. A supply convoy was ambushed as it left the palace. The convoys – usually a dozen tankers and articulated lorries flanked by Warriors – were easy targets and had been routinely attacked during Maciejewski's tour as the violence worsened. AK-47 fire from both sides of the road raked the lead vehicles, killing one of the truck drivers, whose vehicle careered across the road as he slumped against the wheel.

The convoy ground to a halt as the shooting intensified. Jeremy 'Jez' Brookes, recently promoted corporal and from Birmingham, stood up in the top cover of his vehicle laying down suppressing fire with his 7.62mm machine gun, the spent casings trickling down into the cockpit below. He was shot twice in the chest, but by distracting the ambushers for a moment allowed the convoy to get moving again, leaving just the one stricken tanker to the mob. Brookes was confirmed dead on arrival at the airport.

Attacks simultaneously began on the last British outpost in the city other than the palace.[7] Major Paul Harding was in the command room of the four-storey building the day after Qader's death when the first RPGs slammed into its exterior. Rupert Lane, a young second lieutenant who had only received his commission three months before, ran up to the exposed roof. Men from his section were returning fire under a hail of bullets. Dropping on all fours, Lane crawled commando-style to the edge of the rooftop. The street beyond was flanked by shoddy office blocks, with piles of rubbish at their bases like sand drifts. Lane counted a dozen fire points from the roofs opposite. Down the street gunmen were flitting into buildings and behind abandoned cars, their movements traced by muzzle flashes from their AK-47s whenever they stopped to fire.

The insurgents massed to the right of the building and then the left. Lane had to constantly shift his position to marshal his defenders, only ducking downstairs when their ammunition started to run low. 'We need air cover,' he shouted to Harding over the din.[8] Twenty minutes later an F16 fighter jet appeared overhead. Lane had a target. For the past hour he had watched a man at the edge of the battle driving around in a blue Opel just out of range, directing the attacks. The next time the Opel pulled out into the open, Lane gave its coordinates. A few minutes later a 250-pound bomb obliterated the vehicle. The fighting continued for a further half an hour, but it was clear the tide had turned. In the early hours of the morning the last shots rang out. All told, 14,000 rounds had been fired in almost four hours in some of the most intense fighting British forces had faced in Iraq.

It was a heavy price to pay for the death of the latest Jaish al-Mahdi leader, but Shaw was determined to keep the militia under pressure and force them to the negotiating table. Proctor and other officials held several meetings with Fartosi, culminating with a visit by General Shaw himself. He and Fartosi talked for two hours, and Shaw played up Fartosi's credentials as an Iraqi nationalist. He warned him of the threat of Iran taking over southern Iraq. British forces would leave when Maliki asked them, he explained, 'but the Iranians won't'.

Shaw knew he had struck a nerve. Despite Fartosi's dealings with the Iranians to fight the British, he had no desire to cede control of Basra to them or their proxies. 'The Iranians must be driven out from Iraq,' he declared.[9]

'Then you must strike a deal and help secure Basra's future,' Shaw urged.

At first Fartosi resisted, but in a way that clearly left the door open for negotiation. By early June they had agreed to try a three-day ceasefire, a test run for a full deal later in the summer. In exchange for the British staying off the streets, Shaw would release one Jaish al-Mahdi foot soldier and Fartosi's right-hand man Seyyed Sajjad, captured in 2005. In utmost secrecy the two men were driven to the edge of the base, where a phalanx of Jaish al-Mahdi Land Cruisers had gathered. The British knew there would be no way of holding Fartosi to his side of the deal. But that evening there was silence.

Up until then very few people outside his small team and Chief of the Defence Staff Jock Stirrup knew about Shaw's negotiations with the insurgent leader, testimony to the controversial nature of releasing hardened prisoners onto the streets. Richard Dannatt, the army head, was kept out of the decision-making process, and Graeme Lamb, the senior British commander in Iraq, only found out about it weeks into the negotiation from his contacts in the SIS. Lamb was immediately suspicious but only confronted Shaw when he saw his own name on a memo to Stirrup implying he had given his approval. Shaw was all smooth assurance when they spoke.

'Isn't this what you and the Americans are doing in Baghdad?' Shaw asked, referring to the negotiations with insurgent leader Abu Azzam in Baghdad a few months before.

Lamb felt differently. Abu Azzam had been on the defensive when they struck a deal, as most of his associates had been killed or captured, and he himself was hiding from Shia death squads. The Jaish al-Mahdi in Basra was much more powerful, and the offer to Fartosi stank of desperation. The British appeared to be 'cutting a deal to run'. But Lamb stopped short of rejecting the proposal, knowing that withdrawal was imminent and there were few other options.

Britain was about to have a new prime minister. While Gordon Brown had not given any specific withdrawal date, his lack of engagement on Iraq was well known in Whitehall, as was his desire to distance himself from Blair's unpopular reign. Where Blair had kept in touch with his staff, even in his darkest moments, Brown was gripped by a singular, unyielding focus that expressed itself in sullen introspection or aggression. There would be no more delays in getting out of Iraq, even if that meant a decisive break with the Americans.

On Brown's first visit to Baghdad, two weeks before taking office, David Petraeus raised the question of British withdrawal. Brown merely pledged his support to the 'Iraqi cause', a deliberately vague response. 'We all knew that was bullshit,' recalled a member of Petraeus's staff.[10] Even worse, as General McChrystal was briefing on the success of British and American special forces in Baghdad, Brown fell asleep, leaving his aides with the difficult decision of whether to nudge him awake – Brown was a notorious snorer – or stop McChrystal. In the end McChrystal gamely carried talking on as the chancellor slept.[11]

Lamb decided to confront Brown directly about Britain's looming exit and the damage it would cause Anglo-American relations. He gave Brown a broadside at the airport. 'The Americans are winning here, and we're a part of it,' said Lamb. 'If we pull out now all that blood and treasure will have been wasted.' But Brown was unmoved. Iraq had been a poisoned chalice for Blair; it was not going to be his too.

In anticipation of Brown taking charge, Jock Stirrup asked British military headquarters in Northwood to prepare five different planning scenarios to show to Brown. Each showed an early palace handover and the rapid drawdown of troops. The fifth and most radical called for an accelerated withdrawal of troops to Kuwait in a matter of weeks, leaving just 500 behind to guard the airport. On the one hand there were fears that if insurgent attacks continued to mount, anything but a minimal British presence would lead to an unsustainable number of UK casualties. If British forces were withdrawing anyway, better to do it quickly and incisively, went the thinking.

On the other hand, given the critical situation on the ground, a quick withdrawal would leave a power vacuum in Basra, and probably pressure the Americans to move forces to the south. It would certainly be seen as a rebuke of the war and British involvement thus far. However, Number 10, once briefed on the options, saw a way to turn this to political advantage.[12] A precipitous withdrawal from the war that had defined Blair's legacy would demonstrate Brown's independence and willingness to stand up to the Americans.[13]

Using politics to paper over what looked like defeat provoked strong reactions among British officers. 'There was a real concern that Brown was looking for a *Love Actually* moment,' said one senior officer, referring to the 2003 Richard Curtis romantic comedy in which a diffident Blair-like premier stands up to an arrogant American president. Unease ran deep in Northwood. Two of the chief planners, Brigadier Patrick Marriott and Air Vice Marshall Greg Bagwell, ACOS for planning, both confronted Chief of Joint Operations Nick Houghton with their fears for the reputation of the British army. It was not quite a revolt, but feelings ran high. 'We must not rush,' Marriott told Houghton. 'The British public will not forgive us if we squander the sacrifices made by our servicemen.'

Houghton sympathised with them but there was little he could do to overturn a decision taken at Number 10. Besides, there were plenty of officers who believed that Britain had paid its dues and had no further obligation to support America's war – especially as it had been turned into a quagmire by US bungling and arrogance. During the few months that Blair had extended the deadline for withdrawal to appease the Americans, eleven more British soldiers had lost their lives. When would it end? Many back in the UK, including Prime Minister Brown, felt the Brits had sacrificed for four years and it was time to go. An uncomfortable reminder of the tensions created by Iraq came at the end of June, when a British doctor of Iraqi descent attempted to blow up a nightclub in London before driving a SUV filled with gas canisters into the glass doors of Glasgow airport. [14]

Even the accelerated withdrawal was not soon enough for Brown. He instructed Stirrup to prepare a plan that involved the most rapid withdrawal of British troops from Iraq.[15]

When Lamb learned that British forces were preparing to leave, he finally gave his consent to the Fartosi deal, knowing that only a ceasefire in Basra could stave off a disastrous exit. In a carefully worded letter to Stirrup he granted his approval for the ceasefire deal following the earlier, successful test run. The deal was due to start in mid-August, two weeks before the withdrawal from the palace. Lamb stressed that British forces must be ready to resume combat operations against the militia if the deal fell through. Under the agreement, British troops would only enter the city after informing the Jaish al-Mahdi, a proviso intended to reduce the chance of inadvertent firefights.

But that did little to reassure US commanders, who had grown increasingly weary of British attempts – not by Lamb – to portray Basra as a success story. At the last commander's conference in July Jonathan Shaw's description of Basra as ready for handover was met with open disbelief.

'There were rumblings in the room,' recalled one US general present. 'The British withdrawal was a rebuke to the whole US surge strategy.'[16]

To make matters worse, Lamb was also preparing to leave Baghdad, and the next commander was unlikely to have the same ability to mollify the Americans. On the evening of 17 July Lamb was sitting in the ground-floor office of Maude House with a few of his staff officers when Graham Binns, the former commander of 7th Armoured Brigade during the invasion, bounded in. Binns, now a major general, was visiting the country before taking over from Jonathan Shaw the following month. Lamb and Binns had known each other since the 1970s and had a cautious respect for each other's different talents.

'Just got my fucking orders from CDS [chief of the defence staff],' said Binns, slapping the desk and clearly delighted that he would be in charge for the historic moment. 'I'm turning the lights out in Basra palace.'[17]

Lamb gave him a withering look. He felt the US surge was winning over Iraqis, and that a similar strategy in the south would convince them to turn against the Jaish al-Mahdi. 'When will London listen? We've got a golden opportunity in Basra. Al-Qa'eda hoisted a flag over the country, and it was fluttering over a new caliphate until the Iraqis tore it down with our help. We've got to do the same in the south.'

'Well, I've got my orders,' said Binns, taken aback.[18]

Meanwhile, at Basra palace preparations had begun for the withdrawal. The deal had yet to be finalised with Fartosi, and the British enclave, along with the smaller outpost in downtown Basra, was under heavy attack as soldiers attempted to dismantle five years of British occupation: portacabins, computers, desks, even the fleet of battered golf buggies. The idea was to save the taxpayers' assets; the subtext was to give the Iraqi mobs less to steal.

Packing up amid a bombardment was demoralising and dangerous. In mid-June Major Paul Harding, commander of the smaller base, was supervising the unloading of supply trucks from a fortified turret overlooking the compound entrance. Harding knew the position was exposed but in characteristic fashion refused to delegate the job to a junior. When a mortar shell landed in front of him, he was killed instantly. Harding's death was a deep blow to his regiment, as was the inevitable feeling of retreat. Each time they pulled back the mortar and rocket fire seemed to get worse. At times it felt like 4 Rifles was being asked to atone for five years of mistakes.[19] Morale could not have been lower.

By 12 August Shaw was ready to finalise the deal with Fartosi, but when the Iraqi prime minister was informed, Maliki was furious. Since losing the support of Moqtada al-Sadr, Maliki had come to view the Shia cleric's followers as a threat to be defeated not negotiated with. He had appointed a trusted general, Mohan al-Faraji, to wrest control of Basra from the Jaish al-Mahdi. The prospect of a British deal with the militia transformed his calculations. 'Maliki realised he needed to take drastic action in the south to defeat the Jaish al-Mahdi,' said his national security adviser Mowaffak al-Rubaie. But distracted by the battle with Sunni insurgents in Baghdad, Maliki couldn't stop the deal going through.

At the end of August the ceasefire began, and an eerie peace fell over Basra for the following two weeks. Despite British strikes against the Jaish al-Mahdi leadership, the militia's tribal network had proved resilient. SIS station chief Proctor had provided Fartosi with a small office in the detention facility, complete with satellite telephone, fax and television, and enjoyed the small semblance of control he felt, via Fartosi, over the city. 'He's like a tap which I can turn on and off,' Proctor enthused.[20]

The success of the deal with Fartosi so far meant that the order to withdraw troops rapidly to Kuwait would not be given. Brown had had second thoughts about antagonising the Americans, and a slower withdrawal would allow the military to pull back from Basra palace in a measured and dignified fashion. British forces would then be confined to Basra airport and the Shaibah logistics base, where they would remain for the following year training Iraqi security forces. Theoretically, at the request of the Iraqi government, British forces could re-enter the city on security operations, but given the delicate arrangement with the Jaish al-Mahdi that was not likely to happen often, if at all.

By early September the palace had taken on a ghostly air. Everyone bar the military had left the week before. Sanders and his battle group were living out of their rucksacks on battle rations. The offices were empty shells, and the grounds, emptied of portacabins, satellite dishes and generators, had returned to the no-man's-land they had been on the cusp of the British arrival in 2003.

The night before the ceremonial handover several British officers gathered in the small operations room at the palace to sip whisky and steady their nerves about a Jaish al-Mahdi surprise attack. Could the Iraqi forces cope with the handover or would they be instantly overwhelmed by insurgents? General Mohan, the charismatic Iraqi general sent down by the prime minister to oversee the handover, had shown a flair for the dramatic, demanding that the palace handover include a fly-past by British jets and a full parade. Neither was going to happen, but Mohan's bombast and good humour had reassured British officers that the situation might stay under control.

In the early hours of 3 September 2007 Sanders shook hands with his Iraqi counterpart in the grounds of the palace as the British flag was lowered. Within half an hour the entire battle group were in their vehicles. The RSM did a headcount, and Sanders' 600 men were off. The tanks and Warriors moved slowly to the city outskirts, laboriously checking for roadside bombs as they went. When they arrived unharmed at the airport many hours later, they did so with a feeling of triumph. The procession was already getting favourable reports on the news channels, although some media outlets persisted in branding it a defeat. The next morning the *Independent* carried the headline CUT, RUN, and the *Daily Mail* called the withdrawal 'a sad day in our military history'. But the biggest-selling tabloid, the *Sun*, declared JOB DONE by the LIONS OF BASRA.[21]

While the British claimed an orderly handover, a militia video quickly did the rounds, showing the British convoy trailed by Jaish al-Mahdi fighters waving their guns triumphantly in the air. Both sides were happy with appearances.

Chapter 31

ON THE RUN

THE DAY AFTER THE BRITISH WITHDRAWAL Haider attended his brother-in-law's wedding. He had barely left his house in months, fearful of being assassinated because he worked for the British. Since the New Year a dozen interpreters working for the occupation had been hunted down and killed, and many others had received death threats. But the eerie quiet of the city over the past few days coaxed him outside. In Iraqi tradition the bridegroom's family ceremoniously fetches the bride from her father's home to the groom's, and his wife Nora's family had formed a long convoy of vehicles with windows down and Arab music blaring. Haider was sitting in the final van. He had brought his Glock pistol just in case.

As they drove past the Shatt al-Arab Hotel in northern Basra three white pickup trucks with tinted windows and no licence plates overtook the convoy of wedding vehicles, cutting off the lead car. The driver of Haider's vehicle screeched to a halt as gunmen jumped out of the trucks. The gunmen headed straight for the vehicle and pulled open a door. One of his wife's sisters screamed as Haider sat frozen, one hand in his jacket pocket on the gun. A gunman poked his head inside and grabbed Nora's twenty-year-old cousin Ahmed, yanking him out and pushing him into one of the trucks, which sped away.

The wedding party raced back to the house and guests spilled into the driveway. The women were hysterical, and mascara streamed down the bride's face. The men gathered in an angry huddle. Nora's father was furious. 'Is it because we are Sunni?' he yelled. 'Why would they do this?'[1]

Haider was still shaking, imagining militia hands reaching for him as he sat frozen with fear. He had a terrible suspicion that the kidnappers had meant to take him, but he kept quiet. Only Nora and her uncle Ali knew he worked for the British. Talk turned to how the family could contact the militia to try and negotiate a ransom; such kidnappings were common, with families expected to pay £60,000 or more to secure a hostage's freedom. Haider eventually returned home with Nora and their nine-month-old son Husam.

Ali called Haider the next day to tell him Ahmed's body had just been found in a ditch. In a low voice he said, 'Haider, there was a note in his pocket saying they are looking for you. There was a number for you to call for you to give yourself up. Don't worry. I took the note before anyone else saw.'[2]

'What did they do to him?' asked Haider, his face ashen.

Ali wouldn't answer. 'He's dead now, God rest his soul. Now listen. Do you have a passport? Take Nora and Husam to her family,' he advised, 'then you must flee to Iran.'

There was no hiding from Nora what was happening. Haider went through to the bedroom, where she was nursing Husam. He watched her body tense as he told her the news. Husam started crying. Haider told Nora about the threatening text message he had received the previous year. 'Why didn't you stop working for the British sooner?' she screamed.[3]

Nora's anger seemed to hold her together as she packed bags. They discussed their plan. Haider would go alone to Iran; he hated the thought of separation but accepted that Nora's extended family could protect her better than he could. If he disappeared for a while the militia might leave them alone. Haider's brother had gone to the Iranian consulate to collect a visa. He returned with Ali in the early afternoon. Ali handed him $400. 'It's all I can spare at the moment,' he said. 'I'll get you more.' Nora was crying as she got into Ali's car. 'I'm sorry,' she said to Haider. Then they drove away, leaving Haider alone on the street. He hurried down Basra's main thoroughfare. In the early afternoon the street was eerily quiet. He flinched as the shutter of a nearby shop rattled down. A few men in dishdashas lingered in the shade of doorways, watching him. Any of them could belong to the militias; their networks had grown so quickly through the city. Haider took the first cab that drove past.

Driving through the city, he could see scant evidence of four years of British reconstruction. Over £150 million had been spent. Haider spotted a few roughly refurbished schools in otherwise gritty neighbourhoods with sewage pooling in the streets outside and mounds of rubbish between the houses. On the outskirts of the city they passed a grove of recently planted date palm, untended and withering in the desert heat – another British effort that had failed to have much impact.[4]

It wasn't that the British and American money had been completely wasted – there was more fresh water and more electricity to power the consumer goods that Iraqis had rushed to buy after the invasion – but the occupiers' efforts had never met Iraqi expectations or demands, leaving most with a bitter taste in their mouths. In return for these scant rewards, Iraqis had lost their security. None had more willingly embraced the promises of democracy and freedom that the invasion had brought than Haider, but even he was forced to admit that he had felt safer under Saddam. Were the benefits that Britain and America had brought worth the sacrifice? Haider could no longer afford to wait to find out.

He was relieved to find no police checkpoints on the five-mile journey to the border; close to Iran there were often roadblocks. The small passport office at the frontier was also open, and he queued anxiously with the other pedestrians, before making his way to the taxi ranks on the Iranian side. Ali had given him the name of a hotel in Abadan, a port city on the Shatt al-Arab built around a large oil refinery once owned by the British. The hotel owner was a distant relative of Ali, and he had agreed to put Haider up in a basement room without registering his name with the police. Haider's mobile still worked this close to the Iraqi border. Once he had settled in his tiny room he called Nora to check she was safe. Next he took out the number that Ali had found with Ahmed's body and slowly keyed the numbers into his phone.

'We were hoping you would call, Haider. You had some luck yesterday.' The man had a thick Basra accent typical of the city's slum dwellers. He was probably the son of a Marsh Arab, displaced by Saddam Hussein in the 1990s. 'We meant to grab you but Ahmed got it instead. Don't make this difficult,' he said in his gruff voice.[5]

'You'll never find me,' Haider shouted, before hanging up, breathing hard.

As Haider's exile stretched into weeks, he made plans for Nora and Husam to visit. Ali told him that the militia had ransacked his apartment. A few days later they torched a shack he had bought on the Al-Faw peninsula. Haider felt like he had entered an eerie afterlife, watching powerless as the world he'd built disintegrated. He slept late in the mornings before wandering aimlessly around the broad tree-lined streets of Abadan. Nora agreed to come with Husam at the start of November.

Ali collected them early in the morning, hoping to miss the queues at the border. Nora had only left Basra a few times in her life, to visit her family in Baghdad. As they drove through the tired and worn city, she wondered at her wilfulness in pushing to marry Haider. Suddenly two SUVs appeared behind the car, flashing their lights.

'Fuck,' said Ali, who kept driving.

One of the vehicles moved alongside. A gunman leaned out of a window and aimed an AK-47 at Ali, who finally slammed on the brakes. When the vehicles stopped, more gunmen leapt out and hauled Ali from the driver's seat and over to one of the SUVs. Another grabbed Nora roughly by the arm and yanked her out as Husam began screaming. He took her passport and phone. Within moments Nora and Husam were alone on the roadside.

Haider had already begun to suspect things had gone wrong when he received a call from the number he dreaded.

'We've got Ali,' said the familiar voice. 'If you don't believe me, here he is.'

'Haider?'

'I'm here, Ali,' he replied.

'I'm in serious trouble, Haider,' said Ali. He sounded sad, resigned even. He described the kidnap.

'Where's Nora, Ali?'

'I don't know.'

'I'm sorry, Ali.'

The other voice came back on the line: 'Come to us, or we'll kill him.' Then the line went dead.

Haider felt certain that if he returned they would kill both him and Ali, but perhaps he could still save the rest of his family. Packing his small bag, he headed for the border. He tried calling Nora's mobile but there was no response. He tracked her down at her father's house. At first she was reluctant to come to the phone. He could hear her sobbing in the background. Finally she picked up the handset.

'I'm coming back to Basra to get you,' he said.

'And what are you going to do?' she shouted. 'Where will you take us? There's nowhere safe in the city. You're the trouble. Stay away from my family.'

'But you're not safe,' he shouted, but she had already gone.

Next he called Ahmed, a translator at the airport. A plan had started to take shape. After explaining his plight, he asked, 'Will the British take me in?'

Haider's timing was good. Over the summer, as the palace withdrawal loomed, *The Times* reporter Deborah Haynes had written a series of articles on the plight of the interpreters who worked for the British. Initially the British government ruled that no special asylum would be granted to interpreters, for fear of prompting the sight of hundreds of Iraqis clamouring at the gates, but in October Gordon Brown announced a fast-track immigration programme for those who had worked for the British since 2005. Ahmed told Haider he should come to the airport at once to seek asylum.

Haider did not dare go home so arranged to meet his mother, sister and three brothers at the house of a friend. They each brought with them a small suitcase, and stood, a forlorn gaggle, in the walled compound in front of the house. Haider called Nora to tell her his plan. She was reluctant at first but eventually agreed. Half an hour later, with night falling, Haider and his family picked her up in a minibus and headed for the airport.

The Iraqi policemen at the first airport checkpoint interrogated them briefly but let them through. Haider wondered how long it would take for news of his return to reach the militia. At the next checkpoint, manned by former Gurkhas working for a private security firm, Triple Canopy, they slowed to a crawl. A guard appeared with gun raised. Standing orders were to fire on any vehicles after dark which didn't stop. Haider got out with his hands raised. From twenty metres he shouted out why he had come. The man went inside the guardhouse and emerged ten minutes later.

'There's no permission for you to enter,' he shouted back. Haider hurriedly called Ahmed at the airport.[6]

'There's been a problem,' said Ahmed. 'The military are saying you and your family can't enter the base. They say this is not a hotel.'[7]

'We can't go back,' said Haider. Ahmed promised to try again.

By now they had sent away the taxi, and Haider and his family were standing beside the airport road, halfway between the checkpoints. The Iraqi policemen were watching them; Haider could see their cigarettes glowing in the dark.

After an hour and still no word, Haider called William Kearney, the ArmorGroup manager for whom he had worked. 'Billy, is there anything you can do? What am I supposed to do?' he asked. Kearney, at home in England, promised to make some calls.

By now Haider's mood had moved from desperation to a tired despondency. He walked back to his family and took Husam from Nora. 'We've got to go back to Basra,' he said.

'We can't go back,' said Nora, grabbing his arm. 'You said they'd let us in.'

'There's no choice.' He marched back down the road. Past the police checkpoint there were no lights, and the road faded into darkness. Husam went to sleep in his arms. After two miles they reached a junction in the road and a small tea stand lit by a kerosene lamp. A solitary taxi was parked on the roadside. An elderly man in a dirty dishdasha was sipping tea with the proprietor.

'It's late for a walk,' said the taxi driver as Haider approached.

'We're from Amarah,' Haider blurted out. 'The British attacked our neighbourhood. We were looking for our family, but we got lost.'

'You poor things,' said the man as Nora and the others arrived. He insisted they stay with him for the night. They piled into his Oldsmobile, Haider keeping up a constant patter, fearful that if he stopped speaking the spell would break. That night, on a mattress in the man's concrete apartment, he lay awake until the early hours of the morning, too tense to sleep.[8]

They stayed at the taxi driver's house until noon the following day. The answer from the British military was now yes, but he needed to have the most up-to-date Iraqi passport, the so-called G passport, which required extensive documentation, including Ba'athist ID papers

'But Saddam stripped me of my ID papers,' said Haider.

'These are the British conditions,' Ahmed told him.[9]

Haider's quick thinking now saved his family. They couldn't stay with the taxi driver any longer. As the morning wore on, the old man was asking increasingly suspicious questions. To get his paperwork together, Haider needed shelter, and the only place he could think of was a clinic belonging to a friend he had gone to medical school with in the 1990s. Dr Rahim had been a freethinker then, although he had shied away from political talk. The two men had not seen each other since. Now he was Haider's best chance of staying alive.

Rahim greeted him enthusiastically in the clinic's reception area, but Haider's expression brought the flow of courtesies to a halt. In Rahim's office Haider explained the situation and asked, 'Can you check us into your clinic under a false name?'

Rahim paused for a moment as he contemplated the dangers. 'Come,' he said.

Nora and the rest of the family had been waiting in the hall. They now followed Rahim to the end of a poorly lit corridor to a small room that doubled as a store cupboard.

'This is the best I can do,' said Rahim. 'There's a toilet in the corridor. Please don't leave this room during the day. I can bring food in the evenings.'

Haider and his family were safe, for now.

The day after arriving at the clinic Haider received the news he had been dreading. Nora's family called to say that Ali had been found in a ditch, shot four times in the chest but somehow, miraculously, alive. He was in a critical state in Basra hospital. A distraught Haider called up Kearney, overwhelmed by the difficulties of getting together his family's travel documents.

'I'm stuck, Billy.'

'No, you're not,' came the firm reply. Encouraged by Kearney, Haider called round his family's contacts. There was a secular judge in Basra, one of the few left. Haider had met him once, working for Lieutenant Colonel Nicholas Mercer. If he could contact him discreetly, and the judge was prepared to take the risk, he might have a chance. Kearney also called the journalists he knew to encourage them to write about Haider and the plight of Iraqi translators working for the British. He even arranged for Haider to give an interview on the *Today* programme from Rahim's storeroom.

Haider could not move around safely outside the clinic, so he called one of his brother's friends to ask him to deliver a message to the judge. He was called Saddam Hussein, a popular name after the dictator had offered a £150 reward for any baby named after him. At least he was unlikely to be on the militia's side with a name like that, thought Haider. Within a few days of arriving at the clinic, he met with Haider who relayed instructions. He also handed over the last of his money, the equivalent of around £1,500, as a bribe to speed things up in the passport office should the judge agree to help.

Back in the storeroom Husam was screaming and Nora was losing her nerve. After a stoical first few hours, she was now demanding to go back to her family. Part of her knew that was impossible, but that did not stop the panic. Every time they heard footsteps in the hall they froze. At one stage Haider had to restrain her from leaving. His mother and sister did their best to calm her down.

The news from Hussein, however, was good. The judge had agreed to help, and their passports should be ready in a month. 'We haven't got a month,' cried Haider. 'Doesn't anyone understand the situation?' A week later Hussein arrived at the clinic with an envelope of passports. There was no time for rejoicing. Haider got on the phone to Ahmed while Hussein agreed to arrange a minibus to take the entire family to the airport.

Haider felt little emotion as he sped through the city with his family the next day. They arrived half an hour early at the Gurkha checkpoint and had to wait awkwardly outside. Then at the airport they were told that the FCO would not support the claims of Haider's brothers and sister for asylum as they were not technically dependants, being over the age of eighteen. 'What are they supposed to do by themselves in Basra?' shouted Haider, suddenly angry. When the British had needed him he had risked his life, but when he needed their help all he got was red tape. To them he was just another nameless, faceless Iraqi.

The British could not stop his brothers and sister travelling with the rest of the family to Jordan, where they would need to wait for their UK immigration papers to be issued, but at the airport at Amman they were arrested by the Jordanians. Haider's mother clung to her daughter and three sons but they were dragged away by the police and put on the next flight back to Iraq.

Feeling wrenched and torn, Haider and the remainder of his family finally settled down to their first night outside Iraq in a cramped hotel room in downtown Amman. He was free, although his ordeal was far from over.

Chapter 32

CHARGE OF THE KNIGHTS

SINCE THE BRITISH DEAL WITH FARTOSI IN August 2007, the city had fallen fully under the sway of the militia. The palace was not stormed after the British withdrawal, but then the militia hardly needed to hoist a flag over the building to show they were in charge. The fight that the British had long waged in Basra was now left to the Iraqi military and police, and left to stand on their own feet, they quickly stumbled when confronted by the relentless and vicious Jaish al-Mahdi. Its black-shirted operatives patrolled the streets, enforcing a crude form of Sharia law that meant women had to stay indoors or cover themselves in *abayas*. Extortion and kidnapping were common, and the city morgue was full of murdered Iraqis, many bearing the marks of torture.

General Mohan al-Faraji was still leading the fight against the militia. Living in a city where thousands of fighters were plotting his demise, and at least some of his own forces were assisting them, would have overwhelmed a lesser man, but Mohan was most comfortable with his back against the wall. A tank commander during the Iran–Iraq war, he had subsequently plotted a rebellion against Saddam, only to be uncovered and thrown into prison for five years. In his early fifties, he was garrulous, wilful and sly, and he walked with the barrelling gait of a tank commander.

With age came a subtle appreciation of politics. Mohan was under pressure from Prime Minister Maliki to crush the Jaish al-Mahdi. Provincial elections were approaching, and Maliki wanted to undermine Sadr's power base in the south. At the same time Mohan knew the British, with one foot out of the door, were anxious to avoid provoking the militia. The 5,000 British troops remaining were largely confined to their barracks at Basra airport and the Shaibah logistics base. They were still required, under the Overwatch plan, to enter the city, if requested to do so by the Iraqi government, but their main task was to train and mentor Iraqi soldiers before full withdrawal the following year. Mohan had been assigned his own British mentor, Colonel Andy Bristow, and the two had struck up a friendship, with the Iraqi general soon asking for advice in drawing up a security plan for the city and negotiating the political minefield.

The partnership created its own conundrums for Bristow, a quirky, clever officer already selected for promotion when his tour ended in a few months' time. Mohan was effectively asking for his help to defeat the militia although the British military had cut a secret deal with Jaish al-Mahdi to guarantee their peaceful exit. The precise nature of the deal with Fartosi was still a closely guarded secret which few Iraqis knew about, not even Mohan. From the British short-term perspective the deal with Fartosi had been an extraordinary success – after months of incessant bombardment, attacks on the two remaining British bases had all but stopped. To the surprise of some, Fartosi had proved a powerful interlocutor, with his finger still very much on the pulse of the Jaish al-Mahdi. But their Faustian bargain undermined the very institutions that the British had gone to Iraq to create. Talking to Mohan, Bristow heard about the murders, rapes and abuse, and knew that the city was nothing like the peaceful oasis British defence chiefs were trying to portray to the media. Bristow was beginning to realise that the deal might have helped the British army avoid one calamity, but the longer-term consequences for the city, and for the British legacy in Iraq, might be much worse.

The Americans were still pressuring the British not to withdraw too quickly. General Petraeus, on his way back to Baghdad after defending the US surge to Congress, made an unprecedented call on Number 10. Petraeus was growing increasingly alarmed about the scale of the problem developing in the south.[1] He issued Brown with an ultimatum: he was prepared to send American troops to Basra unless Britain retained enough troops there. Brown assured Petraeus that British troops would continue to support the Iraqi security forces in Basra, but told him that Britain's priority lay in Afghanistan, where the mission had swollen to 5,500 troops.

In fact, Brown was doing his best to exploit the British withdrawal from Iraq for his own gain. Riding high in the opinion polls after successfully portraying himself as the antithesis of Blair's shallow spin, the prime minister was considering a snap general election early the following year. The autumn Labour Party conference was seen as a likely venue for an election announcement, but at the last minute Brown backed off; having waited so long for power, he was not prepared to risk it so soon in an election. However, he decided to carry on with what was to have been the first stop on the campaign trail: a visit to Baghdad to announce a troop reduction. The visit, which took place during the Conservative Party conference, was intended to upstage David Cameron, who was fighting for his political life at the time. Instead, the announcement came across as the worst form of politicking. The media soon reported that even the troop withdrawal numbers Brown trumpeted had been fiddled: the figure of 1,000 troops included 500 that Brown had already announced would be leaving back in July. Amid accusations of using the military for political gain, Brown's premiership never recovered.

As the Americans grew increasingly frustrated with the British, so too did Mohan. With Bristow's guidance he had developed a security plan for Basra, modelled on the American success in Anbar and Bristow's experiences in Northern Ireland. Mohan's forces would build military outposts in Basra's most troubled districts to be manned by the newly formed Iraqi 14th Division. Once control had been established around the bases, Mohan's men would push deeper into the city, eventually linking up to create a patchwork of secure zones. Bristow also envisioned a ring of Iraqi army bases around the city to stop militias and Iranian agents entering the city.[2]

The plan would involve some British help in the form of materiel and air support. However, the Iraqi general was left with few illusions about whether the British would deliver a few weeks later when the Jaish al-Mahdi besieged his headquarters at the Shatt al-Arab Hotel. Iraqi police at a vehicle checkpoint opened fire on an unmarked SUV, inadvertently killing a senior insurgent. The militia responded by taking to the streets in force, besieging Mohan and a number of police stations. Under heavy attack, Mohan called up Bristow to ask for British troops to enter the city in his support.

Mohan considered the British response – the overpass of two F15 jets – underwhelming. In fact, the SIS station chief James Proctor had decided to enlist Fartosi's help. Using his contacts, the militia leader arranged for two Sadrist clerics to go to the Shatt al-Arab Hotel to broker a peace deal. Mohan, unaware of the contact with Fartosi, assumed the militia had been compelled to sue for peace by his own brave stand at the Shatt al Arab.[3]

When Mohan next saw Bristow at a dinner, he angrily accused the British of abandoning him, and Bristow revealed the terms of the deal with Fartosi. Mohan was horrified to discover that the British were releasing dangerous Jaish al-Mahdi insurgents into the city without informing him, and that once released no tabs were kept on the men. 'He was gobsmacked,' said Bristow. 'He thought the deal was a bad idea.'

'This must stop at once!' Mohan declared, banging the dining table.

While the deal suited the departing British, they would be leaving behind a city swimming with hardened criminals for Mohan to contend with. It was just the sort of double-dealing the British were infamous for from their colonial days. Back in the 1920s, after control had officially been handed to the first Iraqi monarch, King Faisal, the British had maintained a clandestine network of advisers and Iraqi agents. A rather chagrined Bristow agreed Mohan would be informed of every prisoner released, and each would be obliged to sign a six-point charter pledging to renounce violence. It was a largely symbolic concession.

The British offer of support for his security plan wasn't much better. The British commander in Basra, Graham Binns, who had taken over from Jonathan Shaw in August, was focused on another handover ceremony, this time of the remaining towns and villages around Basra under British control, and had no desire to upset the deal with Fartosi. However, there were stirrings of dissent in Binns's headquarters. The Fartosi deal was increasingly turning into a noose around the neck of the British, and several officers wondered what would happen when their dwindling stock of prisoners, including Fartosi, were all released.

And yet the plan had a high-placed backer, who refused to give it up. SIS Station Chief Proctor had been one of the principal architects of the Fartosi deal and was pushing hard to keep it alive even if that meant overstepping his bounds. An SAS team had been due to conduct an operation in the city, only for Proctor to order them to stop, fearing they would break the terms of the deal with Fartosi. Several officers viewed Proctor's intervention in military affairs as misguided. On another occasion Binns was presented with an opportunity to raid a building in Basra believed to hold Peter Moore, the IT consultant kidnapped in Baghdad in May with his four bodyguards. The intelligence was thin and given the likelihood that a raid would upset Fartosi, Proctor advised Binns to turn down the opportunity.[4]

With the British headquarters seemingly paralysed, Bristow was only able to provide Mohan with some old portacabins for use in outposts. The Iraqi army began collecting and transporting them towards the city, but they both realised that it was an exercise in delusion, and that Mohan didn't have the resources to win a war against the Jaish al-Mahdi without major support from the Americans or more Iraqi forces. Bristow, fearful of the day when the British ran out of prisoners to release, tried to broker a rapprochement between Mohan and Fartosi, but when the two men met in Fartosi's cell there was little disguising the mutual animosity.

'Both men recognised that once the deal with the British ended, the battle between them would begin,' noted Bristow.

They didn't have to wait long. On 31 December 2007 Fartosi was released, and within days the violence had resumed. Rocket fire and mortar attacks on the British base at the airport surged from a few a week to over thirty. Fartosi's response to SIS requests to rein back the violence was lukewarm at best. After six months on the ground, Bristow's tour had recently ended. His replacement Colonel Richard Iron quickly realised that the deal had failed.[5] In London the defence chiefs stuck to the position established by Jonathan Shaw that there was no insurgency in Basra despite overwhelming and grisly evidence to the contrary. Iron understood that this view was politically convenient but blinding Northwood to the reality on the ground: the Jaish al-Mahdi were well equipped, lethally effective and the dominant force in the city. He concluded that the British military needed to jettison plans for an imminent withdrawal and instead support Iraqi forces in a full-scale offensive to clear the militia from the streets. Unlike past operations, when British troops had seized an area, only to withdraw and let the militia back in, this time they would have to stand their ground until local security forces could do the job.

Iron understood the reason for Northwood's wilful blindness: the military was exhausted after four years of fighting in Iraq, and the situation was steadily worsening in Helmand. In December British and American forces had retaken Musa Qala. Since Butler's deal with the local elders and Taliban in 2006 to withdraw from the town, the military had avoided being drawn into manning any more isolated outposts, but under pressure from the Americans they had sent soldiers back in.[6] With violence flaring across southern Afghanistan more troops would be needed soon. But Iron felt that the failure to confront the Jaish al-Mahdi would have implications beyond the fate of an Iraqi city. Upon arriving at the British headquarters in Basra airport Iron had been struck by the air of defeat. The deal with Fartosi had forced the military to make one compromise after another. Judged politically expedient at the time, now that the deal had been exposed as a sham, the military had nothing to fall back on. British troops were glumly hunkered down in the airport waiting for the end. It wasn't just in Basra that the British army was losing its way.

When the US had faced defeat in Baghdad in 2006, Petraeus had spent a year soul-searching and rebuilding the military's mindset, starting with its counter-insurgency doctrine. The resulting surge was now showing tangible results, demonstrating how the application of basic counter-insurgency principles could succeed. Iron, a fastidious and scholarly soldier who had served as a mentor with the Kenyan and Omani armies earlier in his career, knew that history was filled with examples of the British military overcoming an insurgency but feared it now lacked the guts or the brains to do the same in Basra.

From Colonel Alex Alderson, the counter-insurgency expert, Iron learned that British efforts to rewrite their own doctrine had stalled. The two men had been friends since working together on an early lessons-learned report for Iraq. Alderson blamed the usual blend of ego and institutional battling for the failure.[7] He had eventually managed to produce a document that contained the central tenet of the American doctrine: 'Clear, hold, build.' Unfortunately that had provoked a backlash from the army, which viewed the mantra as 'far too kinetic, far too American'. A compromise formulation was reached to better reflect British values: 'Secure, engage, develop.'[8] It was a storm in a teacup and it distracted the army from necessary reform.

Iron didn't have time to wait for British thinking to catch up, but realised that perhaps he didn't have to; the key to transforming Basra lay with General Mohan. It hadn't escaped Iron's notice that as Mohan's mentor he was outside the usual British chain of command with direct access to the Iraqi and American headquarters in Baghdad. The British army would not listen to him, but having for so long paid lip service to the idea that the Iraqis were in charge of security, they were duty bound to listen to them. Using Mohan, he could take on the Jaish al-Mahdi and expose the sham that the deal with Fartosi had become.

Iron was confident that his plan would succeed where almost five years of British efforts had failed because it would be the first offensive to be led by the Iraqis. If there was one lesson he had learned from working in Oman and Kenya and from his reading of the British experience fighting alongside indigenous forces during the colonial era, it was that the Iraqis needed to take ownership. It wasn't a hard sell. Mohan quickly bought into the plan, nicknaming it the Charge of the Knights. In Baghdad Maliki was reportedly eager to be briefed.

Iron next set about bringing British commanders on board. In early February he arranged for the Sandhurst instructor Daniel Marston to visit the headquarters of the latest brigade commander, Julian Free, and brief his staff on counter-insurgency doctrine. Having established the syllabus at Sandhurst, Marston had agreed to teach courses at the American-run counter-insurgency school in Taji, central Iraq, and would be a regular visitor to the country. Iron warned Marston that it might be difficult to convince Free of the need for a new approach, and the instructor did not hold back in the crowded operations room at the airport. 'Unless you start treating this like an insurgency, the British army will fail in Basra,' he said.

The talk ruffled a few feathers, just as Iron had intended, and Free remained sceptical when Iron showed him a one-page summary of Mohan's plan. To make their case, Iron called in Mohan the next day for a full briefing. The Iraqi general gave a bravura performance, and Free was finally won over. However, Free knew that neither the British commanding general, Graham Binns, nor his successor, Barney White-Spunner, nor the British military headquarters in Northwood were likely to accept the plan because it might mean staying longer in Iraq when forces were desperately needed in Afghanistan.

To Iron it was clear the success of the plan hinged on American buy-in. With the help of Marston, he arranged for Mohan to brief Petraeus on the plan in Baghdad. The PowerPoint presentation Iron prepared for Mohan began with a slide declaring that taking on the militia in Basra represented 'the last great urban battle in Iraq'.[9] Arriving in Baghdad, Mohan went straight to meet Maliki in his villa in the Green Zone. Rubaie was present, along with the Iraqi defence minister Abd al-Qadr al-Obaidi. A half-hour presentation turned into an invitation to dinner. An excited Mohan called Iron on his mobile phone as dinner was prepared. 'In Iraqi culture this is a good sign,' he said.[10]

The next afternoon they met Petraeus perched on his chair before a dozen of his commanders in a Green Zone conference room. Iron and the new commander in Basra, Major General White-Spunner sat next to Mohan. The British general had only recently been apprised of the plan. Mohan's presentation was rambling and incoherent, but everyone got the point. A senior Iraqi commander was offering to solve an Iraqi problem. Petraeus told the room, 'This is a brave and courageous plan and let's look into how we can resource it.'

Iron had been sitting poker-faced through the meeting. Inside he exalted. But not everyone was happy. After the meeting had broken up, General Austin took White-Spunner to one side. He could scarcely contain himself. 'What the hell's this?' demanded the six-and-a-half-foot African-American. 'The main effort is meant to be finishing off al-Qa'eda in Mosul.'[11] Implicit in his tone was some of the frustration many commanders had felt building up over the British, especially given the deal with Fartosi and the withdrawal. Hadn't they backed out of the fight? White-Spunner looked uncomfortable.

'And what's this about a British colonel working for Mohan?' Austin continued.

'British policy is to support Mohan,' White-Spunner replied. He recognised the need for action in Basra but insisted that Mohan's plan would not be launched until later in the summer and thus avoid distracting from any American offensives.

Two weeks later the commanders met again in the Green Zone to finalise the plan. Rubaie remained silent until the meal of Iraqi kebabs and hummus was served, although it was clear he could barely contain himself. He had whispered to Iron on entering the room, 'I've got some good news for Basra.' Now he leaned across to Petraeus and asked to have a word outside. 'The Basra plan has been changed. Maliki's sending troops in on Monday,' Rubaie told Petraeus.

It was hard to surprise Petraeus, but the American general was taken aback: 'What do you mean Monday?'[12]

Rubaie informed him that Maliki was so pleased with the Charge of the Knights plan that he had decided to accelerate it. Two battalions from the Iraqi 1st Division had been given top-secret orders. They were already beginning to move south.

Petraeus did not appreciate being ambushed. 'But that's not enough troops to take on the Jaish al-Mahdi. And there's no time to get American mentors in place with the units. This could be a disaster, Mowaffak.' He called over to Dempsey, the American officer in charge of training the Iraqi army, 'Have you heard what Mowaffak is saying?'

The argument went back and forth for twenty minutes. Rubaie took a different view of the Jaish al-Mahdi's capabilities to Petraeus. The 'criminals' in the south would quickly fall into line once the Iraqi army showed up. The Iraqi government's impatience was driven by rapidly approaching provincial elections, which threatened to cement Sadr's grip on the oil-rich south for another five years.[13] Rubaie was also sceptical about involving the British, or any Coalition force, in the sensitive task of putting Iraq's Shia politics in order. Rubaie concluded the argument decisively, 'We're going to do this on our own.'

Iron watched them arguing but was unaware of the dramatic new twist his Basra plan had taken. He was due to head off to spend a week with his family before returning to launch the plan. If this was going to be an Iraqi-led operation, Iron thought, it might take months to get the first forward operation base built and staffed. He left the meeting early. White-Spunner was returning to the UK to brief the incoming brigade and take a holiday. That left Julian Free in Basra to receive the extraordinary news a little later from an irate General Lloyd Austin that for the first time since the invasion Iraqi troops were marching under their own orders. 'I told you we didn't want a second front. We're meant to be winning the war against al-Qa'eda,' Austin shouted down the phone at him.[14] Free placated him as best he could, his mind racing.

If the American general was unhappy, he also wondered how British military headquarters in Northwood would take the news, given the Iraqi prime minister was declaring war on Fartosi and his men, Britain's erstwhile allies in Basra. He tried contacting Northwood that night. It was late and it was Easter Sunday. No one more senior than a captain was on duty. Not for the first time Free wondered how a twenty-four-hour war could be waged with a nine-to-five mentality.

The next morning the scale of the Iraqi enterprise became clear. Some 1,000 troops were on the march from Baghdad to join Mohan's 4,000-strong 14th Division and the remains of the 10th Division. Prime Minister Maliki himself had flown to Basra to take over from Mohan. The speed of the deployment left American teams embedded with the Iraqis struggling to catch up with their units, meaning the local forces were effectively without air and artillery support. Also emerging was just how badly prepared UK forces were. Free was already fielding calls from anxious Americans asking, 'Where are the Brits?'

But despite 5,000 British troops still being stationed in Iraq, no units were set up to embed with the Iraqi army, following a 2005 stipulation from Northwood. Free scrambled to get a detachment of British troops into the joint Iraqi command centre at the Shatt al-Arab Hotel and to prepare men from the Scots Dragoon Guards to take up positions with the 14th Division to help direct air strikes.

Northwood called mid-morning. Free was relieved to hear Brigadier Patrick Marriott, assistant chief of staff for operations, on the other end of the phone. Marriott was a former commander in Basra and understood Iraq. Free wanted to know how much of a risk he could take in putting British forces back onto the streets.

'Given that the British army's reputation is at stake, quite a bit, I would have thought,' said Marriott. 'You crack on and I'll do my best to keep people off your back.'[15]

Both men knew the news was not going to be well received at the MOD, where the change in the status quo would upset the withdrawal plans. Shortly afterwards, James Proctor appeared in the operations room at Basra airport. If British troops were going to enter Basra, then he insisted on informing Fartosi, as the deal with the Jaish al-Mahdi leader stipulated. Free was flabbergasted when he heard. 'I wasn't going to be calling Fartosi to ask him for permission to attack him,' said Free. Proctor was told to back off. Instead Free quickly got his planners to draw up a paper for Northwood outlining how British troops could be safely embedded with Iraqi forces. But the following morning, as Iraqi units began moving around the city, neither US nor British embed teams were in place.

If Maliki thought that a show of force would cow the militia he was in for a rude shock. The Iraqi forces were soon coming under fire from heavily fortified Jaish al-Mahdi positions. Under a withering barrage of fire, the 14th Division's most recently trained brigade, the 52nd, melted away altogether. Other units were blown to oblivion driving down heavily booby-trapped roads. 'The Iraqi strategy seemed to have been to drive around and duff up the Jaish al-Mahdi, but they were being slaughtered everywhere they turned,' commented Free. In areas like Hayaniyah the local police led by the UK's old foe Captain Jaffar predictably sided with the militia, as did some army units. The Iraqi operation appeared on the brink of collapse.

But no one could afford for Maliki to fail. General Austin arrived the next morning with the first of the American embed teams. He also announced he would be sending down the American deputy corps commander, Major General George Flynn, to manage the fighting from the Coalition end. It was an acutely embarrassing moment for Free, who had managed to get a small detachment of British officers over to the Shatt al-Arab Hotel but had not yet received permission from London to let British troops support the 14th Division on the ground.

Further humiliation was to follow when Free accompanied Austin to see Maliki at the palace. The prime minister refused to speak to Free. 'He held us responsible for letting Basra get into such a mess,' said Free. That left Austin to hammer out a more coherent plan with Maliki as Free waited outside. Austin emerged with the news that Mohan had been fired following the Iraqi army's collapse, and General Muhammad, commander of the 14th Division, promoted in his place.

'What the fuck is going on?' said Austin before he jetted back to Baghdad.[16]

The next morning the newly arrived Flynn, a punchy New Yorker, announced to Free's brigade headquarters at the airport that he was there to ensure Britain stopped failing in Basra.[17] It was a damning indictment and Free watched as his officers' heads dropped. Once Flynn had left, Free spoke to his men: 'Look, guys, this is an opportunity to show them what we can do.'

The first thing was to come up with a plan which involved securing some of the city's safer neighbourhoods and isolating the militia stronghold of Hayaniyah. In the Shatt al-Arab Hotel a few hours later Flynn and Free worked with General Muhammad and Mohan – who despite his sacking was still running the show – to develop a plan for retaking the city.

Free also received permission from London for full embedding. He hoped there was time to restore Britain's battered reputation and was bolstered by the return of Richard Iron. The British mentor had spent a single day with his family in the UK before heading straight back to Iraq once he heard about the operation. He was frustrated to have missed the opening of the campaign, when obvious blunders could have been avoided, but was privately delighted to see the Iraqis trying to take back their city. He met Mohan at the Iraqi army HQ outside Basra at dawn. Mohan was all for charging straight into Hayaniyah, and Iron had to deploy every trick he had learned from his time with the general to persuade him to focus on the more limited task of clearing the main routes into Basra.

With American airpower at their disposal, the Iraqi troops overwhelmed insurgent defences along the roads within a few hours, and that evening at the Shatt al-Arab Hotel there was a tangible sense that the tide had turned. Iron was therefore shocked to learn that night that Maliki had declared a ceasefire. All too aware of the need for momentum in a campaign, Iron considered the move a disaster, but he was unaware of the hidden currents within Shia politics. At Maliki's request, a delegation of Iraqi politicians had travelled to the Iranian holy city of Qom to negotiate a ceasefire with the cleric Moqtada al-Sadr and the commander of the Al-Quds Force, Qasem al-Soleimani.[18] The same day Iron returned to Basra the Iraqi prime minister had received assurances from Tehran that they would support a ceasefire. 'The Iranians were as concerned as the Americans about the chaos that would follow if Maliki lost,' noted Iraq's national security adviser Mowaffak al-Rubaie.

The truce had an immediate and dramatic effect. The morning it was declared, the mortars and rockets promptly ceased. Throwing open the gates of Basra palace, Maliki received a steady stream of tribal sheikhs and notables, to whom he handed small pink suitcases. Iron assumed they were stuffed with gold trinkets to buy their support. At the same time the Jaish al-Mahdi fighters were abandoning their defences, first in the outlying districts, then in their heartland Hayaniyah. Fartosi promptly disappeared, only to surface several months later in Beirut. Captain Jaffar, the police officer who had terrorised the city for so long, was spotted in Iran.

Iron greeted the news with shock and then a little ironic reflection. Despite all his efforts to engineer a confrontation between the Iraqi army and the Jaish al-Mahdi, in the end a backroom deal brokered in Tehran had reclaimed the streets of Basra. A few days later, with the ceasefire holding, Iraqi troops began entering Jaish al-Mahdi strongholds. There was sporadic rocket fire, quickly snuffed out by airstrikes, but the real danger was from hundreds of primed roadside bombs. When soldiers entered the Hayaniyah district, former home of the Jamiat police station, they found a warren of boody-trapped houses and underground lairs, but other than the odd potshot, the Jaish al-Mahdi had vanished.

Iron was not disappointed that the Iraqi army had missed the opportunity to prove itself or that the deal appeared to confirm Iranian power in the region. Events had demonstrated that the Iraqis were prepared to confront their own problems and deal with them in their own way. For the first time in weeks Iron saw families on the streets: women in black *abayas* exchanging festive greetings and children running through the rubble. One Iraqi man grabbed hold of Iron's arm. 'Thank God the Jaish al-Mahdi have gone,' the man exclaimed. 'They have no honour. They raped our women and killed our brothers. And they are such terrible drivers!'[19]

Over at the airport Brigadier Free was handling the arrival of a contingent of British troops to embed with the Iraqi army. He was also trying desperately to restore American faith in British competence. That was probably more than he could manage in the remaining weeks of his tour, with Afghanistan and not Iraq likely to be the latest proving ground for British forces, but Free did make one convert. On 25 April General George Flynn was at Basra airport to see off the body of Lieutenant Matthew Vandergrift, a twenty-eight-year-old West Point graduate who had been one of the first into the city with Iraqi troops. He had been killed by an IED along with six Iraqi colleagues early in the fighting.

His coffin, draped in an American flag, was lifted from the back of a truck by half a dozen pall bearers and carried to a waiting plane between an honour guard of British and American soldiers. Suddenly a siren sounded to warn of incoming mortar fire, and half the American soldiers dropped to the ground. Regimental Sergeant Major John McCallum of the Scots Guards bellowed out, 'Stand fast, lads.'[20] The British soldiers remained in place as Vandergrift was carried onto the plane.

It was a touching moment for General Flynn. 'I never doubted the bravery of British soldiers,' he said.[21]

Epilogue

THE NEXT BATTLE

The Charge of the Knights operation in Basra proved a watershed. Within weeks of the Iranian-brokered peace deal, Iraqi forces had secured the city and swept north to Amarah, the capital of Maysan province. Ever since the 2003 murder of the Red Caps in Majar al-Kabir, a few miles outside Amarah, successive British commanders had viewed the province to be an ungovernable morass of tribes and Shia militias but once Iraqi troops appeared, backed by US forces, the Jaish al-Mahdi fled. The militia was similarly routed in Baghdad.

As had been the case with the Sunni insurgency in central Iraq, the key to defeating the Jaish al-Mahdi ultimately lay in the hands of Iraqis themselves, and a combination of political opportunism, war weariness and the rejection of Islamic radicalism. In provincial elections held in January 2009, Maliki's party swept aside local Sadrist politicians who had held sway in Basra and Amarah since 2005. Support for Moqtada al-Sadr did not disappear, suggesting that his appeal to the Shia poor would remain an important voice in Iraqi politics. But against the odds, democracy appeared to have taken hold in Iraq – the most enduring legacy of the Anglo-American occupation.

By the time the bulk of British forces left Iraq over the summer of 2009, peace had brought a quiet prosperity to Basra.[1] Along the main street that curves along Basra's old creek, shoppers packed stores selling mobile phones and satellite dishes, gaudy print dresses and CDs of Arabic pop artists. In the evenings groups of men and women once again drifted from the concourse to the corniche, where they caught the soft breeze, watched the sun set over the Shatt al-Arab, and cast furtive glances at each other. Some of the young men carried beer cans; some of the women wore jeans, not *abayas*. The militias, who once would have punished such offences with death, had disappeared.

The memory of six years of British occupation and the tyranny of the Jaish al-Mahdi has begun to fade. Some Iraqis in Basra are still grateful to the British and Americans for overthrowing Saddam Hussein but remain sceptical about Western intentions, believing that London and Washington planned the chaos of post-war Iraq in order to maintain their military presence. Many Iraqis feel that the modicum of calm they enjoy was achieved *despite* the Coalition, not because of it. More recently, they have eyed the 2011 revolutions in Tunisia, Egypt, Libya, Yemen, Bahrain and Syria and wondered what might have been but for Western intervention in their own affairs. As Haider Samad, the Iraqi translator who once held such high hopes for the British occupation reflected, 'one of the ironies of the war is that it robbed the Iraqis of their own ability to choose their future.' Haider himself lives in a council house in Hull, where he moved as part of a British government resettlement programme in 2008. He and his wife Nora now have two children. They long to return to Iraq but Haider is reluctant, still haunted by the past few years.

Iraq is not without its problems: Maliki's officials have proven as corrupt as their predecessors in siphoning off personal profit from the country's main industry, oil. With the fourth largest reserves in the world, Western oil giants have been eager to start drilling, although US companies have not fared well in early bidding for Iraqi contracts. In apparent recognition of anti-American sentiment in the country, the Iraqi government awarded US oil firms the drilling rights to only one of the country's eleven largest fields – an irony not lost on either the war's supporters or its detractors, who saw the invasion as a US attempt to take control of the country's natural resources. With the Iraqi government seeking to increase production from 2.5 million barrels per day to 12 million by 2017 the benefits of staying in power are huge. Unchecked corruption may yet defeat the fledgling democracy.

Iraq's police service remains as shady and brutal as ever. In September 2008, 200 malnourished Iraq detainees were discovered in a secret prison in Basra. Many of the detainees bore signs of torture. Little consolation was drawn from their discovery by the Iraqi parliament's Human Rights Commission. The Iraq war logs, a trove of leaked US military reports that cover the occupation until the end of 2009, suggest that the police have continued to torture and abuse prisoners, with Coalition forces continuing to turn a blind eye.[2]

Iranian influence also remains strong; its militia proxies might have faded into the background, but its agents are often spotted in Baghdad's Green Zone. In the 2010 national elections, Maliki lost the popular vote to the former scion of the West, Ayad Allawi, now backed by Saudi Arabia and other Sunni Arab regimes. Maliki was only able to muscle Allawi aside after Tehran pushed Moqtada al-Sadr to support the incumbent prime minister. Tehran's ability to forge a partnership between Sadr and Maliki so soon after the Charge of the Knights operation was a reminder of its power in Iraq – and of its ability to disrupt the country should the West confront Iran over its nuclear program.

Despite the lingering concerns, defenders of the war such as Tony Blair have quickly laid claim to this Iraq's modest successes. The reasons for doing so are clear given the costs of the Iraq war. 179 British military personnel died in Iraq; 5,970 were injured, 222 suffering debilitating injuries such as the loss of limbs.[3] There are no accurate figures for the number of Iraqi civilian casualties in southern Iraq. The Associated Press reported 110,600 deaths for the whole country, although many suspect the figure to be much higher. The trauma of war also lives on, in the form of mental health disorders. Veterans under the age of twenty-four are two to three times more likely to commit suicide than their civilian counterparts.[4] Abuse, marriage breakdown, unemployment and homelessness are also more frequent. In total over 120,000 soldiers served in Iraq.[5] Given Vietnam-era levels of post-traumatic stress disorder, 15 per cent of men and 9 per cent of women might be expected to suffer poor mental health.

The financial burden of Iraq has been exorbitant, particularly in the light of the financial crisis. Gordon Brown has revealed that the Iraq war cost Britain about £1 billion a year, with the total bill to the Treasury totaling some £8 billion.[6] The majority of that was spent supporting the British military's presence in Iraq, although it includes £744 million pledged to the Iraqi reconstruction effort. That is a fraction of the estimated $1 trillion spent by the US on the wars in Iraq and Afghanistan, but in the post-recession era of domestic cuts to public services those still seem staggering sums to spend on a distant country of questionable importance to Britain's national interests.[7] To place them in context, in 2010, the new

Coalition government announced plans to axe £1 billion from the Education Department, which included funding for the provision of 715 schools. As schools go unbuilt in the UK, hospitals close, and tens of thousands of teachers, nurses, soldiers and policemen lose their jobs, the Iraq war has become a symbol of the profligacy and waste of the New Labour government.

In Afghanistan, the cost in human lives and money spent make for equally grim reading, with the prospect of a sustainable peace still distant. Britain's failure to provide adequate security or a meaningful reconstruction process in Helmand after three years ultimately forced the US military to intervene in 2009. In a move that would have been anathema just a few years before in Basra, a 30,000-strong American 'surge' in southern Afghanistan rescued the British military from an ascendant Taliban insurgency. Among the bases handed over to the Americans in 2010 was the district centre in Sangin. Four years after Will Pike's men occupied the building against the advice of village elders, 106 British soldiers had died defending it, and the town itself lay in ruins, a sad testimony to how badly a reconstruction mission can go wrong without the consent of the population.

The impact of the American surge in Afghanistan is unclear: security in places like Helmand has improved with the number of attacks against Coalition and local security forces at their lowest since the arrival of the British in Helmand in 2006. There are other encouraging national statistics that reflect the past decade's efforts: 85 per cent of the country now has limited access to basic medical care, compared to 9 per cent under the Taliban; 7 million more children are now in school, more than one third of them girls, up from 1 million in 2001. The Afghan economy is also booming with an annual growth in double digits. Yet despite the progress, the Afghan government's authority is limited to the main cities like Kabul and its security forces considered corrupt and ineffective. The surge in Iraq coincided with Sunni tribes turning against al-Qa'eda and seeking American allies to protect them from growing Shia power in Baghdad. No such dynamic exists in southern Afghanistan. Neither is there an educated class waiting to reassert civilian rule as there was in Iraq. The chances of creating such a class remain a generation away. There may be record numbers of children in Afghan schools but 2 million are still forced by the lack of schools to seek an education in the religious *madrassah* of northern Pakistan, where radical views of Islam are prevalent.

As of early summer 2011, 374 British military personnel have died in the war in Afghanistan, the vast majority since the 2006 occupation of Helmand province. 1,608 have been injured, 493 seriously. Again, there are no accurate figures for Afghan civilian casualties. UN figures suggest there have been over 10,000 Afghan casualties since 2006, although that figure does not include Afghan security forces or Taliban fighters, and is widely suspected to be much higher. Gordon Brown estimated that the war in Afghanistan has cost the British taxpayer £10 billion during its first nine years.[8] The death of Osama bin Laden in May 2011 was a reminder of the reason why Coalition forces originally invaded Afghanistan a decade before – and how all wars take on a momentum and justification of their own.

In seeking to understand what went wrong in Iraq and Afghanistan it is easy to blame Tony Blair, the man who authorised both wars. In his 1999 Chicago speech he displayed a nuanced understanding of the dangers of humanitarian invention. And yet in the run-up to the Iraq war he ignored many of the prescriptions he had set – pre-war planning, clear objectives, a long time frame. 9/11 clearly played a role in releasing him from such constraints, but Blair's flaws as a leader, his preference for grand gestures and sense of moral mission, always made the creed of humanitarian intervention open to abuse. Blair's foreign policy advisers should have tempered their master's enthusiasm but instead they were swept along by it. When there were dissenting voices at Number 10 – Clare Short's faltering rebellion being one – Blair's circle dismissed them too readily.

Blair has paid a heavy price. For some he is the disgraced leader who misled parliament and plunged the country into an illegal war; for others he remains that tantalising figure of a great reforming prime minister whose early potential was consumed by Iraq. In May 2010, the New Labour government he helped to shape was finally voted out of office. Unlike the 2005 election, when Reg Keys stood against Tony Blair in his Sedgefield constituency, Iraq did not polarise the country – although the war still offered moments of political drama thanks to the opening of the Iraq Inquiry a few months before the election.[9] The Inquiry's ostensible aim was to learn the lessons of the conflict, but once it became clear that Blair would be giving evidence in public, the prospect of holding the former prime minister to account gripped the media. In January 2010, Blair appeared before the Inquiry in a packed 700-seat auditorium in central

London. He looked tense and agitated but struck a defiant pose, saying he did not regret the war or its aftermath. As Blair reflected in his memoir, he had no intention of giving the newspapers a headline such as 'BLAIR APOLOGISES FOR WAR.' 'AT LAST HE SAYS SORRY.'[10]

Such recalcitrance did little to mollify those in the audience like Reg Keys, who had hoped to hear Blair admit his mistake and provoke a national dialogue on the lessons to be learned. Seven years after his son Tom's death, the Ministry of Defence has not recognised that the murder of the Red Caps in Majar al-Kabir was preventable.[11]

Yet, Blair is not the only one who should shoulder responsibility for Iraq and Afghanistan. At crucial moments Britain's military leaders – the very men who understood the horrors of war and the vicissitudes of politics – gave poor advice. The individual vanity of senior officers explains some of the missteps. However, there was also an institutional failing that resulted from the military's 'can-do' attitude; a virtue in the heat of battle but not when calm-headed judgments about the feasibility of deployments are necessary. Once the operation had begun, overly optimistic planning fed down the chain of command, creating a culture of positive reporting that further distorted the picture. With each six month tour a new commander would arrive in theatre, declare how bad the situation had become, offer a change of plan, and then insist things had improved by the end of his tour.

Senior officers should have recognised the meandering passage of British strategy and changed it, either by extending tour lengths in the field to enable field commanders to shape a longer-term strategy, or by offering more clear-headed direction from London. Instead they continued to deny the seriousness of the insurgency in southern Iraq, even when events like the hostage crisis at the Jamiat police station clearly indicated the effort was failing. Likewise with the Helmand deployment, military planners lacked understanding of the dangers they faced in Afghanistan and ignored the warnings of experts in the field.

With hindsight some British generals now realise how their advice brought the British Army close to defeat in Iraq and Afghanistan. Lieutenant General Rob Fry, one of the architects of the Helmand operation, concedes that UK forces rushed too quickly out of Iraq without completing the mission. What was meant to be a measured withdrawal from one successful operation to another turned into the dreaded war on two fronts, bringing both campaigns to the verge of defeat. He speculates whether the current British effort in Afghanistan is less about fighting the Taliban, and more about rehabilitating the Army's battered standing. 'We are atoning for the mistakes of Iraq with blood in Afghanistan,' noted Fry.

But such recognition only touches the surface of where the MOD has gone wrong in recent years. The uncomfortable truth is that decision-making has been driven as much by fear of budget cuts as by military expediency. In an era of frugality, senior generals have come to see war as a way of guaranteeing funding. 'Use it or Lose it' has long been the mantra in the MOD. Furthermore, the use of emergency funding to pay for operational requirements during the campaign has become a tacit way of re-equipping ground forces, thus saving money from the main budget.

The MOD's stance is understandable if one accepts the vast range of possible war scenarios they prepare for – everything from nuclear holocaust and ground invasions of Europe to counter-insurgency and cyber war. There is a long-standing debate in the MOD over whether the military should focus on certain types of warfare. So far, no chief of the defence staff has done so for fear of shifting the balance of forces between the army, air force and navy. Instead they have sought to meet every challenge going with often disastrous results, as Future Rapid Effects System demonstrates – expensive equipment designed for a future war that might never be fought. Of course, the MOD is not the only government department to face strained budgets after poor decision-making. The difference is that when the going gets tough, going to war can unlocks billions of pounds worth of fresh funding. With that incentive is it any wonder that Britain has been at war so often in the past fifteen years?

One solution is to separate the military's lines of reporting and planning from the advice given to ministers. A similar change took place in the UK's intelligence services after pre-war reporting on Iraq was judged to not have been thorough enough. However, there are considerable challenges to making this reform. The Post-Conflict Reconstruction Unit was initially set up in 2004 to provide impartial assessment of possible deployments but its independence threatened entrenched government departments and the effort fizzled out. A second attempt to unite Whitehall to evaluate potential threats, this time with a US-styled National Security Council, has so far proven to be a glorified talking shop since its inception in 2008.[12]

Changing how decisions are made pre-deployment should only be the first step in overhauling how Britain goes to war. The military must discard the idea of a 'quick' war. The average length of a counter-insurgency campaign is about fourteen years – a much more realistic frame of reference than the three years envisioned for the southern Afghanistan deployment.[13] Thinking over such a time frame would require planners to address deeper questions about the long-term health and sustainability of the communities in which they are intervening, and help prevent the reactive policy making that has crippled recent strategies.

Six-month rotations for the military should also be abandoned. The idea that a commander or soldier can develop meaningful relations with local communities over that time frame is naïve. Short tours were meant to reduce strain on soldiers – unfortunately they contributed to the failure of the missions in Iraq and Afghanistan, thereby exposing soldiers to greater pressures.[14] There has been some reform in this area. In September 2007, the former commander in Basra, Lieutenant General Jonathon Riley insisted on year-long deployment as deputy head of NATO in Afghanistan, and since then senior officers and their headquarters have been required to stay in country for at least a year. However, the junior and mid-ranking officers who make the vital connections on the ground remain on a merry-go-round of rotations.

The military must also re-examine the nature of its mission. No matter how sensitive an officer might be to a community, there is a limit what British soldiers can achieve. As commanders in Iraq like Riley realised early on, the key to success is in training local forces to take over security. Although lip service has been paid to this idea for years in Iraq and Afghanistan, the military's priority remained offensive operations and self-preservation. In 2005, the military decided not to 'embed' British forces with their Iraqi counterparts below the level of colonel. The concern was twofold: firstly that British soldiers at company level would require additional resources to protect them, which senior generals were not willing to commit to as withdrawal loomed. The second, entirely mistaken reason was that the Iraqi military would develop a 'dependency' culture on their British allies. The result was that for three crucial years, Iraqi 10 division fell under militia influence and collapsed altogether during the Charge of the Knights operation.

Lessons have since been learned, and the decision not to embed was subsequently reversed. Yet the example set by officers like Colonel Richard Iron during the Charge of the Knights is too often the exception. Iron understood that success depended not on teaching military tactics or drill techniques but in understanding how Mohan thought. His engagement was as much diplomatic as it was military – a reminder of the need for an overarching political strategy in the first place.

Even with these changes, the military cannot expect to intervene successfully without a civilian effort to supervise reconstruction and provide political solutions. The FCO and DFID were wholly unprepared for the challenges of Iraq and unable to adapt to adapt to the needs of the military or the Iraqis they found themselves in charge of. From the start, DFID was ideologically opposed to the invasion, which many diplomats viewed as a wholesale rejection of the department's post-colonial philosophy of working with local leaders to empower themselves. Health and safety guidelines further hindered them in the field, and drove a wedge between the civilian and military effort. Relations have improved since the disastrous experience of the Provincial Reconstruction Team in Iraq in 2006, but DFID remains uncomfortable with the implications of nation building. Success in places like Afghanistan requires diplomats to have a long-term immersion in the region, fluency in the language, and the readiness to take risks – a job description more suited to a colonial officer from the days of empire than a modern-day bureaucrat.

THE PROSPECT OF EXPENSIVE, DRAWN-OUT missions of limited ambition is likely to make wars more difficult to sell to the public – and that is precisely the point. In a parliamentary system in which the power to declare war is concentrated in the hands of the prime minister, public opinion provides a vital check. Tony Blair chose to ignore widespread reservations about the war; the result was deep divisions in the country and a fractured war effort.

Since coming to power, Prime Minister David Cameron has brought a greater transparency to the decision-making behind the use of force. When Cameron announced the use of British airpower against Libya in 2011, he made sure that UN backing for the intervention was clear. The attorney general's advice was published as soon as it was given, and the issues were later debated fully in Parliament – in marked contrast to the veil of secrecy that shrouded Iraq.

And yet, the Libya campaign has borne many of the hallmarks of earlier campaigns: hasty planning without clearly articulated long-goals; a military on the defensive from a sceptical public and desperate to stave off further budget cuts.[15] Cameron has insisted on the use of air power to assist the Libyan rebels – marking a rejection of the 'boot-on-the-ground' approach taken in Iraq and Afghanistan. But far from heralding a new approach to humanitarian intervention, the reliance on air power has seen a return to the thinking of the early 1990s, which gave us the desultory effort in the Balkans and Iraq and the belief that ground forces were necessary for a successful intervention.

Cameron's problem is that he shares with his predecessors a fundamental belief in the ability of Western powers to intervene in troubled regions; a conviction that the right war, with sufficient support and legal backing, can succeed. This outlook is based on a classically liberal and humanist view of the world and underpinned by what the philosopher John Gray has labeled the 'myth of progress' that stretches back to the Enlightenment. Yet, the past decade of war provides scant evidence to support such faith in intervention. Until Britain's leaders recognise the limits of intervention, the country is bound to repeat the mistakes of the past.

DRAMATIS PERSONAE

Abu Azzam, the leader of the Sunni insurgent group Jaish al-Islami who renounced violence at British urging during a pivotal moment in the battle for Baghdad in 2007.

Abu Hatem, a tribal leader from Maysan province who shifted allegiances between the British and Shia militia groups in 2004.

Abu Khadr, Wissam, the leader of Jaish al-Mahdi militia 2006–7.

Alderson, Alex, a British army doctrine writer.

Alderson, Andrew, an army reservist and southern Iraq's finance minister 2003–4.

Aldred, Margaret, deputy director of the security and defence committee, one of the overseers of the 2006 deployment to Helmand province.

Allawi, Ayad, Iraq's interim prime minister 2004–5.

Asquith, Dominic, the head of the Iraq desk at the FCO 2004–5 and ambassador to Baghdad 2006–7 who approved the partial evacuation of Basra palace following the Queen's Birthday party.

Baldwin, Gil, a commander of an early British detainment facility in southern Iraq. Later served as duty head of the post-conflict reconstruction unit, the British government's effort to plan and coordinate future overseas operations.

Bashall, James, a brigade commander in southern Iraq in 2007 during the withdrawal from Basra palace.

Bell, Gertrude, a colonial-era British diplomat who played an important role in establishing the state of Iraq.

Binns, Graham, a brigade commander during the 2003 Iraq invasion, and commander of British forces in southern Iraq 2007–8 during the contentious deal with a Shia militia leader.

Blair, Tony, prime minister 1997–2007.

Bourne, John, the governor of Dhi Qar province in 2003.

Bowron, Johnny, a company commander in Maysan in 2004, who later led a battle group in Basra during the British operation that sought to regain control of the city in 2006.

Brackenbury, Alan, a lance corporal who became the first British soldier to be killed in Iraq with a new type of precision-made roadside bomb in 2005.

Bradley, Tobin, American political adviser, Dhi Qar province 2003–4 who oversaw some of the first local elections in Iraq.

Bratcher, Simon, a military bomb disposal expert in Maysan province in 2005 who identified a new type of precision-made roadside bomb.

Bremer, Paul, civilian head of the US administration in Iraq, the Coalition Provisional Authority 2003–4.

Bristow, Andy, the mentor to the Iraqi general commanding southern Iraq in 2007 who helped conceive of the Iraqi-led operation that took on the Shia militia in Basra.

Broome, Christopher, a section commander during the ambush at the Danny Boy checkpoint outside Majar al-Kabir in 2004 who later suffered from post-traumatic stress disorder.

Brown, Gordon, chancellor and prime minister 2007–10.

Browne, Des, defence minister 2006–7.

Bryan, Terry, a sergeant who led his men to safety after being caught in an ambush by Shia militia in Basra in 2004.

Bush, George, US president 2000–8.

Butler, Ed, the commander of British forces in Afghanistan during the 2006 Helmand deployment.

Campbell, Alastair, Tony Blair's communications director 1994–2003.

Carter, Nick, a brigade commander in southern Iraq during the April 2004 uprising against the occupation. In 2002, he had conceived of provincial reconstruction teams to deliver aid and reconstruction in Afghanistan, an idea that was later used in Iraq.

Casey, George, the American commander of Coalition forces in Iraq 2004–6.

Chakrabarti, Suma, the senior civil servant at the Department for International Development, 2002–8.

Clissit, Nick, the British officer who conceived of the post-conflict reconstruction unit, the British government's effort to plan and coordinate future overseas operations.

Cowan, James, the commander of a battle group deployed to Sunni tribal areas during the second battle of Fallujah in 2004, who later served as the chief of staff to the commander of British forces in southern Iraq in 2006.

Cross, Tim, a logistics specialist and member of the British military's planning team sent to US military headquarters. Cross subsequently became deputy head of the Office for Humanitarian Reconstruction and Assistance, Iraq's first post-war administration.

Curry, Charlie, the commander of the British headquarters in Amarah during the later part of a siege by the Jaish al-Mahdi in 2004.

Dannatt, Richard, the commander of the allied rapid reaction corp 2003–4, the UK-led deployable NATO headquarters, then commander-in-chief at Land Command 2005–6. He warned of the dangers of the military being overstretched, before seeking to relieve some of those pressures as the head of the British army, 2006–9.

Daoud, Mohammed, the governor of Helmand province 2006–7.

Ellis, Lee, a soldier killed in Maysan province in 2006 whose sister Karla later became an anti-war campaigner.

Etherington, Mark, the governor of Kut 2003–4, where he survived a siege during the April 2004 uprising. He was the operations officer for the post-conflict reconstruction unit in 2005, and warned Whitehall of the lack of knowledge about Helmand before the British deployment, before becoming head of southern Iraq's first provincial reconstruction team in 2006.

Everard, James, a brigade commander in southern Iraq in 2006 during the British operation that sought to regain control of the city.

Faraji, Mohan, see Mohan

Fartosi, Ahmed, the commander in Basra of the Shia militia called the Jaish al-Mahdi.

Featherstone, Justin, the commander of the British headquarters in Amarah during the early part of a siege by the Jaish al-Mahdi in 2004.

Flynn, George, the US Marine commandant who oversaw Coalition support for the Iraqi-led operation to defeat the Shia militia in Basra in 2008.

Free, Julian, the brigade commander in Basra during the Iraqi-led operation to defeat Shia militias in the city in 2008.

Fry, Rob, the deputy chief of the general staff for operations 2003–6, the key strategy role at the Ministry of Defence, where he planned for the deployment to Helmand and withdrawal from Iraq.

Gharawi, Ahmed Abu Sajad, commander of the Shia militia called the Jaish al-Mahdi in Maysan province.

Goldsmith, Peter, attorney general 1997–2007, the senior legal adviser to the government, whose decision that the war was legal paved the way for the UK's involvement in the Iraq invasion.

Greenstock, Jeremy, the British ambassador to the United Nations during the build-up to the Iraq invasion and the senior UK representative in Baghdad 2003–4.

Haider, Samad, Iraqi translator.

Hamilton-Jewell, Simon, a sergeant in the Royal Military Police who was one of six soldiers to be murdered in Majar al-Kabir in 2003.

Harding, Paul, a company commander in Basra who was killed manning a British outpost in the city shortly before it was handed over to Iraqi control in 2007.

Heatley, Charles, a British spokesman for the US occupation in Baghdad in 2003 and an adviser to Iraq's interim prime minister, Ayad Allawi, in 2004.

Henderson, Nick, a battle group commander in Basra during the 2005 hostage crisis involving two SAS soldiers who were held at the Jamiat police station.

Hickey, Christopher, a sergeant killed checking for roadside bombs in 2005, whose mother Pauline later campaigned against Snatch Land Rovers.

Holmes, Richard, a captain who was killed in Amarah in 2006, sparking a period of violence in southern Iraq that only ended with the British deal with a Shia militia leader.

Hoon, Geoff, defence secretary 1999–2005.

Houghton, Nick, the head of the British military headquarters in Northwood 2005–7, who steered UK forces through the withdrawal of Iraq.

Hughes, Chris, brigade commander in southern Iraq in 2005, and later, senior planner at the MOD before the 2006 Helmand deployment.

Hussein, Saddam, Iraqi dictator 1978–2003.

Iron, Richard, the senior mentor to General Mohan, the commander of Iraqi forces in Basra 2007–8. Iron helped conceive and execute the Iraqi-led operation to take on the Shia militias in Basra in 2008.

Jackson, Mike, the head of the professional army, 2003–6, a period when the military came close to breaking point.

Jadiri, Basima, senior adviser to Iraqi Prime Minister Nouri al-Maliki whose friendship with Emma Sky helped relieve tensions between the Iraqi government and the US military.

Jaffar, commander of the serious crimes unit located at the Jamiat police station 2004–6 who played an instrumental role in the detention of two British SAS soldiers.

Jarvenpaa, Minna, member of the post-conflict reconstruction unit 2004–7.

Jones, Rupert, the chief of staff of the brigade headquarters that took charge of the hostage crisis involving two British SAS soldiers at the Jamiat police station in 2005.

Karzai, Hamid, Afghan president, 2004–

Kearney, William, ArmorGroup manager 2004–6, the private security firm training and mentoring Iraqi police in southern Iraq who tried to raise concerns about the abuse of detainees by local security forces.

Keys, Reg, father of murdered Red Cap Tom Keys, and independent candidate for Tony Blair's Sedgefield seat in the 2005 general election.

Khazali, Qais, commander of the Shia militia called the Jaish al-Mahdi 2003–4, then leader of a splinter group 2004–7 that refused to recognise a ceasefire with Coalition forces.

Labouchere, David, commander of British forces in Maysan province in 2006 during the withdrawal from Camp Abu Naji.

Lamb, Graeme, the commander of British forces in southern Iraq in 2003, then a director at the Land Warfare centre, in charge of re-writing British doctrine 2004–6, before becoming deputy commander of Coalition forces in Baghdad 2007–8, where he played an instrumental role in turning the tide against Sunni insurgents.

Lamb, Robin, senior British diplomat in Basra in 2006 who sought to define the UK's legacy in southern Iraq.

Lorimer, John, brigade commander in southern Iraq during the 2005 hostage crisis involving two SAS soldiers held at the Jamiat police station.

MacCartney, Charlie, a British police mentor to the Jamiat police station 2004–6.

Maciejewski, Justin, a battle group commander in Basra during the British operation to regain control of the city in 2006.

Maer, Matt, commander of Maysan province in 2004 during the first heavy battles with the Shia militia called the Jaish al-Mahdi. Later served as a planner for Iraq and Afghan operations at the British military headquarters in Northwood.

Maliki, Nouri, Iraqi prime minister 2006–

Manning, David, Tony Blair's foreign policy adviser during the build-up to the invasion and ambassador to Washington 2003–7.

Marriott, Patrick, the chief of staff to the commander of British ground forces during the invasion of Iraq, then a brigade commander in southern Iraq in 2006, and subsequently a senior Iraq planner in British military headquarters in Northwood, where he raised concerns about a too-hasty withdrawal.

Marsden, Rosalind, an ambassador to Afghanistan 2004–6, then senior British diplomat in Basra in 2006.

Marston, Daniel, an American academic and instructor at Sandhurst 2004–7 who sought to transform Britain's approach to counter-insurgency.

McChrystal, Stanley, US Special Forces commander 2003–8.

Mercer, Nicholas, the British military's senior lawyer in southern Iraq during the summer of 2003 who first raised concerns about the UK's treatment of detainees.

Messenger, Gordon, the chief planner for the 2006 deployment to Helmand whose findings were later ignored.

Mills, Dan, a sergeant who led a sniper company during the siege of the British headquarters in Amarah by the Jaish al-Mahdi in 2004.

Mohan, the Iraqi general in command of Iraqi forces in Basra 2007–8 during the operation to take on Shia militias in Basra.

Morrison, William, managing director of Adam Smith International, an NGO that trained Iraqi Interim Prime Minister Ayad Allawi's staff how to govern.

Muhammadawi, Abdul Kerim Mahud, see Abu Hatem.

Odierno, Ray, the divisional commander in Iraq in 2003–4, and then put in charge of day-to-day running of Coalition forces in Iraq 2006–8.

Parker, Chris, the chief of staff of the brigade headquarters that took charge of the city after the invasion of Iraq.

Pennett, Miles, army reservist and briefly southern Iraq's minister of culture in 2003.

Petraeus, David, the commander of Combined Arms Centre in Fort Leavenworth, Kansas, 2005–6, where he transformed the US military's thinking about counter-insurgency before becoming commander of American forces in Iraq 2007–9.

Phee, Molly, the American governor of Maysan province 2003–4.

Philby, Harry St John, colonial-era diplomat who resigned from the British administration after disagreeing with the policy of rigging an Iraqi referendum that established a pro-British candidate as the first king of Iraq.

Pike, Will, a company commander during the 2006 Helmand deployment who led the first British troops into the Afghan town of Sangin.

Reid, John, defence minister 2005–6.

Rice, Condoleezza, US National Security Adviser 2000–4, then Secretary of State 2005–8.

Richmond, David, British special representative to Iraq in 2004 who persuaded the UN representative Lakhdar Brahimi not to resign at a critical moment of the April 2004 uprising.

Riley, Jonathon, a brigadier who contributed to early efforts to rebuild the Iraqi army, and was later a commander of British forces in southern Iraq in 2004.

Rogan, Janet, senior British diplomat in Basra during the summer of 2003, who formed the first British administration in southern Iraq since the colonial era.

Rubaie, Mowaffak, Iraqi national security adviser, 2003–

Rumseld, Donald, US Secretary of State for Defense 2000–6.

Sadoon, Ali, the Iraqi translator Haider Samad's uncle-in-law.

al-Sadr, Moqtada, the son of a popular ayatollah murdered by Saddam Hussein, who went on to found a nationalist Iraqi political movement and the Shia militia called the Jaish al-Mahdi.

Sajjad, Seyyed, right hand man of Ahmed al-Fartosi, the Shia militia commander who was released as part of the contentious deal with the British army.

Sanders, Patrick, a battle group commander in Basra who led the last British forces out of the city in 2007.

Sandiford, Tim, a battle group commander in Basra 2006–7, who led the operation that destroyed the Jamiat police station.

Sewan, Ali, the commander of the Jamiat police station 2003–6.

Shaw, Jonathan, the commander of British forces in southern Iraq in 2007 and the officer who brokered a contentious deal with a militia leader that enabled UK forces to safely withdraw from Basra palace.

Shearer, Richard, a second lieutenant who was killed in Iraq in 2005.

Sheinwald, Nigel, Tony Blair's foreign policy adviser 2003–7 who sought to marshal Whitehall to confront the crisis in Iraq, before serving as ambassador to Washington 2007–

Shirreff, Richard, the commander of British forces in southern Iraq in 2006 who led the British operation that sought to regain control of the city.

Short, Clare, minister for the Department for International Development 1997–2003.

Siddique, Rabia, a member of the British military legal team who was sent to the Jamiat police station in 2005 to try and resolve the hostage crisis involving two SAS soldiers.

Sistani, Ali, a Grand Ayatollah and Iraq's most senior Shia cleric.

Sky, Emma, the governor of Kirkuk, 2003–4, and then a political adviser to American General Ray Odierno 2007–10.

Smith, Colin, the senior British police adviser in 2005 who sought to reform the Iraqi police in Basra.

Soleimani, Qasem, the commander of Iranian Al-Qods force who directed operations in Iraq against Coalition forces.

Stewart, Andrew, the commander of the British force in southern Iraq during the April 2004 uprising against the occupation.

Stewart, Rory, deputy governor of Amarah and Nasiriyah 2003–4.

Stirrup, Jock, the chief of the defence staff 2006–10 who succeeded in withdrawing British forces from Iraq and sought to stabilise the growing insurgency in Helmand.

Straw, Jack, foreign secretary 2001–6.

Synnott, Hilary, the senior British diplomat in southern Iraq 2003–4, who moved the British headquarters to Basra palace.

Tansely, James, a career diplomat and senior British diplomat in southern Iraq 2005–6 who hosted the Queen's Birthday Party in 2006 that led to the partial withdrawal from Basra palace.

Tootal, Stuart, 3 Para battle group commander during the 2006 Helmand deployment.

Waeli, Mohammed Musbeh, the governor of Basra 2005–8.

Wakefield, Anthony, a soldier killed on the eve of the 2005 general election.

Walker, Michael, the head of the professional army, in the build-up to the invasion of Iraq and then chief of the defence staff when the decison was taken to send British forces to Helmand.

Wall, Peter, deputy head of the British military headquarters in Northwood 2004–6.

Warrick, Tom, the US State Department official who wrote 'The Future of Iraq Project,' a thirteen-volume plan for post-war Iraq that was discarded by the Pentagon.

Webster, Martin, a corporal who filmed the abuse of Iraqi detainees by British soldiers during violence in Amarah in 2004, which was later published by the *News of the World.*

White-Spunner, Barney, the head of the Joint Forces Headquarters 2003–5, Northwood, which performed the initial planning work for British deployment to Helmand, and then the commander of British forces in Iraq in 2008.

Whitford, Victoria, a press officer for the US administration in Baghdad in 2004, and then for the British Embassy 2004–5.

Williams, Andrew, the military commander of Maysan province in 2005 who first encountered the devastating effect of precision-made roadside bombs.

Williams, Richard, lieutenant colonel and SAS commander.

NOTES

PROLOGUE – THE BIRTHDAY PARTY

1 James Tansley, interview 18 August 2010
2 Private interview
3 *News of the World*, 14 February 2006
4 Dominic Asquith, interview 25 November 2009
5 Private interview
6 James Tansley, interview 18 August 2010
7 Ole Jepsen, interview 1 August 2009
8 Rachel Schmeller, interview 24 September 2009
9 Muzahim al-Tamimi, interview 11 August 2010
10 DFID, 'Eliminating world poverty: A Challenge for the 21st Century,' 1997, cited in Anthony Seldon (ed.), *Blair's Britain:* 1997–2007, Cambridge University Press, 2007, p.555
11 Tony Blair, Iraq Inquiry, 21 January 2011
12 'Iraq: Options Paper Secret UK Eyes Only,' 8 March 2002
13 Prime minister to Powell minute 17 March 2002. He went on: 'A political philosophy that does care about other nations – e.g. Kosovo, Afghanistan, Sierra Leone, and is proud to change regimes on the merits, should be gung-ho on Saddam. So why isn't it? Because people believe we are only doing it to support the US, and they are only doing it to settle an old score. And the immediate WMD problems don't seem obviously worse than three years ago. So we have to reorder our story and message.'
14 Jonathan Powell, Iraq Inquiry, 18 January 2010
15 Lawrence Freedman, *National Interest*, 20 April 2010
16 Ibid.

1 – STRATEGIC COMMUNICATION

1 Tim Cross, interview 19 August 2009
2 Tony Blair, Iraq Inquiry, 29 January 2010
3 Christopher Meyer, interview 9 September 2010
4 Private interview. See also *Iraq: Conditions for Military Action*, Cabinet Office Paper, 21 July 2002, published in *Sunday Times*, 12 June 2005. The British ambassador to the US, Christopher Meyer, believes Blair never set out these conditions to Bush, presenting British positions in a much more equivocal light. Andrew Rawnsley, *The End of the Party: The Rise and Fall of New Labour*, Penguin, 2010, p.96.
5 An August ICM poll quoted in the *Guardian* found that 52 per cent of Britons were opposed to military action and just 33 per cent were in favour. See http://www.icmresearch.co.uk/media-centre-polls.php?year=2002.
6 Jack Straw, Iraq Inquiry memorandum, 08 February 2010. Under Chapter VII of the United Nations Charter, a nation's use of force is authorised in individual or collective self-defence, as outlined in Article 51, or pursuant to a Security Council resolution, as outlined in Article 42. The legality of intervening on humanitarian grounds, as Blair argued for in Chicago, is not widely accepted.
7 On 23 July 2003 Straw made his concerns clear in the famous 'Downing Street Memo'. Leaked to Michael Smith of the *Sunday Times*, the memo revealed for the first time the belief among senior British officials that by the summer of 2002 Bush had already decided to go to war. The memo recorded the head of the Secret Intelligence Service warning that 'intelligence and facts were being fixed around the policy'.
8 Peter Goldsmith, memo to prime minister, Iraq Inquiry, 30 July 2002
9 As the independent inquiry into the UK's use of intelligence, the Butler Report, later revealed, Blair dismissed the system of parliamentary committees and Cabinet Office steering groups as slow and ineffective.
10 Jonathan Powell, Iraq Inquiry, 18 January 2010
11 Andrew Rawnsley, *The End of the Party* p.99. The letter goes on to urge Bush to go down the UN route. Blair's emphasis on unwavering support rather than the important caveats exasperated some of his advisers. Rawnsley writes of the letter, 'When [Sir Christopher] Meyer [Britain's ambassador to the US] learnt of it, he rang [David] Manning [Blair's foreign policy adviser] in horror. "It's a brilliant note except for this bloody opening sentence: 'Whatever you do, I'm with you,'" the ambassador expostulated. "Why in God's name has he said that again? He's handed Bush carte blanche." Manning sighed down the phone: "We tried to stop him. We told him so, but he wouldn't listen."'
12 The Butler Inquiry later revealed that MI6 had received the claim 'third hand' and that there were considerable doubts in the intelligence community about the

material. In preparing the dossier, Joint Intelligence Committee chairman John Scarlett responded to complaints from Alastair Campbell and 'tightened' the language to remove the speculative nature of the evidence. John Scarlett, email to Campbell, 18 September 2002. Major General Michael Laurie, a former director general of the Defence Intelligence Staff, gave evidence to the Iraq Inquiry that the dossier was drawn up to 'make the case for war, rather than setting out the available evidence.' Laurie's statement contradicted Campbell's claim that the dossier was an impartial presentation of the evidence, making the general the first senior official to concede some form of manipulation had taken place.

13 Clare Short, Iraq Inquiry, 2 February 2010. Short did not see the original Cabinet Office paper prepared for Blair's Crawford meeting. Suma Chakrabarti, in his evidence to the inquiry on 8 February 2009, noted the DFID was banned from talking to NGOs or the UN about war preparations. Despite the order, the DFID spoke to the UN anyway.

14 The US commander at Central Command, Tommy Franks, had appointed a one-star general to look at post-war Iraq, although as Tim Cross noted in his testimony to the Iraq Inquiry this was a peripheral member of Franks's headquarters, which saw itself as a 'war-fighting team'. Iraq Inquiry, 6 January 2010

15 Michael Jackson, Iraq Inquiry, 29 July 2010

16 Michael Boyce alleged that Downing Street had forbidden him to talk to the chief of defence (logistics), Iraq Inquiry, 21 December 2009. Geoff Hoon told the inquiry that the enhanced body armour did not arrive into Iraq until 31 March the following year. As Hoon claimed at the Iraq Inquiry, 'When we both [Hoon and Lord Boyce, chief of the defence staff] went to meetings in Downing Street saying, "Look, you need to get on with this", we were told, "Calm down, you can't get on with it while the diplomatic process is under way."' He went on: 'The argument I was given very clearly from the prime minister and the foreign secretary [Jack Straw] was that if we were seen to be overtly preparing for war, that would affect our ability to secure a [UN security council] resolution.' (19 January 2010)

17 Sergeant Steven Roberts, from Shipley, West Yorkshire, was accidentally killed on the night of 23/4 March 2003, when UK troops opened fire during a disturbance near Basra. A MOD inquiry into his death confirmed Roberts would have survived if he had been wearing enhanced combat body armour (ECBA). During the inquest an MOD director, David Williams, said a request for the extra sets of ECBA, sent to Hoon on 13 September 2002, was returned by the minister with the annotation 'Further advice required.' Approaching manufacturers would have had the effect of announcing UK war preparations, Williams said. Roberts was wearing combat body armour (CBA) at the time, after being issued with ECBA only for it to be withdrawn four days before his death owing to shortages.

18 Private interview
19 http://www.gwu.edu/~nsarchiv/NSAEBB/NSAEBB198/index.htm
20 Paul Bremer, interview 3 June 2009
21 Tim Cross, interview 19 August 2009
22 The unit was originally called the Iraq Planning Unit, but its name was changed to Iraqi Policy Unit to reflect the more strategic nature of its thinking and that Iraq was no longer in the planning stage, but grim reality. Iraq Inquiry, Dominic Chilcott, 8 December 2009
23 Private interview. Up until then the only official forum for discussing the invasion was the Ad Hoc Group for Iraq, a Cabinet subcommittee chaired by a senior MOD civil servant, Desmond Bowen.
24 Clare Short, Iraq Inquiry, 2 February 2010
25 Short made the point to Blair in a letter dated 15 February 2003: 'My department has tight budgetary constraints. We have major humanitarian disasters across the world and my resources are stretched. I'm happy to prioritise Iraq from my contingency reserve, but I cannot take resources from other poor and needy people to assist post-conflict Iraq. Without some understanding on finance, I cannot responsibly commit DFID to the exemplary partnership with MOD which we discussed.' Short noted the response: 'Treasury has no answer, nothing and it was this period of stand-off. Gordon Brown was pushed out and marginalised at the time, and having cups of coffee with me and saying, "Tony Blair is obsessed with his legacy and he thinks he can have a quick war and then a reshuffle", et cetera.' The chief secretary to the Treasury, Paul Boateng, finally responded after the invasion began: 'There is no money. Money is very tight, and, therefore, we have got to have a UN Resolution so we can get the World Bank and the IMF and all the others in.'
26 Clare Short, Iraq Inquiry, 2 February 2010. As she later noted to the Iraq Inquiry, 'I was thrown into a tizz by the thought I might be asking them to do illegal things.' Her legal arguments were: 'An occupying power is required to keep order, provide for humanitarian needs, and is not allowed to change the institutions of the country they occupy or its laws. So we knew that if we didn't get another UN Resolution, we were in big trouble. We could do humanitarian, but you can't reconstruct the country.'
27 As an MOD paper at the time noted, 'On public finance grounds there is a strong case for stepping back from military leadership in the aftermath and allowing other countries to take on this role.' John Dodds, *Post Iraq Military Options*, Iraq Inquiry, 4 March 2003. The DFID's permanent undersecretary at the time, Suma Chakrabarti, noted to the Iraq Inquiry, 'We did not think ORHA [Garner's organisation] – because we were still heavily believing in UN leadership of this

post-conflict effort – we didn't think ORHA would be actually the lead in terms of post-conflict work. We thought the UN would take that role on.'

28 The so-called 'dodgy dossier' was released by Alastair Campbell's press team on 3 February 2003. Sections of the dossier were plagiarised from an article originally written by academic Ibrahim al-Marashi in the journal *Middle East Review of International Affairs*.

29 She later resigned after being accused of mismanaging the Iraq's UN-run Oil-For-Food programme.

30 Tim Cross, interview 19 August 2009

31 The FCO's senior legal adviser, Michael Wood, did not think the resolution threatening 'serious consequences' provided legal cover, and had been quick to police comments by Cabinet ministers that suggested otherwise. In January Wood had been involved in a war of words with Jack Straw after he had told US Vice-President Dick Cheney that Britain would 'prefer' a second resolution but it would be 'OK' if they tried and failed to get one 'à la Kosovo'. The NATO-led operation in Kosovo was later justified on humanitarian grounds – a parallel to Iraq Wood was quick to dismiss, adding, 'I hope there is no doubt in anyone's mind that, without a further decision of the [UN Security] Council, and absent extraordinary circumstances of which at present there is no sign, the UK cannot lawfully use force against Iraq to ensure compliance with its Security Council WMD obligations . . . To use force without Security Council authority would amount to a crime of aggression.' Jack Straw gave the extraordinary response, 'I note your advice but I do not accept it.' Jack Straw, Iraq Inquiry, letter 29 January 2002.

32 Simon Walters, 'Proof Blair was told war could be ruled illegal,' the *Daily Mail*, 24 April 2005

33 After the invasion the French ambassador to the US, Jean-David Levitte, claimed he told the Americans, 'Don't ask for a second resolution because you don't need it. You can go without it,' a view that may have been coloured by the passage of time.

34 Peter Goldsmith, Iraq Inquiry, 29 January 2010

35 Hans Blix, Iraq Inquiry, 27 July 2010. In his evidence Straw noted that the draft document was withheld from other members of the Security Council.

36 Chirac's exact wording was: 'Whatever the circumstances, France will vote no because we believe, this evening, that there are no grounds to wage war, to achieve the objective we have set – the disarmament of Iraq.' At the Iraq Inquiry Blair's EU adviser Stephen Wall pointed out that Campbell manipulated Chirac's words. Chirac was only saying he would veto '*ce soir*' any resolution that gave backing for war because those circumstances had yet to be met. Campbell seized on the

words as saying that France would *never* back a resolution. Iraq Inquiry, 20 January 2011

37 Michael Boyce, Iraq Inquiry, 3 December 2009

38 Alastair Campbell, *The Blair Years: the Alastair Campbell Diaries*, Knopf, 2007, p.676

39 Michael Jay, interview 9 September 2010

40 Jack Straw, memorandum, Iraq Inquiry, 21 January 2010. Straw stated that he was also 'fully aware that my support for military action was critical. If I had refused that, the UK's participation in the military action would not in practice have been possible. There almost certainly would have been no majority, either in Cabinet or in the Commons.'

41 President Bush did agree to support the Roadmap a month later, an initiative that was doomed from the start with Ariel Sharon's government in power.

42 As Goldsmith's legal secretary David Brummell noted to the inquiry, 'The evidence had to be confirmed by someone.'

43 Clare Short, Iraq Inquiry, 2 February 2010

44 The paper, entitled *Humanitarian Strategy and Immediate Assistance: A Plan for Iraq*, was the first DFID plan to talk about reconstruction. As Chakrabarti summarised, 'One scenario was: that we didn't get the second resolution and we didn't get the UN mandate, and essentially we had a US viceroy in the field leading. That was the first scenario. The second scenario: that we didn't get a second resolution, but we did get a UN Security Council Resolution that did lead to the UN being in the lead on post-conflict effort. The third scenario: that we got both, the second resolution and the Security Council Resolution that we wanted.' As Chakrabarti concluded, 'So the definition for certainly us in DFID was that the exemplary approach was attributed to that third scenario, one that didn't actually obtain in the end.' Suma Chakrabarti, Iraq Inquiry, 8 December 2009

45 Private interview

46 Tim Cross, interview 19 August 2009

2 – BASRA MON AMOUR

1 Haider Samad, interview 14 May 2008

2 The writer Najem Ali recounts how Basra's first nightclub rose to fame by featuring young male dancers who dressed in women's clothes. The most famous, a Christian from Aleppo called Na'im, was shot and killed by a passionate admirer. http://wordswithoutborders.org/article/basra-stories/#ixzz1AMdz1VYx

3 – OCCUPATION

1 The brigade had pedigree for desert campaigning. As a division, the unit had fought in every major battle in North Africa during the Second World War, earning the nickname the Desert Rats in the process.

2 Chris Parker, interview 12 February 2010

3 Despite widespread jubilation in London and Washington, Majid survived the bombing, but was later captured in Baghdad in a raid led by British special forces. He was sentenced to death in 2007 by an Iraq court and hanged a few weeks later.

4 The three who died were Lance Corporal Ian Malone, Piper Christopher Muzvuru and Fusilier Kelan Turrington, of a total of thirty-four British soldiers who died before the end of major combat operations.

5 Gordon Messenger, interview 25 January 2010

6 Graham Binns, interview 14 April 2003

7 Geoff Hoon, letter 18 March 2003. Jack Straw also signed the letter. On the eve of war Hoon had warned the prime minister that troop levels needed to be lowered by two thirds before the autumn to avoid 'long-term damage to the armed forces'. Kevin Tebbitt, permanent undersecretary at the MOD, saw keeping no more than a battle-group-size force in Iraq – somewhere between 1,500 and 3,000 troops – for no more than two or three years. Kevin Tebbitt, Iraq Inquiry, 3 December 2009

8 Admiral Boyce, Chief of the Defence Staff, later outrageously claimed that as much time was spent planning for post-war Iraq as the initial invasion. Iraq Inquiry, 21 December 2009

9 Chris Parker, interview 12 February 2010

10 Toby Dodge, *Inventing Iraq: The Failure of Nation Building and a History Denied*, Columbia University Press, 2003, p.69

11 Chris Parker, interview 10 February 2010

12 Gil Baldwin, interview 15 April 2010

13 Another argument for not applying the European Convention on Human Rights in Iraq was that the British occupation did not exercise full jurisdiction, a prerequisite under Article 1 of the convention. The counter-argument to this was that the occupation required the army to exercise 'effective control' over territory, and therefore could be said to be exercising jurisdiction. There was also an exception sometimes known as the embassy principle, which treats an embassy as UK territory, where the convention would apply. In 2007, the British government conceded that detention was akin to being held on UK territory.

14 Ian Cobain, 'Humiliate, strip, threaten: UK military interrogation manuals discovered,' *Guardian*, 25 October 2010. A joint services interrogation centre was subsequently located at a military base in Ashford, Kent.

15 Rachel Quick, email 24 March 2003
16 Gil Baldwin, interview 15 April 2010
17 Nicholas Mercer, testimony, R vs Donald Payne court martial, 2006
18 The officer later showed Mercer another document that did reference hooding, but despite repeated requests the MOD has never been able to reproduce it.
19 Nicholas Mercer, testimony, R vs Donald Payne court martial, 2006
20 Lieutenant Colonel Mercer was subsequently prevented by the secretary of state for defence from talking about this incident and was the only officer I was prohibited from interviewing for this book.

4 – THE EMERALD CITY

1 The author was present at scene.
2 Charles Heatley, interview 26 May 2003
3 Condoleezza Rice, interview 12 November 2009
4 Paul Bremer, interview 3 June 2010
5 UN Secretary General Kofi Annan told Clare Short he refused to 'bluewash' the invasion and occupation. Clare Short, Iraq Inquiry, 2 February 2010
6 John Sawers, memo 7 May 2003. He further complained, 'No leadership, no strategy, no co-ordination, no structure, and inaccessible to ordinary Iraqis. Garner and his top team of sixty-year-old retired generals are well-meaning but out of their depth.'
7 David Manning, interview 5 August 2009
8 Paul Bremer, interview 3 June 2010
9 Tim Cross, interview 19 August 2009
10 Ambassador Bremer later claimed that his deputy Walt Slocombe fully briefed British officials on his plans, and that they should not have come as a surprise. Manning says the prime minister and he met Bremer on his way to Baghdad but there was no mention of disbanding the army.
11 David Manning, interview 5 August 2009
12 Anthony Seldon, *Blair Unbound*, Simon and Schuster, 2007, p.191
13 Cross retired from the military in 2007 after serving as general officer commanding for theatre troops, but without gaining further promotion.

5 – BREADBASKET

1 Adrian Bradshaw, interview 16 September 2010
2 Richard Williams, interview 10 October 2009. 'Albertine Jwaideh, 'The Marsh Dwellers of Southern Iraq: Their Habit, Origins, Society and Economy', University of Toronto, 30 October 2007.

3 Michael Walker, Iraq Inquiry, 1 February 2010. As Walker noted, the army had a 'slightly rose-coloured view about – or at least, we did at that stage – about how much better we were at internal security, whatever you like to call counter-insurgency operations, we had a rather better view of ourselves than we thought the Americans were capable of'.

4 Chris Parker, interview 12 February 2010

5 Scott Carpenter, interview 3 August 2009

6 The scale of abuse during the first summer of the occupation is still unclear although a number of cases have been brought to light by Public Interest Lawyers, a UK legal firm that specialises in human rights abuses. One alleged incident took place 12 April 2003, when British soldiers arrested Faisal al-Saadoon, the sixty-seven-year-old chairman of a local branch of the Iraqi Red Crescent. Saadoon was accused of being a member of the Ba'ath party and taken to a British base where he claims he was beaten with electric cables and batons. Saadoon was later transferred to the main British detainment facility where he was interrogated for a further five days. Because of the injuries sustained during the beating, Saadoon says the British flew him to Kuwait for an operation where the International Federation of the Red Crescent visited him. On 27 April 2003 the ICRC requested the British free Saadoon. He and two hundred others were subsequently released on the highway between Basra and Az-Zubayr.

Saadoon's testimony has provided important insight into the death of another British detainee, Tarek Hassan, arrested on 24 April 2003. Five months later his body was found dumped in Samarra, north of Baghdad. His family allege he was held hostage by the British in exchange for the surrender of his brother, Kadhim Hassan, a member of the Ba'ath Party. Saadoon met Hassan while in British custody, the last confirmed sighting of Hassan before his death.

Other instances of alleged abuse include an assault case and two suspicious drownings in May 2003. The military later investigated these cases but subsequently dropped charges against the soldiers accused of being involved. In 2007, the army commissioned Brigadier Robert Aitken, director of Army Personnel Strategy, to examine the scope and nature of the abuse claims. Released the following year, Aiken's report found only isolated acts of criminality and recommended, among other broad suggestions, better education for soldiers.

7 Nicholas Mercer, Testimony R vs Donald Payne court martial 2006. The fifty-year-old Saadoon was held in UK custody until 2008 when he and another prisoner accused of murdering the British soldiers, Khalef Hussain Mufdhi, were handed over to Iraqi authorities. In 2009 they were tried and cleared of all charges in an Iraqi court, but are still being held in custody pending an appeal by the prosecutor.

8 Nicholas Mercer, interview 15 July 2003

9 Public Interest Lawyers, *British Forces in Iraq: The Emerging Picture of Human Rights Violations and the Role of Judicial Review*, 30 June 2009
10 Victims like Ali had no role in these proceedings and their evidence was not taken into account by the Army Prosecuting Authority despite the British legal firm representing the Iraqis, Public Interest Lawyers, submitted statements to the court martial. Perhaps as a consequence of this, the charges were essentially disciplinary in nature and did not encompass the severity of the abuse. No action was taken against Major Taylor. Following the court martial, further allegations of abuse at the camp emerged, including claims by a fourteen-year-old that he was forced to perform oral sex on another prisoner. See http://www.guardian.co.uk/uk/2008/jul/14/military.defence.
11 Private interview
12 Clare Short, Iraq Inquiry, 2 February 2010
13 On 16 June Sally Keeble, Short's former deputy, wrote an angry letter to Blair, describing the DFID's 'disastrous' performance to date in Iraq. She listed the lack of pre-planning, the difficulties in providing humanitarian supplies for the troops and, most recently, the refusal to contribute £6 million towards the dredging of Umm Qasr, a port near Basra, or provide staff for the new administration. Keeble warned, 'My concern is that if there is ever a situation like this again, the government cannot have a repeat of that performance.' Sally Keeble, letter, Iraq Inquiry, 16 June 2003
14 Alastair Campbell, *The Blair Years*, p.698
15 Private interview
16 Private interview

6 – DEATH IN THE MARSHES

1 Jasim Bahadili, interview 6 April 2004
2 Then named Saddam City after the Iraqi dictator.
3 Janet Wallach, *Desert Queen: the Extraordinary Life of Gertrude Bell: Adventurer, Advisor to Kings, Ally of Lawrence of Arabia*, Anchor Books, 2005, p.158
4 H. St John Philby, *Arabian Days: An Autobiography*, Robert Hale Ltd, 1948, p.119, and Janet Wallach, *Desert Queen*, p.189
5 Gavin Young, *Return to the Marshes: Life with the Marsh Arabs of Iraq*, Hutchinson, 1983, p.36
6 Georgina Howell, *Gertrude Bell, Queen of the Desert, Shaper of Nations*, Farrar, Straus and Giroux, 2006, p.265
7 Ibid. p.159
8 Frank Haigh, 'Irrigation Development Commission: Control of the Rivers of Iraq and the Utilization of their Water', 1949

9 http://www.guardian.co.uk/world/2003/jan/06/iraq.rorymccarthy
10 Private interview
11 Abu Hatem, interview 10 May 2004
12 Private interview
13 Patrick Bishop, 3 *Para*, Harper Press, 2007, p.10
14 Mark Nicol, *Iraq: A Tribute to Britain's Fallen Heroes*, Mainstream Publishing, 2009, p. 274. Nicol also provides a fuller account in *Last Round: The Red Caps, the Paras and the Battle of Majar*, Cassell Military Paperbacks, 2007.
15 Stuart Tootal, interview 15 September 2010
16 Ibid.
17 Witness statement, Royal Military Police investigation
18 Mark Nicol, *Iraq: a Tribute*, p.279
19 Witness statement, Royal Military Police investigation
20 Mark Nicol, *Iraq: A Tribute*, p.280
21 Ibid. p.281
22 Hamilton-Jewell's translator later told royal military police investigators that the Red Caps' vehicles were parked in front of the station. Reg Keys believes the platoon commander may have turned a blind eye as he sought to escape ambush.
23 Witness statement, Royal Military Police investigation
24 Mark Nicol, *Iraq: A Tribute*, p.285
25 Ibid. p.288
26 Stuart Tootal, interview 15 September 2010
27 Mark Nicol, *Iraq: a Tribute*, p.293
28 Dan Collins. *In Foreign Fields: Heroes of Iraq and Afghanistan in Their Own Words*, Monday Books, 2007, p. 74. Private John Healy, 1 Para was also shot and injured providing suppressing fire. In 2006 John Dolman was killed working as a private security guard in Iraq.
29 Mark Nicol, *Iraq: A Tribute*, p.302
30 Witness statement, Royal Military Police investigation
31 Reg Keys, interview 15 August 2010

7 – MINISTRY OF CULTURE

1 Miles Pennett, interview 25 August 2009
2 The Danish lead in southern Iraq was one of the few marks of British pre-war efforts to internationalise the country's occupation. Dominic Chilcott, Iraq Inquiry, 8 December 2009
3 Hilary Synnott, *Bad Days in Basra: My Turbulent Time as Britain's Man in Southern Iraq*, I. B. Tauris, 2008, p.26. US contractor Kellog, Brown & Root was meant to

provide 'life support' for the building, but due to a mix-up with the British military for several weeks no escort could be found to accompany its personnel to and from its base in Kuwait.

4 Janet Rogan, interview 7 September 2010

5 Miles Pennett, interview 25 August 2009. In fact Clarke did better than that. Iraq has a long football tradition. In 2004 Clark found himself managing Iraq's national football team at the 2004 Athens Olympics, where the so-called Lions of Mesopotamia came fourth, beating Portugal, Costa Rica and Australia along the way. Udaay Hussein had infamously tortured and abused the football team before the invasion, and Iraq was barred from most international competitions.

6 Janet Rogan, interview 7 September 2010

7 Charles Monk, interview 16 April 2009

8 Entitled *Achieving the Vision to Restore Full Sovereignty to the Iraq People*, the paper had an improbable October 2003 deadline for achieving many of its goals, which included 30,000 trained Iraqi police, re-establishing the Iraqi Border Guard, reopening all courthouses, building eleven new prisons and detention centres, reforming the ministries, improving electricity generation capacity to 4,000 megawatts, restoring basic health care services to prewar levels, rehabilitating 1,000 schools and reopening the airports and railroads.

9 Hilary Synnott, *Bad Days in Basra*, p.13

10 David Manning, interview 5 August 2009

11 This was the verdict of the coroner, although doubts persist about whether the slits to his wrists would have killed him.

12 Hutton Inquiry, 28 January 2004. There is extensive evidence in the Inquiry of Campbell and Hoon's desire to leak Kelly's name – Hutton took the view that his name was likely to have emerged through the Foreign Affairs Committee's inquiry into pre-war intelligence and that if the government did not release the fact that a government official had come forward as the leaker to Gilligan they might be accused of some sort of cover-up.

13 Tony Blair, *A Journey*, Knopf, 2010, p.450

14 Miles Pennett, interview 25 August 2009

15 Ibid.

16 Graeme Lamb, interview 4 September 2009

17 Charles Monk, interview 16 April 2009

8 – OUR MAN IN BAGHDAD

1 The exception was Ryan Crocker, briefly head of the CPA's political team between May and August 2003.

2 L. Paul Bremer, *My Year in Iraq: The Struggle to Build a Future of Hope*, Simon & Schuster, 2006, p.41
3 Richard Armitage, interview 19 April 2010
4 Jeremy Greenstock, interview 21 July 2009
5 Private interview
6 Private interview
7 Private interview
8 Paul Bremer, interview 3 June 2009
9 Scott Carpenter, interview 17 July 2009
10 Ibid.
11 Doug Brand, interview 16 September 2009
12 Jonathan Riley, interview 3 September 2009
13 Eric le Blan, interview 24 July 2009
14 Jeremy Greenstock, Iraq Inquiry, 5 February 2010
15 Charles Heatley, interview 15 October 2003
16 L. Paul Bremer, *My Year in Iraq*, p.225
17 Private interview
18 Private interview

9 – NEOCOLONIALISTS

1 Hilary Synnott, *Bad Days in Basra*, p.62
2 Niall Ferguson, *Empire: The Rise and Demise of the British World Order and the Lessons for Global Power*, Basic Books, 2009. Introduction xi.
3 Anthony Seldon (ed.), *Blair's Britain*, p.555 and Suma Chakrabarti, Iraq Inquiry, 22 January 2010
4 Private interview
5 Private interview
6 Kifah Taha al-Mutari, witness statement 28 July 2004. Mutari, one of the prisoners, described how they were prevented from sleeping and made to remember the names of English football players or risk being severely beaten. He recalled a game the soldiers played involving kick-boxing: 'The soldiers would surround us and compete as to who could kick-box one of us the furthest. The idea was to try and make us crash into the wall.'
7 Shortly after Baha Mousa's death, Northwood banned the hooding of detainees, although the practice continued. Three years later seven members of the regiment, including its commander Lieutenant Colonel Jorge Mendonca, were brought before a court martial on charges of ill-treating detainees. Payne was also charged with war crimes under the International Criminal Court Act 2001. On 19

September 2006 Payne pleaded guilty, becoming the first member of the British military to be convicted of a war crime. He was subsequently jailed for one year and expelled from the army. The charges against the six other soldiers were dropped for lack of evidence. In 2008 the MOD agreed to pay £2.83 million in compensation to the families of Mousa and nine other Iraqi men mistreated by British troops. The following year a public inquiry was launched.

8 Graeme Lamb, interview 2 September 2009. Most of the contract would be awarded to Mott Macdonald, the UK contracting firm that had begun the marsh drainage programme back in the 1960s.

9 Graeme Lamb, Iraq Inquiry, 9 December 2009

10 Suma Chakrabarti, Iraq Inquiry, 22 January 2010

11 Bertram Thomas, *Alarms and Excursion in Arabia*, The Bobbs-Merril Company, 1931, p.85

12 Rory Stewart, *The Prince of the Marshes and Other Occupational Hazards of a Year in Iraq*, Harcourt, 2006, p.58

13 Toby Dodge, *Inventing Iraq*, p.57

14 Abu Hatem, interview 10 May 2004

15 Mark Etherington, *Revolt on the Tigris: The Al-Sadr Uprising and the Governing of Iraq*, Cornell University Press, 2005, p.113

16 The UK's contribution was considerably more modest – £544 million spread over three years – much of it going through UN agencies that now only operated with local staff due to the danger. Disappointingly for British efforts to increase international participation, there had been no pledges from Germany, France or Russia, who felt this was the Americans' war.

17 The presentation was entitled 'Overview of Planning Milestones' and served as an update to Bremer's July *Achieving the Vision* paper. Hilary Synnott, *Bad Days in Basra*, p.158

18 Rory Stewart, *The Prince of the Marshes*, p.112

19 H. St John Philby, *Arabian Days*, p.113

20 Adrian Weale, interview 9 May 2010

21 In the first eight city council elections, when only one vote was allowed per household, less than 2 per cent of the turnout was women. Following the rule change up to 40 per cent of the voters were women.

22 Private interview

23 Patrick Nixon, interview 7 August 2009

24 Molly Phee, interview 15 October 2010

25 Rory Stewart, *The Prince of the Marshes*, p.241

26 Ibid. p. 272

27 Bertram Thomas, *Alarms and Excursion*, p.38

10 – UPRISING

1 Private interview
2 Little did Richmond realise that the previous month Bremer had come close to launching an operation to arrest or kill Moqtada al-Sadr. In early 2004 Bremer revealed to a secret gathering of American officials that the US military had the cleric under surveillance and was preparing to strike. There appeared to be agreement for the operation until Molly Phee, governor of Maysan, exclaimed, 'Are you kidding me? This is the most ridiculous idea I have heard.' Phee's intervention and military doubts scuppered the plan. Molly Phee, interview 15 October 2010
3 Iraq's one television station, al-Iraqiya, was effectively a mouthpiece for CPA press releases. A US company specialising in communication equipment, the Harris Corporation, had won the initial contract to build up Iraq's media operations. It was replaced by Science Applications International Corporation, which proved mildly more effective. Miles Pennett, the former culture minister for southern Iraq, ran its southern Iraq operations.
4 Private interview
5 Condoleezza Rice, interview 12 November 2009
6 Michael Walker, Iraq Inquiry, 1 February 2010
7 Private interview
8 Private interview
9 In December British forces had raided Maliki's offices and removed him from his position, although he continued to claim to be the director general of education and exert considerable control over the school system.
10 Andrew Stewart, interview 26 April 2010
11 David Richmond, interview 15 April 2010
12 Condoleezza Rice, interview 12 November 2009
13 L. Paul Bremer, *My Year in Iraq*, p.327
14 Lakhdar Brahimi, interview 19 September 2009
15 Mark Etherington, interview 13 April 2010
16 Mark Etherington, interview 14 April 2010
17 Victoria Whitford, interview 5 May 2010
18 Patrick French, *Younghusband: The Last Great Imperial Adventurer*, Flamingo, 1995, p.118
19 Private interview
20 Paul Bremer, interview 3 June 2009
21 Private interview
22 Private interview
23 Richard Armitage, interview 19 April 2010

11 – DANNY BOY

1 Piers Morgan, *Daily Mirror* editor, was sacked over the hoax, although he continued to claim the pictures portrayed abuse carried out by the Queen's Lancashire Regiment. In 2005 the Crown Prosecution Service dropped charges against Private Stuart Mackenzie, the member of the Lancashire and Cumbrian Volunteers who had helped fake the photograph. Mackenzie had been deployed in Iraq with the QLR.
2 Molly Phee, interview, 15 October 2010
3 Dan Mills, *Sniper One: On Scope and Under Siege with a Sniper Team in Iraq*, St Martin's Press, 2007, p.53
4 Ibid. p.xxiv
5 Ibid. p.52
6 Private interview
7 Justin Featherstone, interview 15 August 2010
8 Richard Holmes, *Dusty Warriors*, Harper Perennial, 2007, p.125
9 Ibid. p.126
10 Both Williamson and Philips made a full recovery.
11 Molly Phee, interview 25 October 2009
12 Richard Holmes, *Dusty Warriors*, p.218
13 Andrew Stewart, interview 5 May 2010
14 Muhammad al-Fartosi, interview 20 May 2004
15 Khuder al-Sweady, interview 20 May 2004
16 Richard Holmes, *Dusty Warriors*, p.230
17 Mark Keegan, 'On Whose Orders?' *Panorama*, 25 February 2008
18 Public Interest Lawyers, *British Forces in Iraq: The Emerging Picture of Human Rights Violations and the Role of Judicial Review*, 30 June 2009. James Coote denies that any prisoners were abused. Private interview.
19 Andrew Kennett, interview 15 July 2010
20 Richard Holmes, *Dusty Warriors*, p.238
21 Dan Collins, *In Foreign Fields*, p.156
22 Mahdi Jassim Abdullah, 'On Whose Orders?' *Panorama*, 25 February 2008
23 Hussein Jabbari Ali, 'On Whose Orders?' *Panorama*, 25 February 2008
24 Hussein Fadhil Abbas, 'On Whose Orders?' *Panorama*, 25 February 2008
25 Of the nine prisoners, two were acquitted and seven jailed for attacking the British.
26 Ahmed Fausi, interview 20 May 2004
27 The events of Danny Boy are currently the subject of an inquiry, after pressure from the Sweady family and others for the British government to investigate their claims.
28 Khuder al-Sweady, 'On Whose Orders?' *Panorama*, 25 February 2008
29 Khuder al-Sweady, interview 20 May 2004

30 Private interview
31 Stephen Grey, 'Why you don't hear from your brave boys', *New Statesman*, 31 May 2004

12 – OUT OF IRAQ

1 See Butler Review, www.archive2.officialdocuments.co.uk/document/deps/hc/.../898.pdf
2 Robert Peston, *Brown's Britain: How Gordon Runs the Show*, Short Books Ltd 2005, p.338
3 Anthony Seldon, *Blair Unbound*, p.270
4 Condoleezza Rice, interview 12 November 2009
5 The eventual vote on top-up fees was the largest backbench rebellion since 1945. As Alan Johnson, former Labour health secretary, noted, 'People think the UK government was almost brought down by Iraq. No, it wasn't. It was almost brought down by higher education.' Anthony Seldon (ed.), *Blair's Britain*, p.253
6 Anthony Seldon, interview 5 June 2009. See also Anthony Seldon, *Blair Unbound*, p.273.
7 Private interview
8 Private interview
9 Richard Dannatt, Iraq Inquiry, 28 July 2010
10 Geoff Hoon later described the decision as the most difficult of his career. Iraq Inquiry, 19 January 2010
11 Rob Fry, interview 14 January 2010
12 Rob Fry, interview 10 September 2010
13 Private interview
14 L. Paul Bremer, *My Year in Iraq*, p.395
15 Private interview
16 Ayad Allawi, interview 2 July 2009
17 Michael Walker, Iraq Inquiry, 1 February 2010
18 Ibid. At the inquiry Walker admitted two deployments broke defence planning assumptions but offered the caveat, 'We made it absolutely clear, but as we hadn't stuck within our Defence Planning Assumptions for the previous deployment . . . I'm trying to think when we did stick with them since 1998.'
19 As Richard Dannatt, then assistant chief of the general staff, the army's equivalent of managing director, noted, 'Many of the things that were added back in as urgent operational requirements at the start of Iraq were things that had been taken out of the equipment programme, the savings in earlier years.' By the end of the war UORs had risen to £2 billion. (Richard Dannatt, Iraq Inquiry, 28 July 2010)

20 Anthony Seldon, *Blair Unbound*, p.293

21 The MOD believed there was no cash control in the settlement. As Kevin Tebbit pointed out in his Iraq Inquiry testimony, the only reference to a limit came in Annex E to the settlement letter, where it was described as being there for presentational and illustrative purposes, not as a control mechanism. (3 February 2010)

22 Paul Cornish and Andrew Dorman, *Blair's wars and Brown's budgets: from Strategic Defence Review to strategic decay in less than a decade*, Chatham House, 2008

23 The helicopters were not due to come into service until 2010. Air Marshall Jock Stirrup, then head of the RAF, argued to the Iraq Inquiry that the issue of helicopter shortages 'was not a significant issue in our discussion' between the chiefs 2003–6, despite a 2004 MOD study showing a 38 per cent shortfall in helicopter lift. See www.publications.parliament.uk/pa/cm200405/cmselect/. . ./386.pdf.

24 The Ministry of Defence's budget rose from £29.7 billion in 2004 to £33.4 billion by 2007–8. The MOD also regained the power to shift efficiency savings to cash, but this time with a £350 million cap. The military viewed the 2004 settlement as adequate at the time.

25 Anthony Seldon, *Blair Unbound*, p.277

26 Ibid. p.301

13 – POLICE

1 William Kearney, interview 4 March 2009

2 Hilary Synnott, *Bad Days in Basra*, p.223

3 Andrew Alderson, *Bankrolling Basra: The incredible story of a part-time soldier, $1 Billion and the collapse of Iraq*, Robinson, 2007, p.211

4 The brigade headquarters stationed in the palace was withdrawn, although the compound continued to house two battle groups.

5 Private interview

6 Tim Spicer, interview 7 September 2009

7 Private interview

8 Andrew Rathmel, interview 19 October 2009

9 Hilary Synnott, Iraq Inquiry, 9 December 2009

10 Doug Brand, interview 16 September 2009

11 William Kearney, interview 4 March 2009

12 Phil Read, interview 12 July 2009

13 William Kearney, interview 4 March 2009

14 Haider Samad, interview 16 May 2008

15 Haider Samad, interview 12 June 2008. The first recorded incident between British forces and the Jamiat happened within weeks. On 18 April Lance Corporal

Christopher Balmforth and his patrol were ambushed shortly after inspecting the station. No one was injured.

16 Robert Wilson, email 14 June 2004

17 Charlie MacCartney, interview 15 August 2009

14 – FARTOSI

1 Both were subsets of the Albu Muhammad tribe headed by Abu Hatem.

2 Ahmed al-Fartosi, interview 17 October 2010

3 Ibid.

4 Ibid.

5 Terry Bryan, 'The heroes' story: How nine British soldiers fought off gun-blazing 200 Iraqis in a life-or-death siege', *Daily Mail*, 10 November 2007

6 Dan Collins, *In Foreign Fields*, p.188

7 Terry Bryan, 'The heroes' story'

8 Dan Mills, *Sniper One*, p.273

9 Ibid. p.300

10 Ibid. p.305

11 Ibid. p.338

12 Joseph Felter and Brian Fishman, 'Iranian Strategy in Iraq: Politics and "Other Means",' Occasional Paper Series, Combating Terrorism Center at West Point, 13 October 2008, p.21

13 Private interview

14 Ahmed al-Fartosi, interview 17 October 2010. Fartosi denied meeting Sheibani, but through an intermediary Gharawi confirmed the meeting took place.

15 Ahmed al-Fartosi, interview 17 October 2010

15 – GET ELECTED

1 Ayad Allawi, interview 15 August 2010

2 William Morrison, interview 22 July 2009. In fact Morrison's grandmother had passed away a few days before he arrived in Amman for a reconstruction conference about Iraq, on his way to Baghdad. He had his grandmother's obituary in his briefcase. The article recounted how she had encouraged boys from outside the palace compound to come in and play cricket with the young monarch, whom she considered a sickly child. Despite the strictures of the FCO to keep quiet, he found it impossible to resist telling the elderly Iraqi gentleman sat next to him at the conference. The old man looked startled. 'My dear boy, I was one of those children your grandmother asked to play with the king.'

3 Ayad Allawi, interview, 2 July 2009
4 Henry Ensher, interview 24 September 2009
5 Private interview
6 Les Campbell, interview 1 October 2009
7 Elizabeth Monroe, *Philby of Arabia*, Ithaca Press, 1998, p.98. It is worth noting that T. E. Lawrence, a supposed champion of Arab nationalism, was all too happy to hand Faisal the throne. Meanwhile, H. St John Philby felt so strongly that Britain had betrayed its promises for fair elections that he resigned, marking his decisive break with officialdom and the beginning of his journey to becoming a staunch anti-imperialist.
8 Private interview
9 Private interview
10 Al Elsadr, interview 1 December 2009
11 Ayad Allawi, interview 13 August 2009
12 Tom Warrrick, interview 25 September 2009. Tom Warrick later denied he tried to get Allawi elected, a claim greeted with incredulity by Les Campbell. Warrick was forced out of the State Department after returning from Iraq and now works at the Department of Homeland Security.
13 Private interview
14 Seymour Hersh, 'Get Out the Vote', *New Yorker*, 25 July 2005
15 Private interview
16 Private interview
17 Private interview
18 Robin Wright, *Washington Post*, 24 October 2004
19 Azzam Alwash, interview 11 January 2010. The deputy campaign manager, Alwash, worked with Ilham al-Madfaie, a legendary Iraqi folk singer from the 1960s, to rework the song, which went: 'Doctor, Ya, Doctor/ We have a disease/ From within us/ We need an elixir/ Even if it is bitter/ We must drink it.'
20 Azzam Alwash, interview 11 January 2010
21 Private interview. A few months later, on leave, Heatley broke his leg attempting the Paris–Dakar rally on a motorbike. He later left the FCO and became a consultant for William Morrison's organisation, Adam Smith International.
22 Ali Allawi, interview 2 July 2009. Of a military procurement budget in 2003/4 of $1.3 billion, Allawi, the trade minister, claimed less than $200 million may have been spent on usable equipment. Patrick Cockburn, who broke the story in 2005, explained how equipment was frequently overpriced or a low-budget rip-off. In one shipment MP5 machine guns were priced at $3,500 each. In reality, the Egyptian copies that arrived cost $200. *Independent*, 19 September 2005. In May 2007 an Iraqi court convicted Allawi's former defence minister, Hazem

al-Shaalan, in absentia of embezzlement and sentenced him to seven years in prison.

16 – TRANSITION

1 Rudyard Kipling wrote a poem about the Battle of Maiwand, entitled 'That Day', containing the memorable lines: 'There was thirty dead an' wounded on the ground we wouldn't keep/No, there wasn't more than twenty when the front began to go/ But, Christ! along the line o' flight they cut us up like sheep/ An' that was all we gained by doing so.'
2 This campaign was a mild improvement on an earlier invasion of Afghanistan in 1838, called the First Afghan War. Forced to withdraw from Kabul four years later, a 16,000-strong force was almost entirely annihilated as it trekked back over the mountains to India. British Army chaplain G. H. Gleig, one of the few survivors wrote, 'a war begun for no wise purpose, carried on with a strange mixture of rashness and timidity, brought to a close after suffering and disaster, without much glory attached either to the government which directed, or the great body of troops which waged it. Not one benefit, political or military, was acquired with this war. Our eventual evacuation of the country resembled the retreat of an army defeated'. Cited in William Dalrymple, 'Souter Takes a Call', *Outlook India*, 30 August 2010. After the 1880 Battle of Maiwand, British forces rallied at Kandahar and defeated Ayub Khan, marking the end of the Second Afghan War. British forces withdrew from the country having secured control over the country's foreign policy but not much else.
3 Jonathon Riley, interview 3 September 2009. As Chief of the Defence Staff Michael Walker later told the Iraq Inquiry, 'I was always certain that the prime minister himself got it, but I could never be certain that him having got it meant that it was going to happen further down [the line].'
4 Jonathon Riley, interview 3 September 2009
5 Colin Smith, interview 19 October 2009
6 William Kearney, interview 4 March 2009
7 In early June the Danish ambassador Torben Getterman, accompanied by a British diplomat, Tim Torlot, and the US chargé d'affaires David Satterfield, met the interior minister to express their 'grave concern'.
8 Private interview
9 In March 2005 Lieutenant General Richard Dannatt, commander-in-chief at Land Command, was working on the basis there would only be 1,000 troops left in Iraq by the autumn of the following year. One plan circulating in the MOD at the time had British forces pulling out of Muthanna and Dhi Qar province as early

as October, with a Basra withdrawal taking place at the start of the following year ahead of the Helmand deployment.

10 In October 2004 Rob Fry, the MOD operations chief, had learned that the Canadians were pushing for the lead role in the newly coined Regional Command – South. Ottawa's refusal to take part in the invasion of Iraq, and its rejection of the Ballistic Missile Defense Program – Bush's contentious plan to defend the US from rocket attack – had left many Canadian officials with the sense they owed Washington a favour. Janice Gross Stein and Eugene Lanf, *The Unexpected War: Canada in Kandahar*, Penguin, 2007, p.181. Fry was informed that the Canadians would provide a divisional headquarters and 1,000 troops.

17 – A PROPOSAL

1 Haider Samad, interview 8 July 2008

18 – DON'T MENTION IRAQ

1 John Miller, interview 7 January 2011
2 It is standard UK practice to hold an inquest in whichever local jurisdiction a violent or suspicious death occurs. In the case of soldiers killed in Iraq, the bodies were first registered at the Oxfordshire Coroner's Office after arriving at the airbase at Brize Norton. Nicholas Gardiner, the coroner, had also been involved in the investigation into the death of UN Weapons Inspector David Kelly, although the Hutton Inquiry ultimately provided the official verdict on the cause of death.
3 Rose Gentle, interview 8 October 2009
4 Private interview
5 Andrew O'Hagan, 'Iraq May 2nd, 2005', *London Review of Books*, 6 March 2008
6 Anthony Seldon, *Blair Unbound*, p.342
7 Ibid. p.343
8 Ibid. p.342

19 – IRAN

1 Andrew Williams, interview 5 September 2009
2 Ibid.
3 *The Insurrection in Mesopotamia* 1920 (Battery Press Inc. and the Imperial War Museum. 2005), p.314. As a young officer, Haldane had ridden to the rescue of Charles Townshend at the siege of Chitral. Aylmer Haldane.
4 Private interview

5 Bishop informed the divisional commander about the RAF medics' behaviour, which led to the incident being debated in Parliament.
6 Joseph Felter and Brian Fishman, 'Iranian Strategy in Iraq' p.38. Their evidence relies on material provided by the National Council for Resistance in Iran, an exiled Iranian opposition movement.
7 Private interview
8 Henry Ensher, interview 24 September 2009
9 Ali Allawi, interview 2 July 2009
10 Private interview
11 Private interview
12 Richard Dalton, interview 18 August 2010
13 David Satterfield, interview 18 February 2007
14 Condoleezza Rice, interview 12 October 2009
15 Mowaffak al-Rubaie, interview 21 August 2009
16 Private interview
17 Richard North, *Ministry of Defeat: The British War in Iraq* 2003–9, Continuum, 2009, p.70

20 – JAMIAT

1 Private interview
2 Steven Vincent, 'Switched off in Basra', *New York Times*, 31 July 2005
3 Private interview
4 Richard Williams, interview 11 September 2009
5 Nick Henderson, interview 10 October 2009
6 Dan Collins, *In Foreign Fields*, p.258
7 Rabia Siddique, interview 21 October 2009
8 Ibid.
9 Ibid.
10 Karl Hinnett left the army in 2008 to retrain as a gym instuctor. He now spends most of his time working for charity. George Long is still serving.
11 James Tansley, interview 5 August 2009
12 Private interview
13 Nick Henderson, interview 10 October 2009
14 Siddique later brought a race and sex discrimination case against the MOD, alleging that she was unfairly treated by the military after her ordeal. James Woodham received the Military Cross for his heroics that day; but she received little more than 'a pat on the back.' The claim was settled out of court, and Siddique has since left the military.

15 Private interview
16 Richard Williams, interview 11 September 2009
17 Michael Jackson, Report, Iraq Inquiry, 5 October 2005
18 Richard North, *Ministry of Defeat*, p.82

21 – HELMAND

1 Private interview
2 Private interview
3 Eliza Manningham-Buller, MI5 head at the time, later told the Iraq Inquiry she had warned the Joint Intelligence Committee that Iraq would increase the terrorist threat to the UK. She also highlighted the dangers of radicalised British Muslims travelling to Iraq to attack US and British forces. According to intelligence sources, by mid-2005 around seventy had already gone there to fight. (David Leppard and Hala Jabar, '70 British Muslims join Iraq fighters', *The Times*, 26 June 2005) Blair, however, preferred to listen to the assertions of MI6 head Sir Richard Dearlove on the threat Iraq posed and rarely met Manningham-Buller one-to-one. (Eliza Manningham-Buller, Iraq Inquiry, 20 July 2010)
4 Air Commodore Mark Leakey, a planner at Northwood, had recently returned from Helmand with more detailed scenarios for the number of troops that might be needed. They ranged from light forces of 1,500 to a full brigade of 5,000. Andrew Kennett, the former brigade commander in Basra who had recently taken on the top planning job at Northwood, did not think enough planning had been done for the deployment. 'They were all about what we could do. There was next to nothing about how the Afghans might receive us,' said Kennett. He wanted to go back to the drawing board, but deputy chief of joint operations Peter Wall told Kennett, 'the plans were the plans' and there would be no backtracking.
5 Mike Jackson, Iraq Inquiry, 30 July 2010
6 The paper, entitled 'Options for Future UK Force Posture in Iraq', was leaked to the *Mail on Sunday* the following month, forcing John Reid to backpedal. At the time the Bush administration insisted there was no timetable for withdrawal. The plan would enable British forces in Iraq to be reduced from 8,500 to 3,000 by the middle of 2006.
7 Private interview
8 Private interview
9 Private interview
10 Ed Butler, interview 14 January 2010
11 Richard Dannatt, Iraq Inquiry, 28 July 2010 'Should we have revisited that decision, the one taken in 2004, to do more in Afghanistan in 2006; revisited it perhaps

during the latter part of 2005/early 2006? Perhaps we should have done. I could have played a part in saying I think we should revisit it. I didn't say we should revisit it. One accepted it as a policy decision and we got on with it. Maybe that was an error.' But on the contrary, events in Basra only strengthened the argument of Fry and others for exchanging Iraq for Afghanistan. Michael Walker wrote to reassure Reid that the British military could handle a war on two fronts: 'Our ability to fulfill our plan in Afghanistan is not predicated on withdrawal of such capabilities from Iraq and, notwithstanding these qualifications, in the event that our conditions-based plan for progressive disengagement for withdrawal from southern Iraq is delayed, we will still be able to deliver our . . . mandated force levels in Afghanistan.' (Michael Walker, letter, Iraq Inquiry, 19 September 2005)

12 Rosalind Marsden, interview 24 October 2010
13 Private interview. See also James Risen, 'Reports Link Karzai's brother to Afghanistan Heroin Trade', *New York Times*, 4 October 2008
14 Nick Clissit, interview 14 September 2010
15 Gil Baldwin, interview 25 September 2009
16 Private interview
17 Mark Etherington, interview 14 September 2009
18 Private interview
19 Suma Chakrabarti, Iraq Inquiry, 22 January 2010
20 Private interview
21 Mark Etherington, interview 14 September 2009
22 The water in the irrigation canals was often too salty for conventional crops. Poppies, however, are able to thrive in dry and moderately saline conditions. Nick Cullather. *The Hungry World: America's Cold War Battle Against Poverty in Asia* Harvard University Press. 2010, p.131
23 Traditional Pashtu poem
24 Minna Jarvenpaa, interview 12 March 2010. Ghani's contention was later supported by Mullah Abdul Salam Zaeef, Afghan ambassador to Pakistan before the US invasion of Afghanistan, who was detained in Pakistan and sent to the Guantanamo Bay camp. In his autobiography published after his release in 2005 he noted, 'Whatever the reality might be, British troops in southern Afghanistan, in particular in Helmand, will be measured not on their current actions but by the history they have, the battles that were fought in the past. The local population has not forgotten and, many believe, neither have the British. Many of the villages that see heavy fighting and casualties today are the same that did so some ninety years ago.' (Mullah Zaeef, *My Life with the Taliban*, Columbia University Press, 2010, p.242)
25 Memo, Afghan Steering Group, October 2005

26 'Information and intelligence is lacking in all sectors. Period of sustained field research and intelligence gathering by in-country team and targeted studies essential to inform coherent long-term strategy. Lack of knowledge about Pakistan border area a particular concern.' (UK *Plan for Helmand: Joint Interim Report*, 28 November 2005)

27 The 'Next Steps' section of the report noted, 'preliminary study in early 2006 will be required to plan in detail in five main areas: policing, justice, rural livelihoods, governance and information operations. The requisite subject matter experts should be identified as soon as possible to formulate strategy, consistent with existing initiatives and the tenets of this plan. There will be a particular premium in this context in maintaining continuity of knowledge between the assessment and implementation phases.' UK *Joint Plan for Helmand: Final Report*, 12 December 2005

28 Mark Etherington, interview 14 January 2010

29 Private interview

30 After a year in Helmand the British military had fired some four million rounds of ammunition. This statement, justly notorious, has a backstory involving the Dutch. They had agreed to deploy to the neighbouring province of Oruzgun with the Australians but had started to backtrack, casting their own involvement and the fate of the entire southern effort into question. After the massacre at Srebrenica, when Dutch peacekeeping troops had stood by as 15,000 Bosnian Muslims were slaughtered, Holland's cabinet was held directly accountable for any deployment. The Dutch parliament was due to vote on the deployment on 2 February. Reid's statement and its emphasis on peacekeeping was partly intended to reassure the Dutch over the nature of the mission and emphasise that the time for prevarication was over. In early February the Dutch parliament voted in favor of deployment to Afghanistan.

22 – PIKE FORCE

1 Harvey Pynn, interview 15 September 2010

2 Ed Butler, interview 12 January 2010

3 Ibid.

4 James Risen, 'Reports Link Karzai's Brother to Afghanistan Heroin Trade', *New York Times*, 4 October 2008

5 Stuart Tootal, interview 11 September 2010

6 Tom Coghlan, Deborah Haynes, Anthony Loyd, Sam Kiley and Jerome Starkey, 'Cut off, outnumbered and short of kit: how the Army came close to collapse', *The Times*, 9 June 2010

7 Private interview.
8 The incoming NATO commander, British general David Richards, shared the American view but he did not yet have the casting vote.
9 Private interview
10 Stuart Tootal, *Danger Close: The True Story of Helmand from the Leader of 3 Para*, John Murray, 2010, p.97
11 Ed Butler, interview 12 January 2010
12 The Apache helicopter pilot Andy Cash had realised shortly after the drop-off that they had put Pike down in the wrong position and insisted on circling back.
13 Patrick Bishop, *3 Para*, p.126

23 – NEW COIN, OLD ROPE

1 Nigel Aylwin-Foster, 'Changing the Army for Counter-insurgency Operations', *Military Review*, January 2006.
2 David Petraeus, interview 29 October 2010
3 Private interview.
4 Alex Alderson, Interview, 15 September 2009. There is a long line of British thought on counterinsurgency that stretches back to the 1909 Field Service Regulations, Chapter 10, entitled 'War in Uncivilized Countries', has no mention of the enemy, preferring to focus on terrain in keeping with the expansionist philosophy of the empire.
5 Like the American doctrine writer John Nagl, Marston had also studied under the Australian-born Oxford academic Robert O'Neil, who had served in Vietnam with the Australian army.
6 Duncan Anderson, interview 16 September 2010
7 Andrew Kennett interview 15 September 2010
8 William Kearney, the ArmorGroup manager, who regularly visited Camp Smitty in Muthanna, had noticed that more and more greens and ambers were being awarded and there were open conversations among military commanders about 'accelerating' the approval rating.
9 Ben Edwards, interview 4 November 2009
10 The army had spent some £400 million on the Phoenix programme of unmanned surveillance vehicles. Unable to operate in high temperatures, they proved useless in Iraq and were later removed from service.
11 Richard North, *Ministry of Defeat*, p.106
12 Mowaffak al-Rubaie, interview 15 September 2009
13 The issue of extending tours had already been raised among defence chiefs but rejected on the grounds it would overstretch an already fragile army. Michael

Jackson, Iraq Inquiry, 29 July 2010. Jackson made an additional and extraordinary statement contrasting the American attitude to the UK's: 'They were a nation at war. I don't think that has been an attitude that has coloured us.'

14 Richard North, *Ministry of Defeat*, p.234. During early discussion about FRES in 2001 experts at the Defence Evaluation and Research Agency did not expect the vehicle to be ready for another twenty-five years.

15 Richard North, interview 2 March 2011

16 Richard North, *Ministry of Defeat*, p.234

17 Richard Dannatt, *Leading from the Front: the Autobiography*, Bantam Press, 2010, p.210. Specifically, he blamed the Defence Management Board, the MOD's main corporate committee, for second-guessing the Army Board and hiring a firm of external analysts to reassess the army's decision.

18 Drayson's name was raised in the cash-for-honours crisis that had engulfed Tony Blair in 2006. At the time, Drayson insisted he was innocent of any wrongdoing. No criminal charges were ever brought against him.

19 After being up-armoured the SE432 became known as the Bulldog.

20 Richard Dannatt, Iraq Inquiry, 28 July 2010. The difficulty in finding new vehicles for Iraq contrasts with the rapid although no less flawed decision-making over Afghanistan. In April 2006 an order for eighty up-armoured Pinzgauer trucks known as the Vector was placed at a cost of £35 million. They arrived in theatre later that summer. Unfortunately the vehicle was a death trap, with the driver's seat perched over the front wheel and especially vulnerable to roadside bombs.

21 The next brigade sent to Iraq would have the revamped Bulldog and Mastiff. The army's helicopter shortage was solved by converting eight Chinooks, currently going through a slow and expensive refit for use by special forces, back to regular service. Predator drones were also ordered to improve surveillance capabilities, replacing the defective Phoenix system. But Dannatt was not prepared to give up on FRES. Over the summer of 2006 he invited Drayson to come to the training area of Salisbury Plain to demonstrate to him the current fleet of army vehicles and the glaring absence of medium-sized armoured vehicles. Drayson agreed to champion FRES and succeeded in arranging a competition among manufacturers to come up with the best design by November 2007. Unfortunately, the same month that the decision was due to be made, Drayson resigned, effectively delaying FRES by several more years, as his successor struggled to work out the Byzantine world of defence procurement. On 22 March 2010 the MOD announced that General Dynamics UK had been awarded a £500 million development contract to build a FRES specialist vehicle. Trials are expected to start in 2013.

22 See *National Vietnam Veterans Readjustment Study*. In 2010, the King's Centre for Military Health Research published the first major study into the mental health

of UK troops deployed in Iraq. The study found that the earlier research into Vietnam veterans might not be relevant to British troops because of substantial differences between the two countries in how they deploy forces to warzones. US soldiers deploy for longer, with less time between tours; they are typically younger, likely to have less previous deployment experience. The King's Centre researchers concluded that there was no increase in psychiatric problems among forces that took part in the 2003 Iraq invasion. However, it should be noted that the vast majority of soldiers who took part in the initial invasion saw little combat compared to the intense bombardment most British bases experienced between 2006–8. The study makes a similar claim for Afghanistan but likewise its sample was drawn from 2007, a period of reduced fighting compared to later. The researchers still found that about a fifth of 611 UK soldiers studied after a 2009 tour of Iraq – another period of relative calm – showed signs of mental distress. Wessely, Simon and Dandeker, Christopher, 'Kings Centre for Military Health Research: A fifteen year report', September 2010.

23 Louis Lillywhite, MOD press release 25 November 2009. The surgeon general's comments in 2009 were reflective of the MOD's consistent attitude. In 2004 the directorate had closed down the military's only psychiatric in-patient facility, the Duchess of Kent's Psychiatric Hospital, and awarded a private contract to the Priory Group of hospitals to run fifteen small clinics around the country, which were clearly destined to struggle as the wars in Iraq and Afghanistan escalated. The mental health charity Combat Stress, set up to deal with traumatised soldiers after the First World War, was reporting a worrying increase in the number of Iraq veterans referred to the three homes for veterans it ran.

24 Dan Collins, *In Foreign Fields*, p.163

25 Mark Nicol, *Iraq: a Tribute*, p.310

26 http://www.youtube.com/watch?v=3YzH1GQI2Bg&has verified=1. The youths and policeman were released after several hours and subsequently brought legal action against the MOD. After leaving Iraq, Webster became an army instructor at Winchester and only belatedly transferred his footage onto a DVD to share with the other squad members. One of his colleagues, immediately realising its potential, sold the DVD to the *News of the World* for £8,000.

27 Martin Webster, interview 15 October 2010. Webster went on to produce a documentary film entitled *Diary of a Disgraced Soldier* based on a video diary he kept of his own battle with depression. He now works with the UK charity Talking2Minds promoting the issue of PTSD in the military.

24 – THE SHERIFF

1 Private interview
2 Richard Shirreff, Iraq Inquiry, 11 January 2010
3 Private interview
4 Tim Foy, interview 5 August 2009
5 Following his election victory the previous year, he had hired David Bennett, a former management consultant at the US firm McKinsey, to create a coherent 'narrative' to the end of his premiership. Since January Bennett had been probing Blair for a list of his priorities, which Ruth Turner, another strategist, had written up for the Chequers meeting. She had entitled the paper *Aiming High, Right to the End*, which seemed to evoke the stiff upper lip that Blair was going to need to withstand Brownite pressure to force him out sooner. (Anthony Seldon, *Blair Unbound*, p.421)
6 Private interview
7 The deputy head of the committee was Yusef al-Musawi, leader of the Thar'allah militia. His fighters were responsible for many of the killings in Basra, but he was too powerful to exclude. Distasteful as dealing with people like Musawi was, Everard managed to build a working relationship with him.
8 James Everard, interview 26 January 2010. In July the British effort had been badly embarrassed when Casey requested two battalions from the Iraqi 10th Division for his Forward Together operation, only for one of the Iraqi units to mutiny at the prospect of going to Baghdad.

25 – BREAKING POINT

1 Tim Wilson, interview 4 June 2009
2 David Labouchere, interview 3 September 2009
3 Private interview
4 Stuart Tootal, *Danger Close*, p.123
5 Richard Dannatt, interview 29 September 2010
6 Ibid.
7 Rachel Schmeller, interview 24 September 2009
8 David Labouchere, interview 2 December 2009
9 This figure reflects a combination of direct British funding and US military funds given to their British counterparts. It does not include Iraqi spending or contributions from other nations; the true cost of Basra's reconstruction between 2003–9 is likely to be much higher
10 In September 2006 British special forces killed the Jaish al-Mahdi leader, Habib Abu Hayder
11 On 1 September 2006 Ranger Anare Draiva was killed and Corporal Paul

Muirhead seriously wounded. The 3 Para commander Stuart Tootal was left with the desperate choice between risking Muirhead's life and delaying the helicopter rescue until darkness fell, three hours hence, or exposing the helicopter crew to a murderous daylight landing. Following the advice of Jowlett's medic Mike Stacey, he opted to wait. Muirhead later died from his injuries

12 At the end of August the chief of the general staff, General Richard Dannatt, paid his first visit to Camp Bastion and was shocked by what he found. Another mortar strike in Musa Qala had again forced Tootal to choose between sending in the helicopters during the day or waiting. Tootal told Dannatt forcefully that the strategic advantages of manning outposts like Musa Qala were outweighed by the dangers of losing helicopters. Brigadier Ed Butler agreed, saying that if he could no longer guarantee his men timely helicopter support then keeping troops in Musa Qala was an unacceptable risk. (Stuart Tootal, *Danger Close*, p.221)

13 Dan Collins, *In Foreign Fields*, p.381

14 Paul Muirhead, injured the week before in Musa Qala, had finally succumbed to his injuries.

15 Stuart Tootal, *Danger Close*, p.260

16 Ibid., p. 261

17 Over the summer the Finnish expert working for the Post-Conflict Reconstruction Unit, Minna Jarvenpaa, had persuaded the Cabinet Office to review what had gone wrong in Helmand, the first official recognition of the scale of the disaster. At one stage Margaret Aldred, the combative director of the Cabinet's Foreign and Defence Policy Committee, flew to Helmand herself, a rare moment when a senior Whitehall official confronted the reality of policy decisions.

26 – SINBAD

1 Anthony Seldon, *Blair Unbound*, p.467, quoting Condoleezza Rice.

2 Blair was shown a video of that mission, which took place on 16 April 2006. Williams's men had as their target an al-Qa'eda operative called Abu Atiya, known as the admin emir for an insurgent cell in the Abu Ghraib neighbourhood. He also did much of the organisation's media work, posting videos online of attacks against Coalition forces. American intelligence analysts were tracking Abu Atiya's cell phone and had a grid reference for him in Al-Yusifiya, a Sunni town on the southern fringes of Baghdad. Abu Atiya was captured during the raid, although another Iraqi called Abu Haider arrested that day was to prove pivotal to the Coalition's fortunes. After weeks of questioning he revealed he knew Zarqawi's religious mentor in Iraq, a fact that ultimately led to the al-Qa'eda leader's death in a Coalition airstrike. Williams's briefing to Blair was meant to last an hour,

but in the end lasted for several. (Mark Bowden. 'The Ploy', *Atlantic Monthly*, May 2007, and Mark Urban, *Task Force Black: The Explosive True Story of the SAS and the Secret War in Iraq*, Little Brown, 2010, p.148)

3 At a G8 meeting in St Petersburg during the crisis Bush famously hailed the prime minister with 'Yo, Blair,' unaware that the microphones at the conference table were on. The incident seemed to reconfirm the widespread suspicion that Blair was an American poodle.

4 Anthony Seldon, *Blair Unbound*, p.473

5 Private interview

6 Private interview

7 Lieutenant Colonel Jonny Bowron, private interview 10 September 2010

8 Sarah Sands, 'Sir Richard Dannatt: A Very Honest General', *Daily Mail*, 12 October 2006

9 Tom Rawstorne and Danielle Gusmaroli, 'The Terrible Neglect of our Troops', *Daily Mail*, 2 October 2006. This was the second article of a two-part series.

10 One of her emails read, 'The anesthesia had worn off by Sunday afternoon and no-one had organised any pain relief for [the injured soldier] (morphine drip which patients can use themselves). He was in excruciating agony – phoned [military liasion officer] literally screaming into his phone in agony.' Sarah-Jane Shirreff, email, 14 August 2006

11 Stuart Tootal had also alerted Dannatt to the inadequate treatment of injured soldiers at Selly Oak Hospital in June 2006. In July 2008 the MOD announced it would open a dedicated military trauma and orthopaedic ward for injured service personnel. This opened in 2010.

12 Richard Shirreff, interview 6 December 2010

13 Ibid.

14 Tim Evans, interview 22 October 2010

28 – LAMBO

1 Rob Fry, interview 14 January 2010

2 Lamb's predecessor Rob Fry had set up an insurgent engagement cell to focus the intelligence resources in Baghdad on engaging with Sunni insurgents, building on what McMaster and others had achieved on the ground in Anbar.

3 Emma Sky, interview 19 September 2009

4 Private interview

5 Graeme Lamb, interview 4 September 2009

6 Private interview

7 Graeme Lamb, interview 4 September 2009

8 Ibid.
9 Peter Mansoor, interview 12 April 2010
10 David Petraeus, interview 29 October 2010
11 Emma Sky, interview 19 August 2009

29 – PALERMO

1 Justin Maciejewski. interview 8 December 2009. Every time British troops left a district, the militia rushed back in. Local leaders in the next areas of operation warned Maciejewski that the British were only provoking the insurgents and asked him not to come.
2 Martin Dinham. interview 7 January 2010
3 In July 2007 Shirreff and his wife Sarah-Jane met the cartoonist and former officer Bryn Parry, who had recently completed a charity cycle ride with his wife for the MacMillan cancer charity. Parry wanted to run a similar event for the Army Benevolent Fund. Sarah-Jane had recently visited Headley Court, the military facility where injured soldiers are rehabilitated. The commanding officer there had told her that top of his wish list was a rehabilitation swimming pool. The idea for the Help for Heroes charity, which has gone on to raise over £100 million for injured service personnel, was thus born. Sarah-Jane became a trustee of the organisation, along with Richard Dannatt and his wife.
4 Jonathan Shaw, interview 2 February 2010
5 Mark Allen had resigned from M16 in 2004 after losing out on the top job at the spy agency to John Scarlett, presenting an interesting what-might-have-been had Allen, a man steeped in Middle Eastern history, become M16's head in the build-up to the war. Possibly some of the intelligence that Scarlett later approved for inclusion in the government's September dossier would not have been included.
6 3 Commando Brigade in Helmand, which had replaced 14 Air Assault Brigade, had 5,400 troops, and the figure was set to rise to 6,300 over the winter.
7 Private interview
8 Ian Thomas, interview 14 January 2010
9 Lou Bono, interview 16 November 2009
10 Ian Thomas, interview 14 January 2010
11 Stephen Grey, *Operation Snakebite: The Explosive Story of an Afghan Desert Siege*, Penguin, 2010, p.34
12 David Petraeus, interview 29 October 2010
13 James Bashall, interview 14 June 2010
14 Jonathan Shaw, interview 2 February 2010

30 – THE DEAL

1 Ahmed Fartosi, interview 10 October 2010
2 Fartosi later tried to sue the MOD, claiming psychological abuse during his detention. His cell was small, with no ventilation in the summer, he complained. At night he was forced to listen to his guards watching pornography on their laptops, which he believed they did to deliberately shame him. 'If they expected me to give in to my basic instincts they did not realise that I am not that kind of man.'
3 Private interview
4 Private interview
5 Tim Evans, interview 22 October 2010
6 Patrick Sanders, interview 23 November 2009
7 The Shatt al-Arab Hotel had been successfully handed over in April.
8 Rupert Lane, private interview 17 August 2010
9 Johnathan Shaw, interview 2 February 2010
10 Private interview
11 Private interview
12 Private interview
13 Private interview
14 The doctor, Bilal Abdulla, survived the attack and was sentenced to 32 years in prison. His accomplice and driver of the vehicle, Kafeel Ahmed, died of his injuries. Five members of the public were injured in the attack.

15 Jock Stirrup, interview 15 January 2011
16 Private interview
17 Private interview
18 Private interview
19 A week after Harding's death, three more soldiers were killed on a supply run as they dismounted in the Al-Antahiuya district. Corporal Paul Joszko, from Abercynon, Wales, was standing a few feet from the roadside bomb when the device detonated and was killed instantly. Privates Scott Kennedy and James Kerr both died from multiple blast injures.
20 Private interview
21 Tom Newton Dunn, 'Job Done' and 'Lions of Basra', *Sun*, 4 September 2007

31 – ON THE RUN

1 Haider Samad, interview 3 June 2009
2 Ali Sadoon, interview 4 March 2009
3 Haider Samad, interview 3 June 2009

4 This figure reflects a combination of direct British funding, Iraqi money allocated by the CPA and US military funds given to their British counterparts. It does not include post-CPA Iraqi spending or contributions from other nations; the true cost of Basra's reconstruction between 2003–9 is likely to be much higher.
5 Ibid.
6 Haider Samad, interview 3 April 2009
7 Ibid.
8 Ibid.
9 Ibid.
10 Ibid.

32 – CHARGE OF THE KNIGHTS

1 Private interview
2 Andy Bristow, interview 21 November 2009
3 Private interview
4 Private interview. Peter Moore was later released after more than two and a half years in captivity in Iraq. Four bodyguards kidnapped with him had been murdered. The Shia insurgent leader Qais al-Khazali, who had masterminded a concerted campaign against Coalition forces, including the kidnapping of Moore, was released from US custody at the same time, leading to suspicions of a deal.
5 Richard Iron, interview 28 July 2009
6 The offer of a local Taliban leader Mullah Salaam to switch sides also helped change British attitudes. Mullah Salaam claimed to have been among the original group of local Taliban commanders who concluded a deal with Ed Butler in September 2006. He visited President Hamid Karzai over the summer of 2007 and convinced the Afghan leader that he was ready to defect. But as Stephen Grey reveals in his vivid account of the fighting in Musa Qala, the Mullah Salaam who sparked the battle was possibly not a Taliban commander at all, but a local tribal leader of the same name (Stephen Grey, *Operation Snakebite*. p.318)
7 In 2006 the MOD had scrapped the army's doctrine development facility in favour of a new centre at Shrivenham that combined the army, navy and air force's capabilities. Rear Admiral Chris Parry, its new director, was keen to discard the army's old-fashioned view of counter-insurgency, which he considered more appropriate for the battles of colonial disengagement than twenty-first century warfare against criminal gangs and stateless terrorists. Parry developed his own discussion paper on 'irregular warfare' and threatened to stall Alderson's work.
8 The doctrine was still shelved a few weeks later, when Richard Shirreff took

offence at a critical description of the destruction of the Jamiat police station. It wasn't until January 2010 that Major General Andrew Kennett, a brigade commander in Basra in 2004 and current director of the Land Warfare Centre, finally resurrected and published a reworked version.

9 Private interview
10 Richard Iron, interview, 28 July 2009
11 Private interview
12 Mowaffak Rubaie, interview 21 August 2009
13 No date had been set, but the Iraqi parliament was under considerable pressure to approve an electoral law that would set the timetable. The elections eventually took place ten months later, in January 2009.
14 Julian Free, interview 29 November 2009
15 Patrick Marriot, interview 16 September 2010
16 Julian Free, interview 29 November 2009
17 Ibid.
18 Linda Robinson, *Tell Me How This Ends*, Public Affairs, 2008, p.341
19 Richard Iron, interview 28 July 2009
20 Julian Free, interview 29 November 2009
21 George Flynn, interview 26 April 2010

EPILOGUE

1 A small British naval contingent was the last UK unit to leave Iraq in May 2011
2 Nick Davies, Jonathan Steele and David Leigh, 'Iraq war logs: secret files show how US ignored torture,' the Guardian, 22 October 2010
3 http://www.casualty-monitor.org/p/iraq.html
4 Walter Busuttil, 'Veteran Services in the UK: The Role of Combat Stress,' Combat Stress: Ex-Services Mental Welfare Society, 08.10
5 In his memoir, Blair tries to favourably compare the number of Iraqi dead as the result of the invasion, with those killed by Saddam during the repression of the Shia uprising in 1992, and the al-Anfal Campaign against the Kurds – a bizarre and ill-judged attempt to exonerate himself.
6 Gordon Brown, Iraq Inquiry, 5 March 2010
7 Some experts place the true cost of the Iraq and Afghanistan as high as £3 trillion. Joseph Stiglitz and Linda Bilmes, "The true cost of the Iraq war: $3 trillion and beyond," The Washington Post, 5 September 2010.
8 Gordon Brown, Iraq Inquiry, 5 March 2010
9 Brown was forced into convening the Inquiry – a long-standing objective of anti-war protesters – after the Conservative Party announced they would hold one if

they came to power. The Inquiry announcement marked a particular victory for Adam Price, the Plaid Cymru MP, who had fought to have Tony Blair impeached in 2004. In 2006, Price had introduced to Parliament a debate over a possible inquiry, after which the Conservatives announced their own policy change.

10 Tony Blair, A Journey (London: Hutchinson, 2010), p.371. Blair later made a second Inquiry appearance in which he tried to clarify his earlier testimony. 'At the conclusion of the last hearing, you asked me whether I had any regrets. I took that as a question about the decision to go to war, and I answered that I took responsibility. That was taken as my meaning that I had no regrets about the loss of life and that was never my meaning or my intention. I wanted to make it clear that, of course, I regret deeply and profoundly the loss of life, whether from our own armed forces, those of other nations, the civilians who helped people in Iraq or the Iraqis themselves.' 2101.11.

11 Little progress has been made in terms of bringing to justice those who led the mob against the police station in Majar al-Kabir, where the Red Caps were stationed. In February 2010, the Iraqis courts arrested eight local men for instigating and carrying out the murders; only two were charged, and they were later released after the presiding judge cited a lack of evidence.

12 Matt Cavanagh, 'How to fix the National Security Council,' the Spectator, 15 May 2011

13 Jones, Seth, 'Counterinsurgency in Afghanistan,' Rand Counterinsurgency Study Volume 4, 17 August 2009

14 As well as the scandalous state of care for veterans, particularly those injured in combat. Since 2006, there have been marked improvements, not least thanks to the extraordinary effort of Emma and Bryn Parry, a cartoonist and former officer, who, in 2007, launched the Help for Heroes campaign to improve medical facilities for wounded soldiers, raising over £100 million in the process.

15 The 2010 Strategic Defence Review led to sweeping cuts in the military's budget, which led to the axing of the Harrier jump jet program and the Navy's flagship HMS Ark Royal, and the loss of 42,000 MOD jobs by 2015. Overall, defence spending will fall by 8% over four years.

BIBLIOGRAPHY

THIS BOOK IS PRIMARILY BASED ON OVER 400 interviews with serving and retired military personnel, diplomats, contractors and Iraqis. Some 300 interviews were conducted during my time in Baghdad as the *Daily Telegraph*'s correspondent and in Kabul for the *Washington Post* but the majority took place on the phone or in person between 2008–10. An important supplement to my reporting has been the Iraq Inquiry, launched in November 2009. The journalist Chris Ames and the *Guardian*'s Andrew Sparrow teased out many of the details I might have missed during the hours of hearings. Evidence supplied to the Baha Mousa Inquiry and a number of other inquests have revealed the full extent of the British military's abuse of Iraqi detainees. Public Interest Lawyers, a law firm specialising in human rights issues, initially brought to light many of these cases.

When I began research there was no history that covered the full sweep of the British involvement in Iraq. However, there were a number of books without which this account could not have been written. Anthony Seldon's *Blair Unbound* provides the definitive chronicle of the period from Number 10's perspective, followed by Andrew Rawnsley's *The End of the Party*. Richard North captures the bureaucratic ineptitude at the MOD in *Ministry of Defeat*. Over the course of two books, Tom Ricks offers a compelling and influential history of the US military in Iraq. Rajiv Chandrasekaran's dark satire captures the madness of the early days of American rule in Iraq. Other accounts that proved invaluable were the late Richard Holmes' assiduously reported *Dusty Warriors*, Dan Mills' *Sniper One*, Mark Etherington's *Revolt on the Tigris*, Paul Bremer's defence of his administration

My Year in Iraq, and Rory Stewart's *Prince of the Marshes*. Mark Nicol's sobering book *Iraq: A Tribute to Britain's Fallen Heroes* was a constant reminder of the cost of war. In Afghanistan, Patrick Bishop, Stuart Tootal and Stephen Grey proved essential reading for key deployments. To understand Britain's historical role in the Middle East and its eerie echoes of the present, David Fromkin's *A Peace to End All Peace* is masterful.

There have been a number of outstanding films and documentaries. James Longley's *Iraq in Fragments* offers a simple and evocative portrait of the Shia south. The BBC's *Panorama* team did not shirk from covering the most difficult and controversial aspects of the British occupation.

Books

Allen, Mark, *Arabs* (London: Continuum, 2006)

Alderson, Andrew, *Bankrolling Basra: The Incredible Story of a Part-Time Soldier, $1 Billion and the Collapse of Iraq* (London: Robinson, 2007)

Allawi, Ali, *The Occupation of Iraq: Winning the War, Losing the Peace* (New Haven: Yale University Press, 2007)

Amis, Martin, *The Second Plane* (London: Vintage, 2008)

Bishop, Patrick, *3 Para* (London: Harper Press, 2007)

Blair, Tony, *A Journey: My Political Journey* (London: Hutchinson, 2010)

Bremer III, L. Paul, *My Year in Iraq: The Struggle to Build a Future of Hope* (New York: Simon & Schuster, 2006)

Buchan, John, *Greenmantle* (West Valley City: Waking Lion Press, 2006)

Bush, George, *Decision Points* (New York: Crown, 2010)

Campbell, Alastair, *the Blair Years: the Alastair Campbell Diaries* (New York: Knopf, 2007)

Chandler, Edmund, *The Long Road to Baghdad* (London: Cassell and Company, 1919)

Chandrasekaran, Rajiv, *Imperial Life in the Emerald City: Inside Iraq's Green Zone* (New York: Knopf, 2006)

Cockburn, Patrick, *Muqtada al-Sadr and the Fall of Iraq* (London: Faber and Faber, 2008)

Collins, Dan, *In Foreign Fields: Heroes of Iraq and Afghanistan in Their Own Words* (London: Monday Books, 2007)

Cullather, Nick, *The Hungry World: America's Cold War Battle Against Poverty in Asia* (Cambridge: Harvard University Press, 2010)

Dannatt, Richard, *Leading from the Front: An Autobiography* (London: Bantam Press, 2010)

Dickson, Violet, *Forty Years in Kuwait* (London: George Allen & Unwin, 1978)

Dodge, Toby, *Inventing Iraq: The Failure of Nation Building and a History Denied* (New York: Columbia University Press, 2003)

Etherington, Mark, *Revolt on the Tigris: The Al-Sadr Uprising and the Governing of Iraq* (Ithaca: Cornell University Press, 2005)

Ferguson, Niall, *Empire: The Rise and Demise of the British World Order and the Lessons for Global Power* (New York: Basic Books, 2004)

French, Patrick, *Younghusband: The Last Great Imperial Adventurer* (London: Flamingo, 1995),

Fromkin, David, *A Peace to End All Peace: The Fall of the Ottoman Empire and the Creation of the Modern Middle East* (New York: Avon Books, 1990)

Geniesse, Jane Fletcher, *Passionate Nomad: The Life of Freya Stark* (New York: The Modern Library, 2001)

Gordon, Michael and Trainor, Bernard, *Cobra II: The Inside Story of the Invasion and Occupation of Iraq* (New York: Pantheon, 2006)

Gray, John, *Straw Dogs: Thoughts on Humans and Other Animals* (London: Granta Books, 2003)

Gray, John, *Black Mass: Apocalyptic Religion and the Death of Utopia* (New York: Farrar, Straus, Giroux, 2007)

Greene, Graham, *The Quiet American* (London: Vintage, 2001)

Grey, Stephen, *Operation Snakebite: The Explosive Story of an Afghan Desert Siege* (London: Penguin, 2010).

Haldane, Aylmer, *The Insurrection in Mesopotamia* 1920 (London: The Imperial War Museum, 2005)

Hemingway, Ernest, *A Farewell to Arms* (London: Penguin Books, 1964)

Holmes, Richard, *Dusty Warriors* (London: Harper Perennial, 2007)

Hopkirk, Peter, *Like Hidden Fire: The Plot to Bring Down the British Empire* (New York: Kondansha International, 1994)

Hopwood, Derek, *Tales of Empire: The British in the Middle East* (London: I.B Tauris & Co, 1989)

Horne, Alastair, *A Savage War of Peace: Algeria 1954–1962* (New York: New York Review of Books, 2006)

Hourani, Albert, *A History of the Arab Peoples* (New York: Warner Books, 1992)

Howell, Georgina, *Gertrude Bell, Queen of the Desert, Shaper of Nations* (New York: Farrar, Straus and Giroux, 2008)

Irwin, Robert, *The Arabian Nights: A Companion* (London: Penguin, 1995)

Ivison, Kevin, *Red One: A Bomb Disposal Expert on the Front Line* (London: Weidenfeld and Nicolson, 2010)

Jackson, Mike, *Soldier: The Autobiography* (London: Corgi, 2008)

Kampfner, John, *Blair's Wars* (London: Free Press, 2004)

Khudayyir, Muhammed, *Basrayatha: The Story of a City* (London: Verso, 2008)

Klein, Naomi, *The Shock Doctrine: The Rise of Disaster Capitalism* (New York: Picador, 2007)

Mallinson, Allan, *The Making of the British Army: From the English Civil War to the War on Terror* (London: Bantam Press, 2010)

Marston, Daniel, *Phoenix from the Ashes: The Indian Army and the Burma Campaign* (Westport: Praeger, 2003)

Meyer, Christopher, *DC Confidential: The Controversial Memoir of Britain's Ambassador to the US at the Time of 9/11 and the Run-Up to the Iraq War* (London: Phoenix, 2006)

Mills, Dan, *Sniper One: On Scope and Under Siege with a Sniper Team in Iraq* (New York: St Martin's Press, 2007)

Monroe, Elizabeth, *Philby of Arabia* (Reading: Ithaca Press, 1998)

Morrison, Elizabeth, *Jane Penelope's Journal* and *Governess to King Feisal II of Iraq 1940–43* (Cambridge: West Meadow Books, 1995)

Nagl, John, *Counter Insurgency Lessons from Malaya and Vietnam: Learning to Eat Soup with a Knife* (Westport: Praeger, 2002)

Nicol, Mark, *Iraq: A Tribute to Britain's Fallen Heroes* (London: Mainstream Publishing, 2009)

Nicol, Mark, *Last Round: The Red Caps, the Paras and the Battle of Majar* (London: Weidenfeld and Nicolson, 2005)

North, Richard, *Ministry of Defeat: The British War in Iraq 2003–9* (London: Continuum, 2009)

Peston, Robert, *Brown's Britain: How Gordon Runs the Show* (London: Short Books, 2005)

Philby, H. St John, *Arabian Days: An Autobiography* (London: Robert Hale, 1948)

Rawnsley, Andrew, *The End of the Party: The Rise and Fall of New Labour*, (London: Viking, 2010)

Reidar, Visser, *Basra, the Failed Gulf State: Separatism and Nationalism in Southern Iraq* (Berlin: LIT Verlag, 2005)

Ricks, Thomas, *The Gamble: General Petraeus and the Untold Story of the American Surge in Iraq, 2006–8* (London: Allen Lane, 2009)

Ricks, Thomas, *Fiasco: The American Military Adventure in Iraq* (New York: Penguin, 2006)

Robinson, Linda, *Tell Me How This Ends* (New York: Public Affairs, 2008)

Ross, Carne, *Independent Diplomat: Dispatches from an Unaccountable Elite* (Ithaca: Cornell University Press, 2007)

Satia, Priya, *Spies in Arabia, The Great War and The Cultural Foundations of Britain's Covert Empire in the Middle Eas*t (Oxford: Oxford University Press, 2008)

Seldon, Anthony (ed.), *Blair's Britain: 1997–2007* (Cambridge: Cambridge University Press, 2007)

Seldon, Anthony, *Blair Unbound* (London: Simon and Schuster, 2007)

Seldon, Anthony and Lodge, Guy, *Brown at Ten* (London: Biteback, 2010)

Sluglett, Peter, *Britain in Iraq: Contriving King and Country* (London: I.B Tauris, 2007)

Smith, Rupert, *The Utility of Force: The Art of War in the Modern World*, (New York: Vintage, 2008)

Stein, Janice Gross and Lanf, Eugene, *The Unexpected War: Canada in Kandahar* (Toronto: Penguin, 2007)

Stewart, Rory, *The Prince of the Marshes and Other Occupational Hazards of a Year in Iraq* (Orlando: Harcourt, 2006)

Stone, Peter G. and Bajjaly, Joanne Farchakh, *The Destruction of Cultural Heritage in Iraq* (Woodbridge: The Boydell Press, 2008)

Stothard, Peter, *30 Days: A Month at the Heart of Blair's War* (London: HarperCollins Publishers, 2003)

Synnott, Hilary, *Bad Days in Basra: My Turbulent Time as Britain's Man in Southern Iraq* (London: I.B Tauris, 2008)

Thesiger, Wilfred, *The Marsh Arabs* (London: Penguin, 2007)

Thomas, Bertram, *Alarms and Excursion in Arabia* (Indianapolis: The Bobbs-Merril Company, 1931)

Tootal, Stuart, *Danger Close: The True Story of Helmand from the Leader of 3 Para* (London: John Murray, 2010)

Townshend, Charles Vere Ferrers, *My Campaign* (New York: The James A. McCann Company, 1920)

Tripp, Charles, *A History of Iraq* (Cambridge: Cambridge University Press, 2007)

Turner, Barry, *Suez 1956: The Inside Story of the First Oil War* (London: Hodder & Stoughton, 2007)

Urban, Mark, *Task Force Black: The Explosive True Story of the SAS and the Secret War in Iraq* (London: Little, Brown, 2010)

Vincent, Steven, *The Red Zone: A Journey into the Soul of Iraq* (Dallas: Spence Publishing Company, 2004)

Wallach, Janet, *Desert Queen: The Extraordinary Life of Gertrude Bell: Adventurer, Advisor to Kings, Ally of Lawrence of Arabia* (New York: Nan A. Talese, 1996)

Wells, H.G., *The Shape of Things to Come* (London: Penguin Classics, 2005)

Winstone, H.V.F., *Leachman: O.C. Desert* (London: Quartet Books, 1982)

Woolley, Leonard, *Excavations at Ur* (New York: Thomas Y. Crowell Company, 1965)

Wright, Lawrence, *The Looming Tower: Al-Qaeda and the Road to 9/11* (New York, Knopf, 2006)

Young, Gavin, *Return to the Marshes: Life with the Marsh Arabs of Iraq* (London: Hutchinson, 1983)

Zaeef, Abdul Salam, *My Life with the Taliban* (New York: Columbia University Press, 2010)

Articles, Papers, Monographs

Bowden, Mark, 'The Ploy', *Atlantic Monthly*, May 2007

Busuttil, Walter, 'Veteran Services in the UK: The Role of Combat Stress,' Combat Stress: Ex-Services Mental Welfare Society, October 2010

Cochrane Sullivan, Marisa, 'The Fragmentation of the Sadrist Movement,' Institute for the Study of War, 15 January 2009

Cole, Juan, 'Marsh Arab Rebellion: Grievance, Mafias and Militias in Iraq,' Fourth Wadie Jwaideh Memorial Lecture, 15 October 2005

Cornish, Paul and Dorman, Andrew, 'Blair's wars and Brown's budgets: from Strategic Defense Review to strategic decay in less than a decade,' Chatham House, 2008

Dalrymple, William, 'Souter Takes a Call,' *Outlook India*, 30 August 2010

Felter, Joseph and Fishman, Brian, 'Iranian Strategy in Iraq: Politics and "Other Means"', Occasional Paper Series, Combating Terrorism Center at Westpoint, 13 October 2008

Hersh, Seymour, 'Get Out the Vote,' *New Yorker*, 25 July 2005

International Crisis Group, 'Where is Iraq heading? Lessons From Basra,' 25 June 2007

Jwaideh, Albertine, 'The Marsh Dweller of Southern Iraq: Their Habitat, Origins, Society And Economy,' Sixth Wadie Jwaideh Memorial Lecture, 30 October 2007

Knights, Michael and Williams, Ed, 'The Calm Before the Storm: The British Experience in Southern Iraq,' The Washington Institute for Near East Policy, February 2007

Kulka, R. A, et al, 'National Vietnam Veterans Readjustment Study: Table of findings and technical appendices,' New York: Brunner/Mazel, 1990

O'Hagan, Andrew, 'Iraq May 2nd, 2005' *London Review of Books*, 6 March 2008

Public Interest Lawyers, 'British Forces in Iraq: The Emerging Picture of Human Rights Violations and the Role of Judicial Review,' 30 June 2009

Rawstone, Tom and Gusmaroli, Danielle, 'The Terrible Neglect of our Troops,' the *Daily Mail*, 2 October 2006

Sands, Sarah, 'Sir Richard Dannatt: A Very Honest General,' the *Daily Mail*, 12 October 2006

Visser, Reidar, 'Britain in Basra: Past Experiences and Current Challenges,' www.historiae.org, 11 July 2006

Vincent, Stephen, 'Switched off in Basra,' the *New York Times*, 31 July 2005

Documentary/Film

Bigelow, Katherine, *The Hurt Locker*, June 2009

Bowker, Peter, *Occupation*, BBC, June 2009

Broomfield, Nick, *The Battle for Haditha*, Channel 4, 17 March 2008

Corbin, Jane, 'Bringing Our Boys Home?' *Panorama*, 18 March 2006

Ferguson, Charles, *No End in Sight*, January 2007

Longley, James, *Iraq in Fragments*, January 2007

Sweeney, John, 'Sweeney Investigates: The Death of the Red Caps,' Panorama, 10 February 2005

Sweeney, John, 'On Whose Orders?' *Panorama*, 25 February 2008

Ware, John, 'No Plan, No Peace,' BBC, 14 January 2008

ACKNOWLEDGEMENTS

THIS BOOK WOULD NOT HAVE BEEN possible without my wonderful agent Clare Alexander and my editor Dan Franklin, who first recognised the need for a comprehensive history of the Iraq war, and whose eye for detail brought the story to life. I would also like to thank Tom Avery and the rest of the team at Jonathan Cape. Hugh Davis gave the manuscript a deft copy-edit.

The reporting for *A War of Choice* began many years ago in a Devon pub, where I discussed covering war zones with my old friend Luke Walker. That discussion became a reality with the help of Alec Russell, the *Daily Telegraph*'s Foreign Editor in 2002, who guided me to Kuwait and Iraq and has been an inspiring role model for a young war reporter and historian ever since. Francis Harris's unbounded pessimism, sound judgement and wry humour helped steer countless pieces to press, including this book.

I would also like to mention the rest of the *Daily Telegraph* team, now sadly disbanded – Patrick Bishop, who taught me the ropes, David Blair, Peter Foster, Abbie Trayler-Smith, Heathcliffe O'Malley, Julius Strauss, Harry de Quetteville, Alex Spillius, Anton La Guardia, Neil Darbyshire, Alan Philps, Sebastian Berger, Oliver Poole, Joe Jenkins and, of course, that wonderful pair, Paul Hill, who made sure I got paid and stayed safe, and Patsy Dryden, whose dulcet tones always reassured. Rasha al-Sabah welcomed me in Kuwait like a long-lost son and Peter Murtagh of the *Irish Times* paid for my first Satellite phone bill.

The strains of living in Baghdad for two years would have been more acutely felt without the camaraderie of my fellow correspondents, Adam Davidson, Jen Banbury, Matthew McAllester, Rory McCarthy, James Hider,

Tara Sutton, Patrick Graham, Jonlee Anderson, Rajiv Chandrasekaran, Anne Barnard, Lulu Garcia Navarres, Jonathan Steele, Thanassis Cambanis, Ray LaMoine, Jeff Neumann, Maya Alleruzzo, Willis Witter, Dumeetha Luthra, Phil Sherwell, Stephen Grey, Victoria Whitford, and Marla Ruzicka, whose tragic death in Iraq in 2005 while campaigning to protect Iraqi civilians was a reminder of the deep bonds that our little community had formed, and the great hole she left. My dear driver and friend Abu Salah saved my life and risked his own on a number of occasions, and embodied the warmth and courage of so many of his countrymen.

Reporting in Afghanistan was made possible with the support of the *Washington Post*'s Hal Straus and Lauren Keane. Rory Stewart kindly lent me his room and made the suggestion that made this project possible. Mokdar followed me across the mountains.

For my research in the UK, I relied on the generosity of my *alma mater*, Lincoln College, Oxford University, where Paul Langford kindly housed my small family during my long journeys across the country. Stephen Gill and Peter McCullough have provided wise words over the years, and Kantik Ghosh first sent me packing to India with a notebook and camera. Debbie Usher revealed the secrets of the St Anthony's archives. David Booker at Atlantic College has been a constant inspiration. I'd also like to thank Richard North, Chris Ames, Mark Nicol and the late Richard Holmes for kindly agreeing to review sections of the text, and Anthony Seldon, Hew Strachan, Michael Smith, Hala Jabbar and Sam Falle for their advice. Nigel Tillyard, Tim Wilson, Lazlou Szomoru and Alan Cromie were all generous with their time and notes. Natasha Vashisht, Sophie Mann, Ahmed Mehdi and Karolina Maclachlan were excellent researchers.

At Harvard University, Cemal Kafadar welcomed me to the Centre for Middle Eastern Studies, and Roger Owen offered sage advice over several pints. Marissa Cochrane Sullivan at the Institute for the Study of War offered her insights into the Shi'ite insurgency. Joy de Menil taught me how to craft a narrative.

The Solutions team gave me the time and flexibility to tackle the intractable problems of the region. Many thanks to Robert Costanza and Ida Kubiszewski. For their support and love over the years I would also like to thank Adam, Chloe and Sophie Fairweather, Phil and Lynn Asquith, Nikki and David Dubois, Jens Munch, Rune Frederiksen, Yuill Herbert,

Lila Pla, Debbie Davies, Caroline Thomas, Lex Vaughan-Jones, Caroline Dale, Herbie Muller, Anthony Lipmann and Sal and Sherine Khadr.

This book would not have been possible without the generosity and openness of those who served and worked in Iraq. Many spent long hours reliving traumatic events and life-changing moments – a familiar feature of life in Iraq. I would like to single out Major Matthew Botsford and A Squadron, the Queen's Dragoon Guards, who first introduced me to the rigours of life in the field, and the finest aspects of the military. Charles Heath-Saunders at the MOD worked wonders arranging interviews. Others felt compelled to tell their stories and trusted me with their memories. I have not been able to include every story here, but every interview has helped deepen this narrative.

Among the many hardships of Iraq, I somehow managed to find my love, Christina, who fell for the offer of a flak jacket and a cup of tea. In return, she has given me endless love, warmth, patience and wisdom – and of course two beautiful daughters. She also worked tirelessly on this manuscript. Iraq has filled our lives. Christina has shown me how to shape those experiences into this book and so much more.

INDEX

Abadan, 321, 322
Abbas, Hussein Fadhil, 123
Abbas, Mahmoud, 127
Abu Azzam, 293–4, 295–6, 298, 299, 313, 351
Abu Ghraib, 115, 119, 293, 294
Abu Hatem, 53–4, 56, 57, 90, 91, 96, 112, 116, 119, 120–1, 123, 124, 155, 156, 196, 351
Abu Khadr, Wissam, 351
Abu Naji *see* Camp Abu Naji
Adams, Jerry, 310
Adam Smith International, 162
Aegis, 137
Afghanistan, xiii, 3, 7, 10, 83, 129–31, 132, 133, 173–5, 181, 194, 221, 224–7, 229–35, 237–45, 263, 271, 275–7, 280, 303, 305, 328, 331, 333, 339, 343, 344–5, 346, 347, 348, 349
Afghan Steering Group, 229–30, 232
Ahmed (consulate office manager), 1
Ahmed (cousin of Nora al-Sadoon), 319, 320, 321
Ahmed (translator), 323, 324, 326
Ahmedinejad, Mahmoud, 203
Ahmood, Ahmed Jabar, 122
Ahtissari, Martii, 228, 230
Air Assault Brigade, 16th, 225
Akhundzada tribe, 227
Akil, Colonel, 273
Al-Askari shrine, Samara, 252, 262
Albu Muhammad tribe, 51, 52, 53, 54, 96
Al-Dawaya, 94
Alderson, Colonel Alex, 250, 332, 351
Alderson, Andrew, 67, 68, 69, 86, 136, 351
Alderson, John, 141
Aldred, Margaret, 229, 233, 351
Al-Hawzat, 100, 101
Ali, Ra'aid, 46–7
Al Jazeera, 95, 100, 104–5, 150, 155, 170, 273
Allawi, Abbas, 178–9, 180
Allawi, Ali, 202
Allawi, Ayad, 131, 161–3, 164, 165–6, 169–70, 171, 343, 351

Allen, Mark, 303
Alli, Waheed, 170
Allied Rapid Reaction Corps *see* ARRC
Al-Qa'eda, 33, 83, 174, 181, 279, 290, 292, 293, 294, 304, 316, 344
Al-Quds Force, 149, 202, 204, 297, 338
Al-Rifai, 93
Al-Uzayr, 56
Alwash, Azzam, 170
Amarah, xv, 2, 52, 54, 89–90, 96–7, 101, 112, 113, 115, 116–20, 124, 149, 150, 155–8, 192, 195, 196, 198, 202, 205, 251, 269, 270, 272–3, 341
Amiery, Daoud, 244
Amman, 326
Anbar province, 170, 296, 329
Anderson, Duncan, 251
Ansar al-Sunni, 293
Arab Bureau, 89
Arafat, Yasser, 167
Arbuthnot, Felicity, 189, 190
Argyll and Sutherlands, 117, 118
Armitage, Richard, 77, 79, 114
ArmorGroup, 136, 137–8, 139, 140, 145, 177, 183 196, 210, 213, 286, 287
Armoured Brigade, 7th, 27
Armoured Division, 1st, 29
Army Board, 254, 255
Army Training Regiment, 257
ARRC (Allied Rapid Reaction Corps), 129, 130, 132, 301
Asa'ib Ahl al-Haq, 159
Ashmore, Lieutenant Colonel Nick, 30
Asquith, Dominic, 3, 5, 78, 221, 351
Associated Press, 343
Association of Chief Police Officers, 139
Austin, General Lloyd, 334, 335, 337
Australians, 34, 181
Awakening councils, 290, 292, 293, 295, 296–7
Aylwin-Foster, Brigadier Nigel, 247–8
Az Zubayr, 45, 47, 135–6, 139, 140, 209

Ba'ath Party, 23, 24, 25, 40, 41, 68, 202
Bach, William, 255
Bacon, Major Matthew, 209
al-Badran, Nouri, 112, 138
Badr Brigade, 116, 149, 159, 195, 202
Baghdad, 2, 23, 28, 29, 31, 37–9, 40, 44, 52, 69, 70, 73, 75, 76–7, 78, 79, 80, 81–2, 87, 90, 91–2, 95, 99, 101, 104, 106, 107, 108, 113, 129, 138, 148, 159, 161–2, 166, 167, 168, 195, 202, 203, 204, 217, 262, 263, 272, 279, 289, 290, 291, 292, 293, 304, 305, 313, 314, 316, 329, 332, 333–5, 341, 343, 344
Bagram, 130, 238
Bagwell, Air Vice Marshall Greg, 315
al-Bahadili, Abdul Sattar, 102, 103, 148–9

al-Bahadili, Ali, 51–2
Bahadili clan, 147
Bahrain, 342
Baldwin, Lieutenant Colonel Gil, 32, 34, 35, 228, 229, 230, 234, 351
Balkans, 6, 7, 11, 32, 65, 66, 129, 228, 350
Balmoral private security firm, 137
Banda Aceh, 229
Barbados, 73
Barlow, Fusilier Andy, 276
Bartlam, Fusilier Gary, 47, 48
Bashall, Brigadier James, 305, 351
Basra, xiv, 1–5, 9, 14, 23–5, 27, 28, 29, 30–1, 43–50, 52, 66–9, 70, 71–4, 85, 86, 87, 92, 95, 102, 103, 112, 124, 136–7, 138, 140, 141–5, 147, 148–9, 150–5, 156, 158, 159, 175, 177, 178–9, 180, 183, 184, 207–16, 217–20, 221, 248, 251, 252, 261–2, 262–3, 264, 265–8, 271, 272, 274, 277, 279, 280–1, 282, 283, 286, 301, 302–3, 304, 305–7, 309, 311–13, 314, 315–18, 319–21, 322, 324, 325, 326,327–8, 329–33, 334, 335, 336–8, 341–2
Basra River Force, 44, 48
Basra Security Committee, 267
Bathurst, Lieutenant Colonel Ben, 195
Baugh, Matt, 228–9
Bayley, Gareth, 100
BBC, 70, 85, 127, 191, 282
  Radio 4, 50
  World Service, 23
Beckett, Margaret, 282
Beckett, Lieutenant Colonel Tom, 54
Beharry, Private Johnson, 119
Beirut, 338
Bell, Gertrude, 52–3, 89, 351
Bell, Martin, 190, 191
Bellinger, John, 18
Beni Lam tribe, 96
Benn, Hilary, 86, 136, 235
Berlusconi, Silvio, 95
Bigley, Kenneth, 166–7
bin Laden, Osama, 345
Binns, Major General (formerly Brigadier) Graham, 28, 29, 31, 316, 330, 333, 351
Bishop, Lieutenant Ben, 198–9, 200
Bishop, Patrick, 55
Blackie, Emma, 47–8
Black Watch, 168
Blackwater, 100, 137
Blackwill, Robert, 82, 105, 114
Blair, Tony, 6, 7, 8, 9, 11, 12, 13, 16, 17, 19, 20–1, 22, 39, 40, 41, 48–50, 63, 70, 71, 73, 75, 78, 83, 100, 104, 127–9, 130, 131–2, 133, 134, 138, 162, 164, 167, 168, 176, 181, 188, 189–90, 192, 193–4, 202, 220–1, 223–4, 227, 230, 235, 265, 279, 280, 290, 301, 303–4, 304–5, 310–11, 313, 315, 343, 345–6, 349, 352
Blix, Hans, 14, 19, 20
Bono, Lou, 304
Booth, Cherie, 194
Bosnia, 65, 88, 225, 250
Botsford, Major Matthew, 31

Bourne, John, 88–9, 92, 93, 94–5, 352
Bowron, Major Johnny, 97, 280, 281, 283, 352
Boxer armoured vehicle, 254
Boyce, Admiral Michael, 8, 13, 20
Brackenbury, Lance Corporal Alan, 199, 200, 201, 202, 204, 205, 352
Bradford, Captain James, 215, 216
Bradley, Major David, 153
Bradley, Tobin, 93–4, 95, 101, 102, 103, 110, 111, 113, 352
Bradshaw, Brigadier Adrian, 43, 48
Brahimi, Lakhdar, 104–5, 111, 114
Brand, Doug, 79, 138, 139, 144
Bratcher, Captain Simon, 200–1, 202, 352
Brealey, Bruce, 71, 72
Bremer, Paul, 39–41, 44, 45, 66, 67, 68, 69, 70, 72, 73, 75, 76, 77, 78–9, 81–2, 90, 92, 94, 95, 99, 100, 101, 102, 105, 106, 111, 112, 113, 114, 131, 352
Bright, Mark, 276
Brims, Major General Robin, 29, 30, 35
Bristow, Colonel Andy, 328, 329, 330, 331, 352
British Council, 168
Brize Norton, 65, 189, 198
Brookes, Jeremy (Jez), 311
Broome, Sergeant Christopher, 121–2, 122–3, 257, 352
Broome, Lynsey, 257
Brown, Gordon, 127, 128, 132–3, 134, 192, 280, 313–14, 315, 317, 323, 328–9, 343, 345, 352
Browne, Des, 235, 266, 304, 307, 352
Brussels, 21
Bryan, Sergeant Terry, 150–4, 352
Bulford, 187, 262
Bullock, Brigadier Gavin, 249
Burton, John, 193
Bush, George W., vi, 7, 11, 12, 13, 14, 33, 39, 40, 41, 75, 77, 78, 82, 86, 105, 111–12, 131, 162, 165, 166, 168, 202, 289, 290, 352
Butler, Brigadier Ed, 225–6, 239–40, 241, 242, 243, 245, 275, 276, 277, 305, 331, 352
Butler Inquiry, 134, 193
Byles, Corporal Mark, 120, 122

Cabinet, 12, 13, 16, 20, 232, 235
Cabinet Office, 8, 50, 89, 164, 165, 207, 216–17, 228, 229, 233, 239, 264, 265, 266
Cairo, 89
Cameron, David, 329, 349, 350
Camp Abu Naji, 117, 118, 121, 122, 123, 155, 192, 193, 195 ,196, 198, 200, 252, 269–70, 271, 272, 273–4, 277, 283, 291, 292, 306
Camp Bastion, 225, 237, 238, 239, 242, 245, 275, 276
Campbell, Alastair, 19, 20, 21–2, 41, 49, 50, 70, 73, 75, 78, 352
Campbell, Les, 163, 164
Camp Breadbasket, 46–8
Camp Cherokee, 151

Camp Dogwood, 168, 169
Camp Victory, 80
Canadians, 129, 131, 181, 225, 241
Care International, 168
Carpenter, Scott, 79, 82, 114
Carter, Brigadier Nick, 102, 103, 263, 352
Casey, General George, 161, 248, 250, 252, 262, 270, 272, 274, 289, 290, 291, 293, 303, 304, 352
Chakrabarti, Suma, 86, 87, 352
Chalabi, Ahmed, 39
Channel 4 News, 192, 193
Chaplin, Edward, 229–30
Charge of the Knights, 333–8, 341, 343, 348, 349
Chequers, 8, 133, 265, 310
Chiarelli, General Peter, 272, 274, 283
Chiarini, Brigadier Gian Marco, 102, 103, 110
Chicago speech, 6, 8, 9, 345
Chicksands, 33, 34
Chilcott, Dominic, 16
China, 8
Chirac, Jacques, 20
Churchill, Winston, 164
CIA, 161, 163, 166, 203, 293, 298
Cimic House, Amarah, 89, 90, 96, 101, 112, 115, 117, 118, 155–8, 195, 251
City of London, 194
Civilian Police Advisor Training Team, 138
Clark, Mark, 66, 67, 68
Clay, Bob, 190
Clinton, William, 161
Clissitt, Nick, 227, 228, 352
CNN, 37
Coalition Provisional Authority *see* CPA
Cobra, 217
Cocup, Major Matthew, 273
Coldstream Guards, 209, 211, 212, 221
Cole, Mike, 140, 141
Collins, Colonel Tim, 28
Colombia, 137
Combined Arms Centre, 247
Commons Defence Committee, 255
Comprehensive Spending Review, 133, 139
Congo, the, 137
Congress, US, 69, 70, 78, 163, 165, 289, 290
Conservative Party, 191, 193, 194
Contini, Barbara, 95, 101, 103, 110–11
Control Risks Group (CRG), 106, 107, 110
Cook, Robin, 6, 21
Cooley, Lance Corporal Mark, 47
Cooper, Major General John, 261–2, 267
Coote, Major James, 122
Corbin, Jane, 252
Cornwallis, Kinahan, 165
Corriere della Sera, 110
Council of Guardians (Iran), 204
Cowan, Lieutenant Colonel James, 168, 169, 353
Cox, Percy, 52, 164

CPA (Coalition Provisional Authority), 40, 41, 66, 67, 69, 72, 76, 81, 82, 90, 91, 93, 95, 96, 99, 100, 101, 103, 104, 106, 107, 109, 110, 111, 113, 131, 136, 139, 144, 150, 230
Crawford, Texas, 8, 11–12
CRG *see* Control Risks Group
Crook, Captain Jonny, 54
Cross, Major General Tim, 11, 13, 15–16, 16–17, 21, 22, 37, 40–1, 353
Curry, Captain Charlie, 155–6, 156–7, 353
Cyprus, 267, 271
Czechs, the, 131

Daily Mail, 281–2, 318
Daily Mirror, 115
Daily Telegraph, 38–9, 41, 220
Dalton, Richard, 203–4
Danes, 66, 131, 178, 179
Dannatt, Lieutenant General Richard, 129–30, 226, 255, 256, 271, 281–2, 313, 253
Danny Boy incident, 121–3, 205, 257
Daoud, Muhammad, 227, 233, 239, 240, 241, 243, 245, 353
Daqduq, Ali Mousa, 297
Dasht-e Margo desert, 174
Davies, Gavyn, 127
Dawa, 163, 202, 252, 297
Defence Intelligence and Security Centre, 33
Defence Medical Services Directorate, 256, 257
Defence Ministry (Iraq), 171
Defence Systems Limited, 137
Democratic Party (US), 289
Dempsey, Martin, 292, 335
Department of Defense (US), 15, 41, 227
Department for Environment, Food and Rural Affairs (Britain), 89, 95
Department for International Development (Britain) *see* DFID
Desert Fox campaign, 6
de Stacpoole, David, 138, 140
DFID (Department for International Development), 6–7, 13, 16, 49, 66, 78, 85–6, 87, 136, 139, 161, 171, 175, 228, 229–30, 238, 239, 263, 265–6, 301, 349
Dhi Qar province, 70, 89, 92–5
Dickson, Harold, 89, 92–3
Dinham, Martin, 301
Directorate General Doctrine and Development, 250
Dolman, Corporal John, 58
Drayson, Paul, 255
Drummond, Jim, 228
Dubai, 139
Dutch, the, 70, 130, 181
Dyke, Greg, 127, 190
Dyncorp, 138, 139

Eaton, Major General Paul, 80, 138
Education Department (Britain), 344
Edwards, Lieutenant Colonel Ben, 252
Egypt, 9, 342

Eikenberry, Karl, 227
Elliot, Major Brian, 30
Ellis, Courtney, 258
Ellis, Private Freddy, 61
Ellis, Karla, 258–9
Ellis, Private Lee, 252, 258, 269, 353
England, Lynndie, 115
Eno, Brian, 189, 190
Etherington, Mark, 88, 91–2, 101, 103–4, 106–8, 109, 110, 112, 113, 230, 231, 232, 233–4, 264–5, 266, 267, 282, 353
Euphrates, River, 52, 53, 97, 101
European Community Monitoring Commission in Bosnia, 88
European Convention on Human Rights, 32, 33, 35, 45
Everard, Brigadier James, 251, 252, 266, 353

Fadhillah, 48, 180
Faisal I, 164, 165, 171, 330
Faisal II, 162
Fallujah, 100–1, 102, 104–5, 108, 111–12, 112–13, 162, 166, 167, 168, 169, 293
al-Faraji, General Mohan *see* Mohan, General
Farhadi, Mahmoud, 203
al-Fartosi, Ahmed, 147, 148–9, 150, 151, 154–5, 156, 159–60, 180, 208–9, 209–10, 211, 214, 307, 309–10, 312–13, 315, 316, 317, 327, 328, 329, 330, 331, 332, 334, 335, 336, 338, 353
al-Fartosi, Muhammad, 120–1
Fartosi clan, 147, 148
Fasal, Firas, 62
Fatabend, Major General David, 295
Fausi, Ahmed, 123
FCO (Foreign and Commonwealth Office), 2, 3, 6, 7, 16, 20, 40, 48, 66, 67, 78, 87–8, 89, 92, 95, 136, 137, 139–40, 162, 209, 217, 221, 228, 229, 230, 279, 301, 326, 349
Featherstone, Major Justin, 117, 251, 353
Feith, Douglas, 40
Fenmore, Captain James, 28
Fergusson, George, 232–3
Feyzabad, 131
Figgures, Major General Andrew, 101, 113
First Gulf War, 8, 14, 16, 23, 261
First World War, 29, 91
Flanagan, Ronnie, 221
Fleming, Roy, 232
Flynn, Major General George, 337, 339, 353
Foley, Tom, 80, 81
Forces of Darkness list, 208
Foreign Affairs Select Committee, 71
Foreign and Commonwealth Office *see* FCO
Forsyth, Frederick, 189, 190
Fort Hood, 303
Fort Leavenworth, 247, 249
Fort McNair, 15
Forward Together, Operation, 262, 289
Foster, Peter, 38–9, 41

Foy, Tim, 263
France, 8, 9, 18, 20, 203
Franks, General Tommy, 40
Frechette, Louise, 17
Free, Brigadier Julian, 333, 335, 336, 337, 339, 353
Freedman, Lawrence, 9
FRES (Future Rapid Effects System), 254–5, 256, 347
Front de Liberation National, 159
Fry, Lieutenant General Rob, 112, 130, 132, 134, 224–5, 229, 289, 291, 346–7, 353
Fulton, Lieutenant General Robert, 255
Future of Iraq Project, The, 15, 163
Future Rapid Effects System (FRES), 254–5, 256, 347

Gaddafi, Colonel Muammar, 128
Gaddafi, Saif, 167
Gardiner, Nicholas, 258
Garner, Jay, 14–15, 17, 21, 22, 37, 38, 39, 41
Gates, Robert, 303, 304
Gaza, 128
Geneva Conventions, 32–3, 46
Gentle, Gordon, 189
Gentle, Rose, 189, 253
George Bush Senior Presidential Library, 12
Gereshk, 231, 238, 239, 240
Germany, 9, 29, 40, 129, 131, 203
Ghani, Ashraf, 231
al-Gharawi, Ahmed Abu Sajad, 118–19, 149, 155, 197, 205, 354
Gharawi clan, 271
Gilligan, Andrew, 50, 70, 71
Glasgow, 315
Glover Webb patrol vehicles, 253, 254
Goddard, Captain Dave, 201
Goldsmith, Peter, 13, 18–19, 20–1, 32, 75, 192, 194, 354
Gould, Philip, 194
Gray, John, vi, 350
Gray, Sergeant Stuart, 168
Green, Lieutenant Colonel Jeremy, 187–8
Greene, Graham, 94
Greenstock, Jeremy, 18, 69–70, 72, 75–7, 78, 79, 81, 82, 83, 95, 99, 114, 354
Greenwood, Christopher, 32
Green Zone, 38, 39, 81, 112, 166, 291, 294, 333, 334, 343
Grey, Stephen, 124
Guardian, 190, 207, 220

Hadfield, Major Andy, 212, 213, 215, 216, 217
Haider, Fakher, 208
Haider Samad, 23–5, 31, 46, 141, 142–4, 178, 183–5, 285–7, 319–26, 342, 354
Haigh, Frank, 53
Halabja, 28
Haldane, General Aylmer, 197
Hale, Lance Corporal Stu, 275
Haman, Gunner Frank, 151, 152, 153
Hamilton, Neil, 190
Hamilton-Jewell, Sergeant Simon, 56–7, 58, 59, 60, 61, 62, 354

Hammond, Major Mark, 275–6
Hamra Hotel, Baghdad, 38, 81
Harding, Major Paul, 312, 316, 354
Hardwick Arms, Sedgefield, 191
Harry, Prince, 88
Harris, Lee, 210, 211, 212, 215, 219, 220
Hart Security Limited, 108
Hashmi, Lance Corporal Jabron, 244
Hasiba, Madam, 23
Hassan (translator), 184
Hassan, General Ahmed, 207
Hassan, Margaret, 168, 169, 189
Hayaniya district, Basra, 52, 144, 210–11, 214, 216, 337, 338, 339
Haynes, Deborah, 323
Heath government, 32
Heatley, Charles, 22, 38–9, 41, 81, 100, 164–5, 166, 169, 171, 354
Helmand, 3, 173–4, 181, 224–7, 229–35, 237–45, 248, 271, 275–7, 280, 305, 331, 344, 345, 346
Helmand river, 173–4, 231, 232, 238
Henderson, Lieutenant Colonel Nick, 211–12, 213, 215, 216, 217–18, 220, 221–2, 253, 354
Hewett, Private Philip, 205
Hezbollah, 128, 148, 202, 210, 279
Hickey, Sergeant Christian, 221, 354
Hickey, Pauline, 253, 254
Hill, Rupert, 67
Hillsborough Castle, 39
Hindu Kush, 174
Hinnett, Private Karl, 216
Holmes, Captain Richard, 252, 269, 354
Home Office, 139
Hoon, Geoff, 29, 70, 129, 130, 132, 133, 188, 189, 217, 354
Houghton, Lieutenant General Nick, 261, 272, 315, 354
Houston, 12
Hughes, Brigadier Chris, 209, 225, 354
Hull, 342
Human Rights Act, 32
Human Rights Commission (Iraq), 342
Hunter, Major Nick, 198–9, 200, 201
Hurley, Kevin, 178
Hussein, Saddam, 2, 3, 4, 6, 7–8, 12, 13, 14, 17, 18, 19, 20, 21, 24, 25, 28, 31, 38, 40, 48, 51–2, 54, 131, 140–1, 147, 148, 149, 159, 342, 354
Hutton Inquiry, 127, 193

Ibn Saud, 197
ICRC (International Committee of the Red Cross), 32, 35
Imam Ali Shrine, Najaf, 113, 120, 150, 155, 170
Independent, 318
India, 11, 173
Indian Ocean tsunami, 229
Indonesia, 229
Ingrams, Adam, 189
Interior Ministry (Iraq) *see* Ministry of Interior

International Atomic Energy Agency, 203
International Committee of the Red Cross (ICRC), 32, 35
International Republican Institute (IRI), 163, 169
IRA, 200, 296
Iran, 149, 159, 163, 199, 202, 203–4, 269, 272, 274, 290, 297, 306–7, 312, 320, 321, 338, 343
Iran–Iraq war, 2, 24, 149
Iraq, xii, 1–128, 129, 130, 131, 132, 135–71, 175–22, 224, 226, 227, 237, 247, 248, 250–74, 277, 279, 280–344, 345–6, 346–7, 348–9
Iraqi Divisions
1st, 334
10th, 267, 280–1, 306, 335, 348
14th, 329, 335, 336, 337
Iraqi Governing Council, 82–3, 112
Iraqi National Accord, 161
Iraq Inquiry, 19, 20, 87, 226, 345–6
Iraq Planning Unit, 16
Iraq Study Group, 289–90
IRI (International Republican Institute), 163, 169
Iron, Colonel Richard, 331–3, 334, 335, 337–8, 339, 349, 354
Ishakzai tribe, 240, 241
Israel, 9, 12, 128, 202, 279
Istanbul: NATO conference, 128, 130–1, 174
Itais, Nouriya, 208
Italians, the, 70, 92, 95, 101, 102, 103, 110, 111, 112, 113, 239
ITV, 307
Jaafari, Ibrahim, 202, 204
Jabr, Bayan, 202, 208
Jackson, Dennis, 136
Jackson, General Mike, 44, 221, 224, 254, 354
al-Jadiri, Basima, 298–9, 354–5
Jaffar, Captain, 145, 178, 179, 180, 208, 210, 211, 212, 214, 218, 219, 283, 337, 338, 355
Jaggard-Hawkins, Major Ian, 30, 43
Jaish al-Islami, 295
Jaish al-Mahdi, 68–9, 87, 100, 101, 104, 106, 108, 116, 118, 119, 143, 147, 148, 149, 155, 157, 158, 159, 180, 197, 205, 208–9, 251, 262, 263, 266, 268, 270, 273, 274, 283, 293, 294, 295, 297, 304, 306, 307, 309, 311, 312, 313, 315, 316, 317, 318, 327, 328, 329, 330, 331, 332, 335, 336, 338, 339, 341, 342
Jamariya neighbourhood, 272
James, Corporal Ryan, 154
Jamiat police station, 144–5, 178, 179, 180, 207–8, 210, 211–16, 217, 218–19, 220, 221, 226, 283, 301, 309, 346
Jarvenpaa, Minna, 228, 230, 231, 232, 233, 234, 238, 240, 355
Jawad, Rear Admiral Muhammad, 43, 44
Jay, Michael, 20
Jeffries, Stuart, 190
Jenkins, Paul, 210, 211, 212, 215, 219, 220
Jepson, Ole, 4, 5
Johnson, Private Phil, 60

Joint Service Interrogation Team, 34
Joint Services Command and Staff College, 249–50
Jones, Corporal Chris, 156
Jones, Lieutenant Colonel Herbert (H), 207
Jones, Major Rupert, 207, 208, 209, 210, 211, 212, 213, 214, 215, 216, 217, 218, 219, 355
Jordan, 138, 290, 326
Jowell, Tessa, 127

Kabul, 128, 173, 174, 226, 230, 238, 239, 344
Kajaki Dam, 242, 275
Kandahar, 173, 174, 181, 225, 226, 231, 232
Karbala, 91, 295, 297
Karzai, Ahmed Wali, 240
Karzai, Hamid, 132, 227, 240, 242, 355
Katyusha rockets, 4–5
KBR, 107
Keane, General Jack, 290
Kearney, William, 135–6, 137, 140, 141–2, 143–4, 145, 177–8, 179, 180, 181, 183–4, 208, 213, 286–7, 324, 325, 355
Keegan, Lance Corporal Mark, 122
Kelly, David, 70–1, 73, 127
Kemp, Major Chris, 56, 57, 62
Kennedy, Lieutenant Ross, 55, 56, 57, 58–9, 60, 62
Kennett, Brigadier Andrew, 122, 251
Kenyon, Corporal Daniel, 47
Kenyon, Lieutenant Colonel Mark, 311
Kerick, Bernie, 79, 138
Kernaghan, Paul, 139
Kerry, John, 168
Keys, Reg, 62–3, 187, 188–9, 190–92, 193, 194, 253, 254, 258, 289, 345, 346, 355
Keys, Sally, 62, 63
Keys, Lance Corporal Thomas (Tom), 56, 62, 63
Khadeem estate, Amarah, 119
al-Khafaji, Aus, 101, 103
Khameini, Ayatollah Ali, 204
Khan, Ayub, 173
Khan, Dos Muhammad, 241
Khatami, Muhammad, 203
al-Khazali, Laith, 297
al-Khazali, Qais, 159, 297, 306, 307, 355
Killick, Corporal Stephen, 272
King's Royal Hussars, 198
Kirkuk, 89, 92
Kissinger, Henry, 39
Kiszely, Lieutenant General John, 169
Knaggs, Colonel Charlie, 240, 241
Kosovo, 6, 21, 27, 65, 230
Kunduz, 128
Kurds, the, 28
Kursk, 190
Kut, 29, 91, 101, 103, 104, 106–10, 112, 113, 119, 230
Kuwait, 2, 14, 19, 21, 22, 37, 88, 282, 301, 314, 317

Labouchere, Lieutenant Colonel David, 269–70, 272, 273, 355
Labour Party, 6, 128, 164, 170, 191, 192, 193, 194, 255, 280, 344, 345
Lamb, Christina, 245
Lamb, Major General Graeme, 72–3, 86, 87, 248–9, 250, 290, 291–2, 293, 294, 295–6, 313, 314 315, 316, 355
Lamb, Robin, 265, 266, 355
al-Lami, Heider, 124
Landers, Kevin, 180
Land Warfare Centre, 249, 250, 291
Lane, Second Lieutenant Rupert, 312
Larkin, Lance Corporal Darren, 47
Lashkar Gar, 225, 231, 232, 232, 233, 238, 240
Lawrence, T. E., 89, 248
Lebanese Broadcasting Corporation, 169
Lebanon, 148, 279, 280
le Blan, Eric, 80–1
Leong, Sergeant Zak, 244
Libya, 342, 349, 350
Light Infantry, 96, 97, 115, 116
Lockwood, Anthony, 191–2
London, 2, 3, 16, 21, 83, 95, 100, 112, 171, 176, 190, 315
bombings, 223–4
Long, Sergeant George, 216
Lorimer, Brigadier John, 209, 211, 212, 213, 216, 217, 218, 219, 220, 355
Lowe, Private Paul, 168
Lyndhurst, Staff Sergeant Christopher, 273

McArdle, Private Scott, 168
McCallum, Sergeant Major John, 339
MacCartney, Charlie, 144, 145, 178, 179, 355
McChrystal, General Stanley, 292, 293, 295, 314, 356
McColl, Lieutenant General John, 113, 168, 169, 225
McDonagh, Margaret, 170
MacDonald, Simon, 264, 301
MacGuiness, Martin, 310
Maciejewski, Lieutenant Colonel Justin, 283, 306, 311, 355
MacKenzie, John, 188–9
McMaster, Colonel Herbert, 290
McNeill, General Dan, 305
Madrid Donors' Conference for Iraq, 91
Maer, Lieutenant Colonel Matt, 117–18, 119–20, 121, 122, 123, 124, 155, 156, 158, 195, 196, 355
Mail on Sunday, 71
Maimonides, vi, 9
Maiwand, 173, 174, 181, 231
Majar al-Kabir, 51, 52, 54–62, 63, 120–1, 123–4, 159, 187, 205, 341, 346
al-Majid, Ali (Chemical Ali), 28
Malalai, 231
Malaysia, 88, 248, 249, 250, 251
al-Maliki, Adel Muhoder, 197, 198
al-Maliki, Ahmed, 68, 69, 73–4, 102

al-Maliki, Nouri, 252, 267, 274, 296, 297, 298, 299, 306, 312, 317, 327, 333, 334, 335, 336, 337, 338, 341, 342, 343, 356
Mandelson, Peter, 41
Manning, David, 8, 41, 63, 69–70, 71, 77, 78, 105, 356
Manning, Fusilier Stephen, 209
Manning, Major Steve, 219
Mansoor, Colonel Pete, 297
Mansoor district, Baghdad, 166
Marriott, Brigadier Patrick, 280, 305, 315, 336, 356
Marsden, Rosalind, 227, 230, 266, 267, 301, 356
Marsh Arabs, the, 51, 52, 53–4, 70
Marston, Daniel, 250–1, 253, 333, 356
Mastiff armoured vehicles, 256
Maude House, Baghdad, 291, 292, 293, 294, 295
Maysan province, 54–63, 70, 88, 89–91, 95–7, 112, 115–25, 140, 147, 175, 224, 251–2, 257, 264, 268, 269–71, 272–4, 283, 287, 341 *see also* names of places
Mazar-e-Sharif, 129, 131
Meade, Fusilier Donal, 209
Mechanised Brigade, 12th, 207, 210, 251
Mello, Sergio de, 76
Mercer, Lieutenant Colonel Nicholas, 33–5, 36, 45–6, 86, 141, 325, 356
Merchant Bridge, 81
Mesopotamia, 53, 89
Messenger, Brigadier Gordon, 226–7, 231, 232, 233, 234–5, 356
MFAW *see* Military Families Against the War
Middle East peace process, 11, 12, 20
Military Families Against the War (MFAW), 189, 253–4
Military Review, 247–8
Miller, John, 188
Miller, Corporal Simon, 56, 62, 188
Mills, Sergeant Dan, 116–17, 118, 119, 156, 157, 158, 356
Ministry of Defence (Britain) *see* MOD
Ministry of Interior (Iraq), 79, 138, 139, 179, 202, 207, 212, 217, 306
MI6, 131, 164
MOD (Ministry of Defence), 7, 14, 29, 30, 33, 48, 78, 112, 115, 130, 132, 133, 180–1, 220, 221, 222, 224–5, 228, 239, 242, 252, 253, 255, 256, 258, 274, 280, 307, 336, 346, 347
  Defence Medical Services Directorate, 256, 257
Mohan, General, 317, 318, 327, 328, 329–30, 331, 332, 333, 334, 335, 337, 338, 349, 356
Monk, Charles, 67, 68, 69, 72, 73–4
Moore, Peter, 330
Morrison, William, 162, 171, 356
Mott MacDonald, 53
Mousa, Baha, 86–7
al-Mudhaffar, Raghib *see* Raghib, Judge
Mufotiyah district, Basra, 280–1

Muhammad, General, 337
al-Muhammadawi, Abdul Kerim Mahud *see* Abu Hatem
Murdoch MacDonald, 53
Musa Qala, 275, 276–7, 305, 331
al-Musawi, Imam Ali, 3, 4
Muthanna, 70, 224, 251, 261, 264, 270, 287
Mutheneb district, 141

Nad-e Ali, 231
Nagl, Lieutenant Colonel John, 248
Najaf, 91, 101, 113, 120, 149–50, 155, 158, 162, 166, 169, 170
Nasiriyah, 89, 92, 93, 95, 101, 102, 103, 110–11, 113, 119
Nathan, Jeremy, 103
National Democratic Institute (NDI), 163
National Security Council (Britain), 347
National Security Council (US), 82, 105, 111, 112, 114, 166
NATO, 6, 27, 128, 129, 130, 134, 305, 348
  Istanbul conference, 128, 130–1, 174
  Prague conference, 128
Naw Zad, 240–1, 242, 243
NDI (National Democratic Institute), 163
Nepal, 230
New Labour *see* Labour Party
News at Ten, 307
News of the World, 2, 252
Newton, Lance Sergeant Craig, 192
New York, 7, 17, 19
New York Times, 208
Nicholson, Baroness Emma, 54
Noorzai tribe, 240
North, Richard, 254, 255
Northern Ireland, 7, 32, 135, 140, 200, 228, 229, 231, 247, 251, 253, 296, 310, 329
North West Frontier Province, 181
Northwood, 21, 33, 34, 35, 41, 176, 195, 226, 231, 242, 245, 251, 261, 263, 267, 272, 280, 303, 314, 315, 331, 333, 335, 336
Norway, 131
Numaniya, 110

al-Obaidi, Abd al-Qadr, 333
O'Callaghan, Private Lee, 153
Odierno, General Ray, 92, 292, 293, 295, 296, 303, 356
Office of Policy Planning and Analysis, 91
Office for Post-War Iraq, 11, 14
Office of Reconstruction and Humanitarian Aid (ORHA), 37, 38, 40
Office of Special Plans, 40
Oil-for-Food programme, 8
Old Electricity Building, Basra, 66–7, 69, 73, 85
Oliver, Jonathan, 71
Olmert, Ehud, 279
Olsen, Mona, 66, 67
Olsen, Ole Woehler, 66–7
Olympic Committee (Iraq), 68
ORHA *see* Office of Reconstruction and Humanitarian Aid

Ostrovskiy, Brigadier Sergey, 103, 106
Overwatch, 176, 195, 196, 207, 261, 328

Pachachi, Adnan, 112
Pakistan, 11, 130, 171, 181, 224, 225, 344
Palestinian Liberation Organisation, 159
Palestinians, 12, 127
Palmerston, Lord, 190
Papua New Guinea, 137
Parachute Regiment, 44, 55, 225,
1 Para, 54, 55, 56
3 Para, 187, 237, 238, 239, 240–1, 242, 275
Parker, Major Chris, 27–8, 30, 31, 43, 44–5, 48, 49, 356
Parker, Corporal Lee, 276
Parliament, 29, 189, 190, 349
Parry-Jones, Lieutenant Colonel Bryn, 258
Pashtu tribesmen, 173
Passmore, Captain James, 121
Patey, William, 212, 217
Pattison, James, 30
Payne, Corporal Donald, 86
PCRU (Post-Conflict Reconstruction Unit), 228–9, 230, 231, 232, 233, 234, 238, 264, 266, 347
Pearson, Corporal Stu, 275
Pelosi, Nancy, 166
Pennell, Lieutenant Ian, 153
Pennett, Miles, 65–6, 67–8, 71–2, 356
Pentagon, 15, 37, 39, 78, 82
Pepper, Corporal Andre, 153, 154
Petraeus, General David, 247, 248, 249, 290, 291, 295, 296, 297, 299, 303, 304, 305–6, 314, 328, 332, 333, 334, 335, 356
Phee, Molly, 95, 96, 115, 116, 117, 118, 119, 120, 150, 356
Philby, Harry St John, 52–3, 89, 197, 356
Philby, Kim, 52
Philips, Lance Corporal, 118
Philips, Lieutenant Richard, 56, 60, 258
Pike, Lieutenant General Hew, 237
Pike, Major Will, 237, 238–9, 240–1, 242, 243–4, 244–5, 357
Pinter, Harold, 189
Plesch, Dan, 190
Pointing, Lieutenant Colonel Bill, 96
Poland, 171
Post-Conflict Reconstruction Unit *see* PCRU
Powell, Colin, 17, 76–7, 82
Powell, Jonathan, 8, 13
Prague: NATO conference, 128
Price, Adam, 189, 190
Princess of Wales's Royal Regiment, 115, 116, 117, 149, 257
Proctor, James, 303, 307, 309, 310, 312, 317, 329, 330, 336
Provincial Reconstruction Teams (PRTs), 232, 263–5, 282, 349
Provincial Security Committee, 196
PRTs *see* Provincial Reconstruction Teams

PTSD (post-traumatic stress disorder), 256–7, 258, 343
Pynn, Harvey, 239, 243, 244

Qader, Wissam Abu, 301, 311
Qalit Salih, 52, 54
Qom, 338
Queen's Birthday Party, 1–5, 265
Queen's Dragoon Guards, 32
Queen's Lancashire Regiment, 67, 86
Queen's Own Yeomanry, 65, 66
Queen's Royal Hussars, 272
Quetta, 225
Quick, Rachel, 33, 34, 45
Qurnah, 29

Raffaelli, Rear Admiral Philip, 281
Raghib, Judge, 213, 214, 215
Rahim, Basil, 81
Rahim, Dr, 324–5
Rahman, Babu, 229, 232
Ramadi, 169, 293, 296
Rangwala, Glen, 190
Rathmel, Andrew, 139
Rawlinson, Lance Corporal Liz, 199
Rawnsley, Andrew, 19
Read, Phil, 139, 141
Red Caps/Royal Military Police, 55, 56, 57, 59, 60, 62, 90, 159, 187, 188, 189, 258, 341, 346
Reg vs Blair documentary, 191
Reid, John, 220, 224, 225, 235, 357
Reuters, 216, 220
Revolutionary Guard, 149, 202, 204
Rheindahlen, 129
Rice, Condoleezza, 18, 78, 82, 100, 105, 111, 114, 128, 166, 167–8, 204, 263, 274, 357
Richmond, David, 72, 73, 99, 100, 101, 102, 104, 105, 106, 108–9, 110, 112, 114, 115, 357
Ricketts, Peter, 3, 4, 5
Ricks, Tom, 248, 292
Rigestan desert, 174
Riley, Lieutenant General (formerly Brigadier, then Major General) Jonathon, 79–80, 175–6, 177, 180, 181, 195, 196, 207, 348, 357
Riyadh, Governor, 96, 97, 123, 124
Robertson, George, 128
Robertson, Sergeant John (Jock), 57–8, 59–60, 61, 62
Rogan, Janet, 66, 67, 68, 71, 72, 73, 357
Rollo, Major General Bill, 150, 156, 158
Romania, 181
Route 6, 54
Rowhani, Hassan, 204
Royal Engineers, 86
Royal Fusiliers, 46
Royal Green Jackets, 135
Royal Gurkhas, 240–1
Royal Hampshire Regiment, 27
Royal Logistics Corps, 200
Royal Marines, 30
Royal Military Police *see* Red Caps/Royal Military Police
Royal Welsh, 219

al-Rubaie, Mowaffak, 204, 296, 317, 333, 334, 335, 338, 357
Rumsfeld, Donald, 15, 20, 39, 80, 82, 137, 357
Russert, Tim, vi
Russia, 173, 190
Rwanda, 6

al-Sadoon, Ali, 142–3, 320, 321, 322, 325, 357
Sadoon, Faisal, 45
al-Sadoon, Muhammad, 183, 184, 185
al-Sadoon, Nora, 24, 142, 183, 184, 185, 285, 286, 319, 320, 321, 322, 323, 324, 325, 326, 342
al-Sadr, Moqtada, 68–9, 72, 73, 87, 90, 91, 99–100, 101, 105, 113, 116, 119, 120, 125, 148, 149, 150, 155, 158–9, 161, 166, 175, 195, 207, 209, 297, 317, 327, 335, 338, 341, 343, 357
Sadr, Ayatollah Muhammad Sadeq, 25, 48, 68, 148
Sadr City, Baghdad, 52
Sadrists, 74, 96, 101, 102, 103–4, 110, 112, 116, 118–19, 120, 148, 180, 196, 197, 202, 252, 297–8, 341 *see also* Jaish al-Mahdi
Sahwah see Awakening councils
Sajjad, Seyed, 210, 313, 357
Salamanca, Operation, 266, 267, 272
Salih, Barhim, 204
Salmond, Alex, 189
Samara, 252
Sanchez, Lieutenant General Ricardo, 296
Sanders, Lieutenant Colonel Patrick, 311, 317, 318, 357
Sandhurst, 250, 251, 333
Sandiford, Lieutenant Colonel Tim, 357
Sandline, 6, 137
Sangin, 240, 241, 242–5, 271, 275, 344
SAS, 167, 209, 210, 211, 213, 215, 217, 219, 220, 221, 225, 272, 279, 309, 311, 330
Satterfield, David, 204
Saudi Arabia, 70, 89, 290, 343
Sawadi, Hassan, 208
Sawers, John, 40, 167
al-Sayeed, Sajjad Badr, 266–7
Schmeller, Rachel, 5
Schulte, Paul, 228, 230
Schwekit, Ahsen, 144
Scots Dragoon Guards, 28, 336
Sea King helicopters, 215–16, 219, 220
Second World War, 9
Secret Intelligence Service *see* SIS
Sedgefield, 128, 190, 191, 193, 345
SE432 personnel carrier, 256
Seldon, Anthony, 127
Selly Oak hospital, 256, 281
Senor, Dan, 100
September 11 terrorist attacks, 7
al-Sewan, Ali, 145, 178, 212, 214, 219, 357
Shaibah logistics base, 137, 210, 214, 237, 309, 317, 328
Shaqir, Major Mohsen, 140, 178, 185, 208

Shatt al-Arab, 4, 23, 28, 29, 150, 307, 321
Shatt al-Arab Hotel, Basra, 220, 306, 319, 329–30, 336, 337, 338
Shatt al-Basra, 28
Shaw, Major General Jonathan, 302–3, 304, 305, 306, 307, 310, 312, 313, 315, 316, 317, 330, 331, 357
Shearer, Second Lieutenant Richard, 201, 205, 357
al-Sheibani, Abu Mustafa, 159–60, 202
Sheinwald, Nigel, 77–8, 82, 83, 104, 105, 114, 138, 165, 229, 230, 239, 357
Sher Muhammad, 227
Shia, 3, 24, 28, 51, 54, 68, 73, 76, 83, 87, 93, 99, 121, 149, 150, 159, 163, 164, 167, 169, 170, 171, 195, 197, 202, 204, 207, 252, 262, 267, 269, 274, 279, 285, 290, 291, 293 294, 295, 297, 302, 317, 338, 341, 344
Shirreff, Major General Richard, 261, 262–3, 267, 269, 270–1, 271–2, 273, 274, 279, 280, 281, 282, 283, 301, 302, 310, 357
Shirreff, Sarah-Jane, 281
Short, Clare, 7, 16, 19, 20, 21, 49, 86, 345, 358
Shrivenham, 250
Siddiq, Irfan, 113–14
Siddique, Major Rabia, 213–14, 215, 218, 219, 220, 358
Sierra Leone, 6, 7, 137
Signals Corps, 244
Simcock, Lance Corporal David, 199, 200
Simon, Sion, 280
Simonds, James, 293–4, 294–5, 296
Sinbad, Operation, 274, 280–1, 282, 283, 301, 302, 304, 305
SIS (Secret Intelligence Service), 28, 31, 159, 167, 170, 180, 214, 225, 228, 303, 307, 309, 313, 317, 331
al-Sistani, Grand Ayatollah Ali, 76, 82, 83, 91, 99, 131, 149, 158, 159, 162, 170, 358
Sky, Emma, 89, 92, 292, 296–7, 298–9, 358
Sky News, 63
Smiles, Trooper Darren, 199
Smith, Colin, 176–7, 178, 179–80, 207, 208, 358
Snatch Land Rovers, 253, 254
al-Soleimani, Qasem, 204, 338, 358
Southern Iraq Steering Group, 282
Soviet Union, 231, 275
Special Branch, 135
Spectre, 254
Spicer, Private Leon, 205
Spicer, Tim, 137
Srebrenica, 130
Stabilisation Unit, 234
Staffordshire Regiment, 195, 306
State Department (US), 15, 79, 93, 163, 164, 166, 274
Stewart, Major General Andrew, 95, 102, 103, 112, 113, 120, 358
Stewart, Rory, 88, 89, 90, 91, 95–7, 101, 358

Stirrup, Air Chief Marshall Jock, 280, 313, 314, 315, 358
'Stop the War' rally, 21
Strachan, Neil, 106, 107
Strategic Defence Review, 7, 14, 132, 133, 175, 254
Straw, Jack, 8, 12–13, 19, 20, 21, 82, 167–8, 203, 204, 220, 235, 358
Suez crisis, 9
Sun, 307, 318
Sunday Times, 245
Sunnis, 31, 68, 87, 93, 163, 167, 168, 170, 175, 202, 204, 247, 252, 262, 279, 290, 291, 292, 293, 296, 297, 298, 302, 343, 344
Sunni Triangle, 168
Supreme Council for Islamic Revolution in Iraq, 163
Suq al-Shuyukh, 89, 93
al-Sweady, Hamid, 121, 123, 124
al-Sweady, Khuder, 121, 124
Sutcliffe, Roger, 167, 180
Synnott, Hilary, 70, 85, 139, 358
Syria, 290, 342

Tactical Support Unit (TSU), 140, 141
Taji, 333
Talafar, 290
Taliban, 7, 128, 129, 181, 225, 227, 229, 231, 238, 239, 240, 241, 242, 243, 244, 245, 271, 275, 276–7, 305, 331, 344, 345, 347
al-Tamimi, Muzahim, 4, 5, 31
Tamworth, 47–8
Tansley, James, 1, 2–3, 4, 5, 139, 140, 216, 358
Taylor, Lieutenant Colonel Alex, 214
Taylor, Major Dan, 46
Taylor, Captain Martin, 244
Taylor, Private, 123
Tehran, 204
Territorials, 65
Thar'allah, 124
Thesiger, Wilfred, 147
Thomas, Bertram, 89, 90, 93, 97
Thomas, Colonel Ian, 303
Thompson, Corporal Terry, 153
Thorpe, Corporal Peter, 244
Tigris, River, 52, 53, 109
The Times, 323
Timmons, Timm, 107, 110, 112
Titanium, Operation, 197–8
Today programme, 50, 282, 325
Tokyo, 71
Tootal, Lieutenant Colonel Stuart, 241, 242, 243, 245, 275, 276, 358
Torpy, Glen, 176
Toward, Ann, 193
Townshend, Major General Charles, 29, 109
Tracer armoured vehicle, 254
Treasury, 16, 29, 132, 133, 139, 224, 343
Triple Canopy, 323
TSU (Tactical Support Unit), 140, 141
Tukutukuwaqa, Private Pita, 169
Tunisia, 342
Turkey, 13, 14, 109
al-Turki, Abdul Basit, 112
Tyson, Captain Amber, 151, 152

Ukrainians, the, 103–4, 106, 107, 109, 110, 112
Umm Qasr, 143, 148
United Nations (UN), 6, 7–8, 12, 13, 14, 16, 17, 18, 19, 20, 21, 22, 39, 40, 41, 49, 75, 86, 93, 99, 105, 345, 349
General Assembly, 13, 17
headquarters in Baghdad, 76
Security Council, 12, 17, 18, 19, 20, 40
United States/Americans, 6, 7, 8, 11–12, 13, 14–15, 17, 18, 19, 20, 21, 28, 33, 39–40, 41, 44–5, 70, 75, 76–7, 78, 79, 80, 82, 91–2, 99, 100–1, 102, 104–5, 107–8, 111–12, 112–13, 114, 115, 129, 136, 138, 149–50, 161, 163, 164, 165–6, 168, 169, 174, 175, 176, 203, 204, 227, 231, 232, 242, 247–8, 257, 262, 272, 289–90, 291–2, 293, 296–7, 298–9, 303–5, 314, 315–16, 328, 332, 342 *see also* names of individuals
USAID, 239
US Marines, 28, 101, 168, 169, 256

Vandergrift, Lieutenant Matthew, 339
Vauxhall Cross, 228
Vernon, Colonel Chris, 35
Vickers, Captain Steve, 192
Vincent, Stephen, 208

al-Waeli, Muhammad, 180, 208, 220, 303, 306, 358
Wakefield, Guardsman Anthony, 192, 193, 194, 196, 198, 358
Walker, Hugh, 228, 232, 233
Walker, General Michael, 44, 63, 130, 132, 133, 134, 358
Wall, Major General the Lieutenant General Peter, 67, 195, 226, 242, 358
Warrick, Tom, 15, 163, 164, 166, 358
Washington, 11, 14, 15, 18, 76, 77, 78, 82, 113, 114, 165, 289, 290
Washington Post, 76, 248
Wasit province, 88, 91
Watson, Tom, 280
Weadon, Lance Corporal Mark, 55
Weale, Adrian, 93, 94
Webster, Corporal Martin, 258–9, 358–9
Welsh Guards, 205
Westmacott, Peter, 14
White House, 39, 40, 75, 78, 82, 100, 111, 114, 165, 166, 289
White-Spunner, Major General Barney, 174, 181, 226, 333, 334, 335, 359
Whitford, Victoria, 108, 109, 167, 359
William, Prince, 88
Williams, Lieutenant Colonel Andrew, 195–7, 197–8, 205, 359
Williams, Lieutenant Colonel Richard, 44, 210, 217, 279, 310–11, 359
Williamson, Corporal Darren (Daz), 116–17
Wilmhurst, Elizabeth, 21

Wilson, Captain Arnold Talbot, 53
Wilson, Robert, 144
Wilson, Tim, 269
Windham, John, 140
WMDs (weapons of mass destruction), 12–13
Woodham, Major James, 211, 212, 213, 214–15, 218, 219, 220
Woolley, Trevor, 133
World Trade Center, 7
Wright, Frank Lloyd, 38
Wright, Private Mark, 275

al-Yacoubi, Muhammad, 101
Yemen, 342

al-Zarqawi, Abu Musab, 166, 279